AF518252

Environmental Studies and Disaster Management

THE AUTHORS

Dr. Ashwani Kumar Dubey is a leading scientist, in the beginning of 21st century in India. He is currently working as Executive Director, Environment & Social Welfare Society, Khajuraho, India. He has completed his Ph.D. in Zoology in 1995 from Vikram University, Ujjain, MP. He has devoted his life in Academic and Scientific research. He has published many books of Zoology, research papers in International and National peer reviewed journals and academic and scientific articles in standard magazines. Scientific talk broadcasted by All India Radio. His expertise has been requisitioned by several National & International organizations. He is widely regarded as one of India's foremost expert with a scientific temper on Zoology & Environmental Sciences. He awarded many prestigious awards by National and International institutions. He is a regular member and distinguished fellow of the learned scientific societies including Indian Science Congress, Kolkota. He is editor/reviewer in research journals published from India, America, United Kingdom, Egypt, France, Syria, Nepal, Iraq, Sudan, Romania, Malaysia and Japan.

Dr. Shiv Ji Malviya Currently working as Deputy Secretary, Uttar Pradesh Higher Education Commission. He is a great naturalist. Previously he was posted in Hemvati Nandan Bahuguna Degree College, Naini, Prayagraj. He has published many research papers in journals of repute. He has many awards in his name. He had performed various duties in Government sectors. He is a nature lover, conservationist and nation builder.

Er. Priyansha Kushwaha FSAJMS, FIJIRG, FESW currently working as a Lecturer in Government Polytechnic Faizabad, (Ayodhaya) Uttar Pradesh in the Civil Engineering Department. She has done her Bachelor's in Technology (B.Tech) in Civil Engineering from Dr. A.P.J. Abdul Kalam Technical University (APJAKTU), Lucknow, and Masters in Technology (M. Tech) in Environmental Engineering. She is an environmentalist, worked in ESW Society West Bengal Chapter also life member in various other societies, working with South Asia Management Association Sunnyvale California, USA, have many awards like a prestigious Young Engineer award in 2020 by Indian Council of Social Science Research, New Delhi; and being an active environmentalist she is Nature lover.

Environmental Studies and Disaster Management

– Authors –

Ashwani Kumar Dubey

Shiv Ji Malviya

Priyansha Kushwaha

2022

Daya Publishing House®
A Division of
Astral International Pvt. Ltd.
New Delhi – 110 002

ISBN: 9789354614606

Published by : **Daya Publishing House®**
A Division of
Astral International Pvt. Ltd.
– ISO 9001:2015 Certified Company –
4736/23, Ansari Road, Darya Ganj,
New Delhi-110 002
Ph. 011-43549197, 23278134
E-mail: info@astralint.com
Website: www.astralint.com

Dedicated to My Late Parents

Mother Godavari Dubey
Father Tulsidas Dubey

Preface

Environment studies are all about learning the way we should live and how we can develop sustainable strategies to protect the environment. It helps individuals to develop an understanding of living and physical environment and how to resolve challenging environmental issues affecting nature. In addition to studying the physical aspects of the environment, it also emphasizes the need to conserve biodiversity and adopt a more sustainable lifestyle and utilize resources in a responsible way. To create awareness among today's generation on pressing environmental problems, the University Grants Commission (UGC) has made it mandatory for the universities to introduce a course in environmental studies and teach students about the eco-system, pollution and problems concerned with the environment.

Disruption on a massive scale, either natural or man-made, occurring in short or long periods is termed as Disaster. Disaster management in India has been an important point of discussion owing to frequent natural disasters ranging from earthquakes, floods, drought, etc. Loss of life and property due to these disasters have been steadily mounting throughout the world due to inadequate technology to combat disasters, rise in population, climate change, and continuing ecological degradation. The global efforts to manage disasters have proven to be insufficient to match the frequency and magnitude of natural disasters.

The present book is divided into two parts. The first part of the book covers environmental studies, covering the multidisciplinary subject, natural resources, ecosystems, biodiversity and its conservation, environmental pollution, social issues and the environment and human population and the environment.

In the second part disaster management have been covered with emphasis on - introduction to disasters, man-made disasters, disaster management and COVID-19 and its impact on environment.

This book will be highly useful for the students undergoing the course of environmental studies and disaster management. This book is intended to give easy readability, and it covers almost all the basic aspects of Environmental Science and Disaster Management. We hope that this book will be of greater use for the students of undergraduate courses of Indian Universities, researchers and policy makers interested in improving the environmental quality.

We are grateful to all those persons as well as various books, manuals, periodicals, magazines, journals etc. that helped in the preparation of this book. In spite of the best efforts, it is possible that some errors may have occurred into the compilation and editing of the book. Further queries, constructive suggestions and criticisms for the improvement of the book are always welcome and shall be thankfully acknowledged.

Authors

Contents

Disaster Management

Environmental Studies

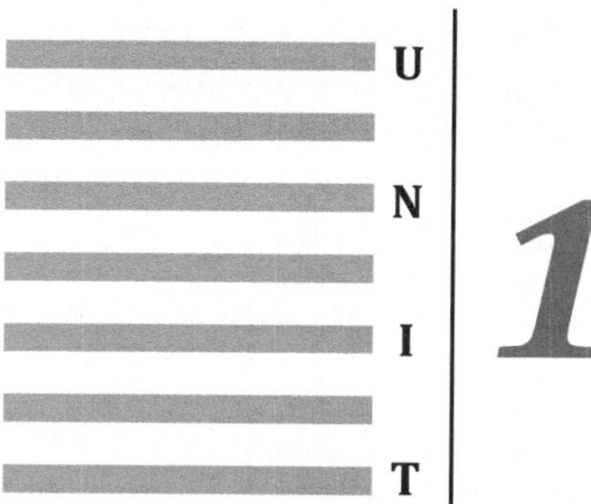

A Multidisciplinary Subject

Definition

'Environment' factually means the surrounding in which we are existing. Environment comprises of all those things on which we are directly or indirectly reliant on for our existence, whether it is a living constituent like animals, plants or non-living constituent like soil, air water.

Definitions of Environment

Some important definitions of environment are as under:

1. Boring: 'A person's environment consists of the sum total of the stimulation which he receives from his conception until his death.' It can be concluded from the above definition that Environment comprises various types of forces such as physical, intellectual, economic, political, cultural, social, moral and emotional. Environment is the sum total of all the external forces, influences and conditions, which affect the life, nature, behaviour and the growth, development and maturation of living organisms.
2. Douglas and Holland: 'The term environment is used to describe, in the aggregate, all the external forces, influences and conditions, which affect the life, nature, behaviour and the growth, development and maturity of living organisms.'

Environmental studies pacts with each subject that affects an entity. It is fundamentally a multidisciplinary approach that brings about an appreciation of our natural world and human influences on its reliability. It is an applied science as it pursues applied responses to making human civilization defensible on the earth's finite resources.

The discipline of Environment studies is a multi-disciplinary subject because it contains numerous branches of studies like chemistry, physics, medical science, life science, agriculture, public health, sanitary engineering *etc.* It is the discipline of physical phenomena in the environment. It studies the sources, responses, transport, effect and fortune of physical a biological species in the air, water and soil and the effect of from human activity upon these.

Scope

As environmental studies is considered to be multidisciplinary in nature, it is generally and broadly considered to be a subject of great scope. The present age of environment is not limited to issues of sanitation and health, rather, is now concerned with pollution control, biodiversity conservation, waste management and conservation of natural resources. The opportunities in this field are enormous not only for scientists but also for engineers, biologists. There is a good chance of opportunity to find a job in this field as environmental journalists. Environmental science can be applied in the following spheres:

Ecosystem Structure and Function

The study of ecosystems mainly entails of the study of the processes that link the leaving organism or in other words biotic component to the non-living organism or abiotic component. So, for the study of environment we should be aware with biotic and abiotic components.

Natural Resource Conservation

For managing and maintenance of forests which are natural resources and for the maintenance of wildlife forms task under natural resource conservation.

Environmental Pollution Control

With the knowledge of environmental science everybody can control the pollution. He/she can handle the waste management and also look for ways to control pollution on the aspect of pollution control.

Environmental Management

There are several independent environmental consultants who are working with Central and State pollution control Board. They offer advice to solve the problems of environment the optimum solution for the upcoming problems. They give direction for controlling pollution due to industrial development.

The Scope of Environmental Studies in Industry

Environmentalists work towards the upkeep of ecological balance, they also work towards the conservation of biodiversity and regulation of natural resources as well as on preservation of natural resources. Most of the industries have a separate environmental research and development section. These sections govern the impact that their industry has on the environment. Our environment is being degraded by the rapid industrialization. To combat this menace there is a growing trend towards manufacture of "green" goods and products. So, we can say that there is a good scope in the field of industry from environmental studies.

Social Development

NGO (non-governmental organizations) help in creating awareness concerning the protection of the environment and making the masses aware of various environmental issues. They also generate a public opinion in this field. They work towards propagating information and in bringing about changes in political policies that are personally effect the environment. The societal dimension of this profession includes controlling population explosion through organizing advisory awareness camps.

Importance

The environment studies make us conscious about the significance of fortification and conservation of our mother Earth and about the annihilation caused by us due to the discharge of pollution into the environment. The upsurge in human and animal population, industries and other issues make the survival cumbrous. A great number of environment issues have grown in size and make the system more complex day by day, threatening the survival of mankind on Earth. Environment studies have, thus, become substantial for the following reasons:

1. Environment Issues are being of Universal Significance

It has been well recognised that environment issues like global warming and ozone depletion, acid rain, marine pollution and biodiversity are not merely national issues but are global issues and hence require international efforts and cooperation to solve them.

2. Development and Environment

Development leads to Urbanization, Industrial Growth, Telecommunication and Transportation Systems, Hi-tech Agriculture and Housing *etc.* However, it has become phased out in the developed world. The North intentionally moves their dirty factories to South to cleanse their own environment. When the West developed, it did so perhaps in ignorance of the environmental impact of its activities. Development of the rich countries of the world has undesirable effects on the environment of the entire world.

3. Volatile Increase in Pollution

World census reflects that one in every seven persons in this planet lives in India. Evidently with 16 per cent of the world's population and only 2.4 per cent of its land area, there is a heavy pressure on the natural resources including land. Agricultural experts have recognized soil health problems like deficiency of micro-nutrients and organic matter, soil salinity and damage of soil structure.

4. Need for an Alternative Solution

It is vital, particularly for developing countries to find alternative paths to an alternative goal. We need a goal as under:

- ☆ A true goal of development with an environmentally sound and sustainable development.
- ☆ A goal common to all citizens of our planet earth.
- ☆ A goal distant from the developing world in the manner it is from the over-consuming wasteful societies of the "developed" world.
- ☆ It is utmost important for us to save the humanity from extinction because of our activities constricting the environment and depleting the biosphere, in the name of development.

5. Need for Prudent Planning of Development

Our survival and sustenance depend on resources availability. Hence Resources withdraw, processing and use of the products have all to be synchronised with the ecological cycle. In any plan of development our actions should be planned ecologically for the sustenance of the environment and development.

6. Misra's Report

Misra (1991) recognized four basic principles of ecology, as under:

- ☆ Holism
- ☆ Ecosystem
- ☆ Succession
- ☆ Conversation

Holism has been considered as the real base of ecology. In hierarchical levels at which interacting units of ecology are discussed, are as under:

Misra (1991) has recognised four basic requirements of environmental management as under:

- ☆ Impact of human activities on the environment,
- ☆ Value system,
- ☆ Plan and design for sustainable development,
- ☆ Environment education.

Keeping in view of the goal of planning for environmentally sustainable development, India contributed to the United Nations Conference on Environment and Development (UNCED), also referred to as "Earth Summit" held at Rio de Janeiro, the Capital of Brazil, 3rd-14th June, 1992.

Environmental studies helps to maintain ecological balance by providing a basic knowledge of environmental systems and their processes. By giving information regarding the changes that take place due to anthropogenic factors environmental study helps us.

It also helps to gain a skill in using techniques to analyse various environmental systems and the effect of human activities on that system.

Environmental studies applies economical methods and concepts to issues of the environment, management, environmental policy analysis.

Need for Public Awareness

In today's world because of industrialization and growing population, the natural resources has been swiftly used and our environment is being progressively besmirched by human actions, so we need to protect the environment.

Environmental problems at local, national and international levels mostly occur due to lack of awareness. Environmental science aims to educate and equip learners with necessary environmental skills to pass to the community in order to create awareness. Environmental awareness can be created through social media, creating a blog dedicated to creating awareness, community centred green clubs, women forums, and religious podiums.

It is not only the responsibility of government but also the individuals to take active part for shielding the environment, so protecting our environment is economically more feasible than cleaning it up once, it is damaged.

The role of mass media such as newspapers, radio, television, *etc.* is also very significant to make people aware concerning the environment. There are various organisations, which are playing confident role towards environment to make people aware regarding environment like BSI (Botanical Survey of India, 1890), ZSI (Zoological Survey of India, 1916), WII (Wild Life Institute of India, 1982) *etc.*

We are facing various environmental challenges. It is essential to get the country acquainted with these challenges so that their acts may be ecofriendly. Some of these challenges are as under:

Growing Population

A population of over thousands of millions is growing at 2.11 per cent every year. It puts considerable pressure on its natural resources and reduces the gains of development. Although population control does automatically lead to development, yet the development leads to a cut in population growth rates. For this development of the women is essential.

Poverty

India has often been described a rich land with poor people. About 40 per cent of our people are still below the poverty line. Environment degradation has adversely affected the poor who depend upon the resources of their immediate surroundings. Thus, the test of poverty and the challenge environment degradation are two facets of the same challenge. The population growth is essentially a function of poverty. Because, to the very poor, every child is an earner and helper and global concerns have little relevance for him.

Need to Groundwater

It is essential of rationalizing the use of groundwater. Factors like community wastes, industrial effluents and chemical fertilizers and pesticides have polluted our surface water and affected quality of the groundwater. It is essential to restore the water quality of our rivers and other water bodies as lakes is an important challenge. It so finding our suitable strategies for consecration of water, provision of safe drinking water and keeping water bodies clean which are difficult challenges is essential.

Development and Forests

Forests serve catchments for the rivers. With increasing demand of water, plan to harness the mighty river through large irrigation projects were made. Certainly, these would plunge forests; displace local people, damage flora and fauna. Forests in India have been shrinking for several centuries owing to pressures of agriculture and other uses. Vast areas that were once green, stand today as wastelands. These areas are to be brought back under vegetative cover. The tribal communities inhabiting forests respects the trees and birds and animal that gives them sustenance. We must recognise the role of these people in restoring and conserving forests. The modern knowledge and skills of the forest dept. should be integrated with the traditional knowledge and experience of the local communities. The strategies for the joint management of forests should be evolved in a well-planned way.

Reorientation of Institutions

The people should be roused to orient institutions, attitudes and infrastructures, to suit conditions and needs today. The change has to be brought in keeping in view India's traditions for resources use managements and education *etc.* Change should be brought in education, in attitudes, in administrative procedures and in institutions. Because it affects way people view technology resources and development.

Evil Consequences of Urbanisation

Nearly 27 per cent Indians live in urban areas. Urbanisation and industrialisation has given birth to a great number of environmental problem that need urgent attention. Over 30 per cent of urban Indians live in slums. Out of India's 3,245 towns and cities, only 21 have partial or full sewerage and treatment facilities. Hence, coping with rapid urbanization is a major challenge.

Air and Water Population

Majority of our industrial plants are using outdated and population technologies and makeshift facilities devoid of any provision of treating their wastes. A great number of cities and industrial areas that have been identified as the worst in terms of air and water pollution. Acts are enforced in the country, but their implement is not so easy. The reason is their implementation needs great resources, technical expertise, political and social will. Again the people are to be made aware of these rules. Their support is indispensable to implement these rules.

Types of Environment

"'Environment', as the term itself indicates, is anything immediately surrounding an object and exerting a direct influence on it". – P. Gisbert. According to Kimball Young, environment refers to "those forces, situations or stimuli which influence the organism from outside".

Geographical Environment

This can be called 'natural environment' for it consists of things that are provided by nature, this can also be called 'physical environment' for it includes the physical conditions of life. The geographic or physical conditions exist independently of man's existence. Man has limited and sometimes no control over them.

This environment includes; the surface of the earth, natural resources, land and water, mountains and plains, fertile lands and deserts, oceans, storms and cyclones, weather and climatic factors, seasons, *etc.* It also includes biological conditions such as plants, animals with all their complexities.

Man-Made Environment

In order to control the conditions of his life man has created a new environment which can be called 'man-made environment' and some have called it 'social-cultural environment'. It can be sub-divided into two types: (a) outer environment, and (b) inner environment.

Psychological Environment

Although physical and social environment are common to the individual in a specific situation. Yet every individual has his own psychological environment, in which he lives. Kurt Lewin has used the term 'life space' for explaining psychological environment. The Psychological environment enables us to understand the personality of an individual. If a person is unable to overcome the barriers, he can either get frustrated or completed to change his goal for a new psychological environment. But adopting this mechanism, the individual is helped in his adjustment to the environment.

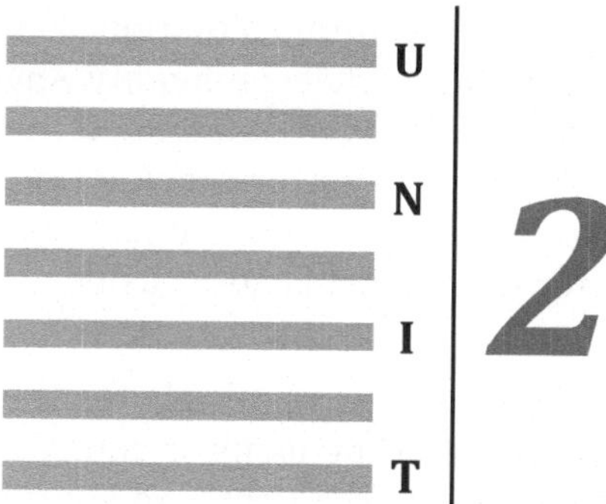

Natural Resources

2.1 Introduction

The term "Natural resources" was first coined and popularized by E.F. Schumacher in the 1970s by his famous book "Small is Beautiful." Basically, 'natural resources' is a term applied to all naturally occurring substances which are considered valuable in their relatively unmodified, *i.e.* natural form.

Natural resources are resources from the Earth that are used to provide life and meet people's needs. Some of the natural resources include air, water, soil, minerals, along with the climate and solar energy, which form the non-living or 'abiotic' part of nature. Earth's biosphere is gifted with tremendously varied kinds of environments that deliver countless goods and services to humankind. Any component of the natural environment that can be used by man to sanction his welfare is measured as a natural resource. The natural resource can be a material, an energy constituent or a natural process or phenomenon. Land, soil, water, forests, grasslands, *etc.* are examples of important natural resources. Some of the resources (*e.g.* soil, water) are significant mechanisms of the life-supporting system. Besides being a source of food, fodder, and shelter, natural resources also deliver recreational opportunities, solace, and even inspiration to mankind. Natural resources have been exploited by humans since the commencement of development or even before. However, since the resources were plentiful then comparative to the human population, no noteworthy depletion occurred. During the last millennia, the human population has increased considerably causing serious damage or destruction of natural resources. In this chapter, we will study major kinds of natural resources, causes of their degradation, and their conservation.

The value of any resource is resolute by its demand-supply situation. The demand-supply again is determined by the utilitarian value of that commodity. Any commodity is a natural resource only if the primary activities of obtaining it are extraction and purification, not creation. Due to this reason, petroleum, hunting, fishing, forestry, and mining are classified as natural resource industries. Since agriculture is an occupation/industry related to the creation, it is not considered a natural resource industry.

Natural resources are used to make food, fuel, and raw materials for the production of goods. All of the food that people eat comes from plants or animals. Natural resources such as coal, natural gas, and oil provide heat, light, and power. Natural resources also are the raw materials for making products that we use every day from our toothbrush and lunch box to our clothes, cars, televisions, computers, and refrigerators.

2.1.1 Earth's Resources and Man

The resources on which mankind is dependent are provided by various sources or 'spheres'.

The troposphere is the layer closest to Earth's surface. It is 4 to 12 miles (7 to 20 km) thick and covers half of Earth's atmosphere. Air is warmer near the ground and gets colder higher up. Nearly all of the water vapour and dust in the atmosphere are in this layer and that is why clouds are found here. It's where the weather happens and comprises the air humans breathe. The air of our planet is 79 per cent nitrogen and just under 21 per cent oxygen; the small amount remaining is composed of carbon dioxide and other gases. The temperature of the troposphere decreases with height.

The stratosphere is the second layer. It starts above the troposphere and ends about 31 miles (50 km) above ground. Ozone is abundant here and it heats the atmosphere while also enthralling harmful radiation from the sun. The air here is very dry, and it is about a thousand times turpentine here than it is at sea level. Because of that, this is where jet aircraft and weather balloons fly.

The mesosphere starts at 31 miles (50 km) and ranges to 53 miles (85 km) high. The top of the mesosphere, called the mesopause, is the unkindest part of Earth's atmosphere, with temperatures averaging about minus 130 degrees F (minus 90 C). This layer is hard to study. Jets and balloons don't go high enough, and satellites and space shuttles orbit too tall. Scientists do know that meteors burn up in this layer.

The thermosphere outspreads from about 56 miles (90 km) to between 310 and 620 miles (500 and 1,000 km). Temperatures can get up to 2,700 degrees F (1,500 C) at this elevation. The thermosphere is considered part of Earth's atmosphere, but air density is so low that most of this layer is what is generally thought of as external space. This is where the space shuttles hovered and where the International Space Station orbits Earth. This is also the layer where the auroras befall. Charged subdivisions from space strike with atoms and molecules in the thermosphere, thrilling them into higher states of energy. The atoms shed this excess energy by emitting photons of light, which we see as the colourful Aurora Borealis and Aurora Australis.

The exosphere, the uppermost layer, is extremely thin and is where the atmosphere combines into outer space. It is composed of very extensively discrete particles of hydrogen and helium. Spreading from the top of the thermosphere to 6,200 miles (10,000 km) above Earth is the exosphere, where weather satellites are. This layer has very few atmospheric molecules, which can escape into space. Some scientists differ that the exosphere is a part of the atmosphere and in its place classify it actually as a part of outer space. There is no clear upper boundary, as in other layers.

Between each layer of the atmosphere is a frontier. Above the troposphere is the tropopause, above the stratosphere is the stratopause, above the mesosphere is the mesopause, and above the thermosphere is the thermopause. At these "pauses," maximum change between the "spheres" occurs.

2.2 Classification of Natural Resources

Natural resources can be confidential in a variety of ways. First of all, there are renewables and non-renewables. Secondly, one could differentiate definite resources, potential resources and stock resources. Furthermore, there are omnipresent resources and contained resources. In addition, one could say there are biotic and abiotic resources.

A large number of natural resources are exhaustible, *i.e.* they have limited supply on the earth and can be bushed if used arbitrarily. Broadly, the exhaustible resources can be either renewable or non-renewable. Renewability is a very popular topic and many natural resources can be considered as either renewable resources or non-renewable resources. One way to name renewables and non-renewables are by describing them as exhaustible and inexhaustible resources. Inexhaustible resources will not be depleted or run out in the foreseeable future. Exhaustible resources have a finite quantity, and will be exhausted in case of overproduction, as a consequence of overconsumption.

Renewable Resources

Renewables concern elements that can be refilled naturally. Some of these elements are repeatedly available and their amount is not strikingly affected by human consumption. Moreover, their retrieval surpasses human consumption. Examples are sunlight, wind and geothermal energy. Renewable resources are those natural resources such as trees, water, sun and wind that can be refilled at about the same rate at which they are used. Renewable resources, however, can be depleted if not properly managed or conserved. Renewable resources are natural resources that can be renewed or replaced by ecological processes on a relevant time-scale. They include biological resources such as biomass, plants, and animals.

They also include freshwater, fresh air, soil fertility, and elements like carbon and nitrogen. Renewable resources also include inexhaustible solar energy.

Water

Annually, each person on Earth consumes about 650 m3 of freshwater for industrial, agricultural, and domestic purposes. Globally, around 3800 km3 of freshwater is used each year. This accounts for 9 per cent of the total environmental space for natural renewable water resources.

With this consumption rate, by 2050, the world will need approximately 1500 km3 of additional freshwater to supply the expanding population. By that time, about 15 per cent of the total natural renewable water resources will be consumed each year.

Although the global water resources are sufficient for the increasing population, the distribution of these resources and their quality is of concern. Canada has the main resources of freshwater. On the other hand, the water poverty index in 2002 shows that 37 countries out of the 147 assessed ones has a value below 50 (on the scale ranging from 0 to 100 in which lower scores indicate water scarcity and poor water provision). Most of these countries at risk are in the sub-Saharan region, where only 58 per cent of the population has access to safe water.

Land and Soil

The environmental space for arable land reductions not only in quantity but also in excellence. Around 30 million hectares of irrigated land are harshly tainted by salinization and around 80 million hectares more are pretentious by water logging.

The uneven distribution of environmental space when it comes to land resources is also reproduced by other types of land use. In urban areas, housing and living space became rarer as the urban population augmented fast during recent decades. It is projected that in Asia, 40 per cent of the urban population lives in slums; in the meantime, in industrial countries, around 7 per cent of people do not have satisfactory quality of shelters.

Wind

Humans have yoked the power of wind for centuries, from windmills to grind grain to yachts used to travel across the oceans. These days large wind turbines can be used to create electricity.

Clean Source

Using wind power does not generate any air pollution like fossil fuels do.

Renewable

You can't deplete the earth's source of wind. It will continue to blow whether or not we yoke it, and harnessing it does not reduce its supply in any way

Biological Resources

The biosphere is a closed system that contains of biotic and abiotic elements, such as air, water, soil, and organisms. The bio-regions vary by their climate, plants, and animals. These are biomes on land and aquatic life zones for aquatic systems.

Non-Renewable Resources

Non-renewable resources are resources that form tremendously slowly and those that do not unsurprisingly form in the environment. Examples are minerals such as petroleum and uranium. Although it must be noted that uranium reduces in a natural way, since it alters into heavy metals. These metallic minerals can be re-used by recycling, while fossil fuels cannot be recycled. Non-renewable resources are those natural resources that are exhausted more quickly than they can renew. Fossil fuels like oil and natural gas were formed over millions of years. Once mined and used entirely, non-renewable resources are gone forever.

Non-renewable resources are used often throughout the world in construction and manufacturing processes. This is due to several reasons such as, lack of consciousness and suitability of renewable bases by producers and consumers, the technical difficulties and immaturities as well as the cost of technology for renewable-based production, and an unpredictable supply of renewable resources due to natural factors. However, biomass-based production has been projected as a part of the explanation to the increasing worldwide resource depletion, climate change, and environmental pollution caused by the use of non-renewable- based systems.

Non-renewable resources are natural resources that are not certainly refilled once they have been used. Non-renewable resources can be used up entirely or spent to such a grade that they become thriftily unreachable. Fossil fuels, minerals, and metals are examples of non-renewable resources.

Fossil Fuels

Incineration of fossil fuels causes most of the climatic problems today, notably the emissions of CO_2 into the atmosphere. Currently, the worldwide average per capita emissions of CO_2 is 3.85 t yr^{-1}. Meanwhile, the world average environmental space for CO_2 is 1.7 t – the level of 1990 which IPCC suggested to prevent global warming and dramatic climate changes. This means, to achieve a sustainable development, much effort will be needed to reduce the consumption of fossil fuel, especially in developed countries.

Minerals

The Earth crust comprises a diversity of both fuel and non-fuel minerals. Fuel minerals comprise coal and oil which have been conversed above. Non-fuel minerals comprise metals (iron, aluminium, copper, *etc.*), industrial minerals (lime, phosphate rock, gypsum, *etc.*), and building minerals (sand, gravel, rock, *etc.*)

2.2.1 Natural Resources and Associated Problems

Extraction

Extracting, dispensation and using natural resources can reason ecological problems such as: air, land and water pollution; disturbance or obliteration of ecosystems; and a reduction in biodiversity. For example, carbon dioxide – which is shaped from scorching coal, oil and natural gas (fossil fuels) – is a critical greenhouse gas. Greenhouse gases engross and recollect heat from the sun. Examples of greenhouse gases comprise methane, ammonia, sulphur dioxide and certain chlorinated hydrocarbons. Many scientists believe that the build-up of greenhouse gases in the atmosphere can cause global climate change (change in the average global temperature of the atmosphere near the Earth's surface). Over time, this disorder could pose hazards around the world such as flooding, drought and disease. Removing and using natural resources also can disturb relationships within ecosystems.

Over-exploitation

Human beings are reducing the planet's natural resources and values of living will instigate to failure by 2030 unless immediate action is taken. The World Wide Fund for Nature (WWF) cautions that the current overexploitation of natural resources is producing an enormous deficit, as 20 per cent more than can be renewed is disbursed each year and this percentage is mounting progressively.

Thus, if we continue at this rate, we would need 2.5 planets to supply ourselves in 2050, according to the latest *Living Planet* report (2016). This publication displays that the world's population of fish, birds, mammals, amphibians and reptiles weakened by 58 per cent between 1970 and 2012 due to human activities and forecasts that by 2020 this percentage will ascend to 67 per cent.

Need for Judicial Use of Resources

The reserves of established non-renewable resources have to be conserved over time. The consumption of non-renewable resources like fossil energy hauliers or convinced materials such as rare metals calls for a predominantly close link to technology and technological progress. The ingesting of non-renewable resources may only be called maintainable if the chronological supply of the resource does not decline in the future.

2.2.2 Forest Resources

In India, forests form 23 per cent of the total land area. The word 'forest' is resultant from the Latin word '*foris*' means 'outside' (may be the reference was to a village frontier or fence unravelling the village and the forest land).

A forest is a natural, self-sustaining community characterized by vertical structure created by presence of trees. Trees are large, usually single-stemmed, woody plants. Forest can happen in many dissimilar regions under a wide range of circumstances, but all true forests share these physical appearances.

Because a forest is a natural community, no forest is static in time. That is, because forest communities reply to outside influences, most forests are in a state of continuous flux. Depending upon the systems within which forest commu-nities exist, such factors might include rainfall, fire, wind, glaciation, seismic activity, flooding, animal activity, lining, and so on.

At any time, a forest is a group of past replies to outside effects and internal competitive interactions. Consequently, the contemporary status of any forest, certainly of any natural community, reproduces what has gone on before. One of the important renewable natural resource on earth. 1/3rd of the world's land surface is enclosed with forest. Forest covering the earth like a green blanket. Forest not only yield uncountable material goods, but also provide several environmental services.

2.2.2.1 Use and Over-exploitation

A forest is a biotic community mainly of trees, shrubs and other forested vegetation, characteristically with a closed canopy. This invaluable renewable natural resource is beneficial to man in many ways.

The direct benefits from forests are:

(a) *Fuel Wood*: Wood is used as a source of energy for cooking purpose and for keeping warm.

(b) *Timber*: Wood is used for making furniture, tool-handles, railway sleepers, matches, ploughs, bridges, boats *etc.*

(c) *Bamboos*: These are used for matting, flooring, baskets, ropes, rafts, cots *etc.*

(d) *Food*: Fruits, leaves, roots and tubers of plants and meat of forest animals form the food of forest tribes.

(e) *Shelter*: Mosses, ferns, insects, birds, reptiles, mammals and micro-organisms are provided shelter by forests.

(f) *Paper*: Wood and Bamboo pulp are used for manufacturing paper (Newsprint, stationery, packing paper, sanitary paper)

(g) *Rayon*: Bamboo and wood are used in the manufacture of rayon (yarns, artificial silk-fibres)

(h *Forest Products*: Tannins, gums, drugs, spices, insecticides, waxes, honey, horns, musk, ivory, hides *etc.* are all provided by the flora and fauna of forests.

The indirect benefits from forests are:

(a) *Preservation of Soil*: Forests prevent soil erosion by binding the soil with the network of roots of the different plants and reduce the velocity of wind and rain – which are the chief agents causing erosion.

(b) *Soil-improvement*: The fertility of the soil increases due to the humus which is formed by the decay of forest litter.

(c) *Lessening of Atmospheric Pollution*: By using up carbon dioxide and giving off oxygen during the process of photosynthesis, forests reduce pollution and purify the environment.

(d) *Control of Climate*: Transpiration of plants increases the atmospheric humidity which affects rainfall and cools the atmosphere.

(e) *Control of Water Flow*: In the forests, the thick layer of humus acts like a big sponge and soaks rain water preventing run-off, thereby preventing flash-floods. Humus prevents quick evaporation of water, thereby ensuring a perennial supply of water to streams, springs and wells.

Forest Area in India

At the beginning of the 20th century about 30 per cent of land in India was covered with forests. But by the end of the 20th century the forest cover was reduced to 19.4 per cent. This is considerably less than the optimum 33 per cent forest area recommended by the National Forest Policy (1988) for the plains and at least 67 per cent for the hills. Of the existing forests, less than two-third are dense forests, and the rest are open degraded forest. Today, per capita forest area available in India is 0.06 ha, which is much below the average for the world (0.64 ha per person).

Sr.No.	*Particulars*	*Functions*
1	Productive	They provide timer, fuel, charcoal, *beedi*, leaves, was and resins, fruits, tanning, materials, manure leaves, grass, bamboo, gums, lac *etc.*
2	Protective	Forests protect water sheds, catchments of rivers and streams against erosion.
3	Aesthetic	Forests add good appearance, landscaping and a thrilling atmosphere to the locality.
4	Recreational	Forest provides picnic resorts and opportunities for sport like hiking, trekking, wild life watching, bird watching.
5	Scientific	Study of ecological process can be made
6	Ameliorative	Forests improve climate and reduce pollution
7	Hygienic	Forests improve the environment and help in reduction of noise, purify the air and give out oxygen to the atmosphere.
8	Industrial developments	Forest meet the need for raw material for industrial development such as Paper pups, rayon grade pulp, saw milk ply wood, hard board *etc.*

Due to over population, there is a mandate fort medicine, shelter, wood and fuel upsurges. Hence corruption of forest material going on increases. Reason for deficit of forest: In India the smallest area of forest required to maintain good ecological balance is about 33 per cent of total area. But at present it is only about 22 per cent. So over utilisation of forest material occur.

Cause of Over Exploitation

1. Increasing agricultural production.
2. Increasing agricultural activities.
3. Increase in demand of wood resources.

Function of Forests

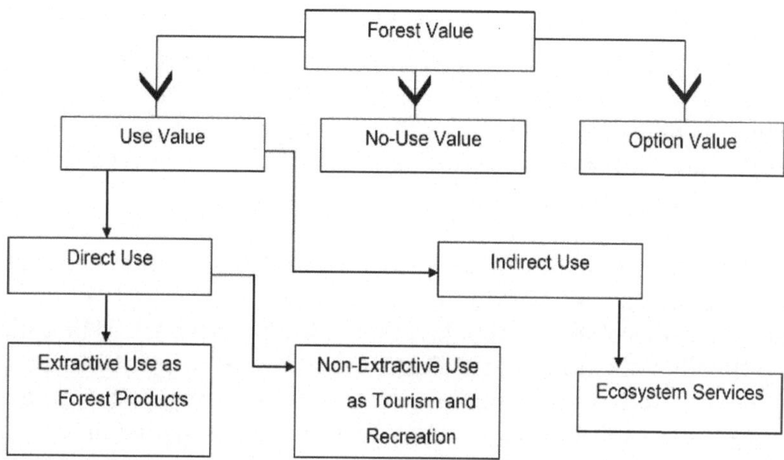

2.2.2.2 Deforestation

Deforestation perhaps created with the use of fire, and approximations are that 40–50 per cent of the Earth's original forest area has been lost. Some of that loss happened before established agriculture began, about 10,000 years ago, but only in latest decades is there dependable information on rates of deforestation. Deforestation is the focussed clearing of woody land. Throughout history and into modern times, forests have been destroyed to make space for agriculture and animal grazing, and to get wood for fuel, manufacturing, and building.

Deforestation has greatly altered landscapes around the world. About 2,000 years ago, 80 per cent of Western Europe was forested; today the figure is 34 percent. In North America, about half of the forests in the eastern part of the continent were cut down from the 1600s to the 1870s for timber and agriculture. China has lost great spreads of its forests over the past 4,000 years and now just over 20 per cent of it is forested. Much of Earth's farmland was once forests.

Deforestation also looms the world's biodiversity. Tropical forests are home to great numbers of animal and plant species. When forests are charted or scorched, it can drive many of those species into destruction. Some scientists say we are already in the middle of a mass-extinction episode.

More immediately, the loss of trees from a forest can leave soil more disposed to to erosion. This causes the remaining plants to become more susceptible to fire as the woodland shifts from being a closed, moist environment to an open, dry one.

Deforestation is one of the most persistent environmental issues that the world is opposite currently. It is the conversion of forested land to non-forested land by humans. Deforestation happens when a land conquered by naturally occurring trees is rehabilitated to provide certain services in response to the human demand. The undiscerning felling of trees has caused in a decrease of 3.16 per cent in the global

Case Studies of Forest Disturbances in Madhya Pradesh

Madhya Pradesh is among one of the states of India which is gifted with rich and diverse forests and comprises the largest forest cover in the country. This is mainly because of the efforts that the state has put in to conserve and harness this invaluable resource through innovative measures like community participation and decentralization (MP). Even after these continuous efforts, there are certain regions within the state where the occurrence of several types of disturbances in the forest has been found. The common disturbances that were being observed during the field survey (2017–2019) include fire, grazing, fuel wood collection, forage removal, litter collection, NTFP collection, lopping, thatch collection, root collection, soil removal, *etc.* Among all these fire has been found as the major recurrent type of disturbance in the forests of Madhya Pradesh. Other major disturbances were grazing followed by fuel wood collection. Since every type of deforestation is not intentional but some which are the results of amalgamation of anthropogenic and natural factors like wildfires and subsequent overgrazing can prevent the growth of young trees and thus eventually degrade the quality and productivity of the forest.

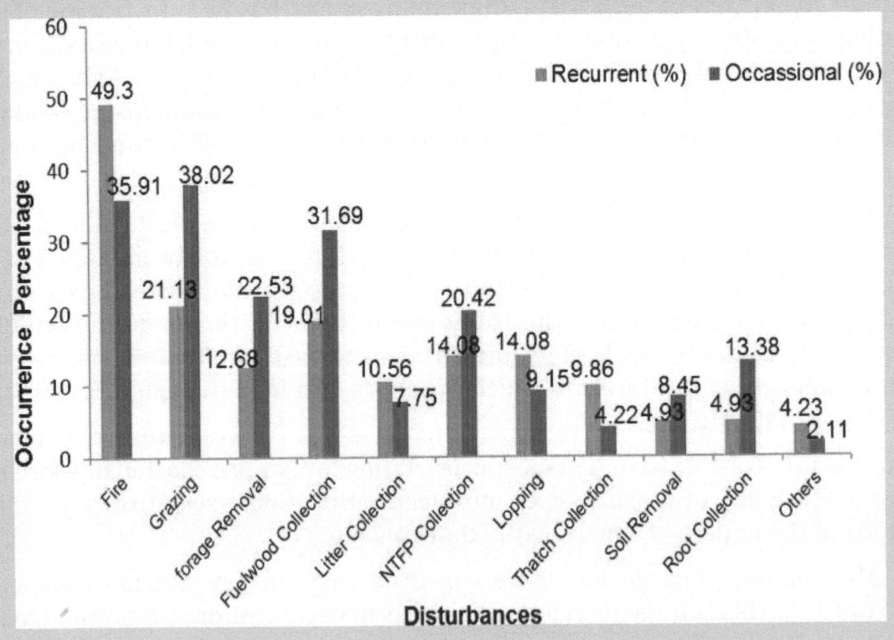

Joint Forest Management in India: A Case Study

Forest management and protection by the local communities is an age-old practice in India which can be traced back to the protective nature of the Bishnoi Community of Rajasthan towards the local forest and animals as the black buck. The idea of community-based forest management arose in an administrative level in the 1970s and 1980s. The declaration of the Government of India in June 1990 marked the establishment of Joint Forest Communities in different India states as per the National Forest Policy of 1988. The Earth Summit of 1992 provided with a clear objective of Sustainable Forest Management to which India responded in a positive way. From an upsurge in the forest cover, non-timber forest product (NTFP) to conservation of native flora and fauna, a whole new realm of forest management strategy by the collaboration of forest departments and local communities aided in decelerating the degradation of natural forest in India. The Participatory Forest Management (PFM) is corresponding to an informal contract in which the local communities are allowed to consume a portion of harvest and NTFP if they protect and conserve it for 5–10 years. In India there is no legal authority of the local community on the forest resource where as in other countries as in Nepal the Community Forest User Groups (CFUG) are registered under their Forest Act, 1993. It is the positive effort of the local communities of India that the area under Joint Forest Management increased from 22,017,583 ha to 2,144,000 ha in March 2006 with 106,482 recognized Joint Forest Management committees countrywide.

forest cover from 1990 to 2015. Although India has seen an addition in the total forest cover of ca. 1 per cent, still there are certain regions in the country that have sought a decrease in the forest cover. The main details credited to the lessening in forest cover are instable cultivation, rotational felling, other biotic pressures, diversion of forest lands for developing activities, *etc.*

Deforestation occurs when a land initially conquered by logically happening tree species is converted to deliver certain services in reply to the human demand. The UN Food and Agricultural Organization (FAO) has distinct deforestation as the conversion of forest to another land use or the long-term decrease of tree awning cover below the 10 per cent threshold.

Preventive Measures to Curtail Deforestation

Deforestation is a main environmental challenge which has been tenacious from the past, and the condition is more deteriorated at present. Therefore, there is a crucial need to focus on the defensive events in order to prevent the distressing effects of deforestation in the near future. In order to alleviate the tricky of deforestation, the approaches should be based on the fundamental causes of the same. Also, the plans for justifying the problematic of deforestation need its actual implementation that needs the acknowledgement of the parts of national, state and municipal governments along with the pro-active role of the public society and private society. The incessant upsurge in the human population particularly in the developing countries has resulted in enhanced pressure on the forests for human settlements and other land use practices. A decrease in the growth rate of human population plays a vital role in reducing the practice of deforestation. Replacements to slash-and-burn agriculture can be adopted as a strategy to mitigate the deforestation by boosting the agricultural and forestry productions for the shifting cultivators. The sustainable alternatives for slash-and-burn cultivation will assist the poor farmers in leading a better life without destroying additional forests. Public policies and laws with greater security, convenience to the minimal inputs required to maintain or enhance food production and a chance to the cultivators to market their products will aid in the sustainability of shifting cultivation systems.

Timber Extraction, Mining and Dams

Extraction is the process of transporting cut timber from the place where it was rising to a point where it can be removed from site. There are a wide range of dissimilar approaches of timber extraction and they vary in their fortes and weaknesses, and in the places where they are most suitable. Unless the site is predominantly vulnerable to damage, machinery will nearly always be used to extract timber.

Mining has always been an environmentally troublesome activity, but modern extractive industries are located in some of the most environmentally subtle forests in the boreal and the tropics. Forests are valuable resources. They provide raw materials for industries, timber for buildings, furniture's and many other uses. The forest ecosystem is conquered by various species of trees.

The principal product that forests supply is wood like timbers. Chief forest products contain of timber small wood and fuel wood. Indian forests produce about 5,000 species of wood, of which about 450 are commercially valued. Hard woods include significant species such as teak, iron-wood, mahogany *etc.*

These woods are used for constructional purposes. Population explosion had its marvellous pressure on request for timber and other wood. Ingesting of wood for industrial uses is more in developed countries than the developing countries. India has plentiful timber. It accounts half of the total biomass shaped by a forest. But demands may become still higher in future.

Dams and their Effects on Forests and Tribal People

More than 40,000 large dams – those that measure more than 15 metres in height – are presently hindering the world's rivers, whose reservoirs cover more than 400,000 square kilometres of land. These reservoirs have inundated millions of hectares of forests – particularly in the tropics – many of which were not even charted and trees were left to slowly rot.

They have also caused in deforestation elsewhere, as growers displaced by the dams have had to clear forests in other areas in order to grow their crops and build their homes. Additionally, dams imply road building, thus allowing admission to beforehand distant areas by loggers and "developers", resulting in further deforestation processes.

However, the dams' effects have included much more than forest loss and the major environmental changes have impacted on local people, at both the dam site and in the entire river basin. Not only are the best agricultural soils flooded by the reservoir, but main changes occur in the environment, where the river's flora and fauna begins to vanish, with strong impacts on people reliant on those resources.

At the similar time, dams imply a number of health hazards, preliminary with diseases introduced by the thousands of workers that are brought in to build the dam (including AIDS, syphilis, tuberculosis, measles and others) and ending with illnesses connected to the reservoir itself (malaria, schistosomiasis, river blindness, *etc.*).

In far too many cases, dam-building has resulted in widespread human rights violations. As most of us would, local peoples have obstinately resisted the obliteration of their homelands and their forced "resettlement." As a result, they have had to face different types of repression, ranging from physical and legal threats to mass murders.

But confrontation, consciousness and unity have grown. Local people have increasingly been able to establish themselves and to establish local, national and international alliances with other worried organizations.

Major examples are the *Narmada Bachao Andolan movement* in India, the Bio Action Group in Chile, the Coalition of Concerned NGOs on Bakun in Malaysia, the People Affected by Dams movement in Brazil among many others. It has now become possible to stop large hydro dams. They are definitely not a symbol of development but one of economic and political power resulting in social and environmental degradation.

2.2.3 Water Resources

Water is one of the plentifully available substance in the nature which men have abusing more than any other resources for the nourishment of life. Water of good quality is required for living organisms. "Water is the elixir of life. Water is obligatory for various domestic purposes like irrigation, shipping, sanitation, power generation, drinking, industries *etc.* About 73 per cent of earth is enclosed with marine and fresh water which is present in rivers, lakes, ponds, glaciers, mountains with ice caps *etc.* The fresh water content is only 2.76 per cent of the total quantity of global water out of which only 2.4 per cent fresh water is available. From this fresh water about 0.3 per cent to 0.5 per cent is available for drinking drive of the total world water. Human population rises day by day and it increases ingesting of quantity of fresh water.

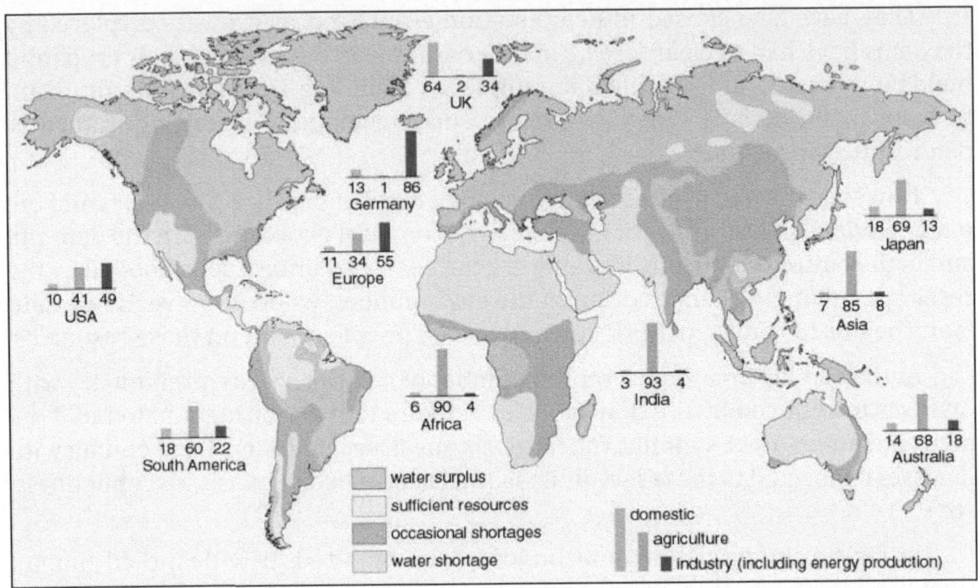

About three-fourth of the earth's surface is busy by oceans which cover about 97.5 per cent of the earth's water in powerfully saline condition. The rest 2.5 per cent is 9 fresh water, and all of this is not available for straight human use. Most of the fresh water is frozen as polar or glacial ice (1.97 per cent). Residual fresh water occurs as groundwater (0.5 per cent) and water in lakes and rivers (0.02 per cent), soil (0.01 per cent) and atmosphere (0.0001 per cent). Thus, only a small fraction of fresh water is available for human consumption. More so the circulation of fresh water is physically uneven varying greatly from country to country and even within a country from one region to another. About 84 per cent of the total global evaporation occurs from ocean surface and 16 per cent from land surface. At any given time, the amount of moisture in the air is only enough to meet a total rainfall requirement of 10 days. Thus, there is very fast movement of water from ocean and land into the atmosphere and an average residence time of water in the air is only about 10 days. About 77 per cent of the total rainfall on earth is received on the sea surface (as against 84 per cent evaporation from this segment) and 23 per cent on land (16 per cent share of total evaporation to the atmosphere). There is a net gain of 7 per cent rainfall water on land and this excess is returned to the oceans by surface runoff through rivers and sub-surface water flows. On global basis, the hydrological cycle is perfectly balanced as the total annual evaporation matches with annual precipitation.

Use and Over-utilization

Water can appear in three states; it is one of the very few materials to be found obviously in all three states on earth. Water takes many different forms on Earth: water vapour and clouds in the sky; seawater and infrequently icebergs in the ocean; in facts, humans and other animals have developed senses which are, to a degree, able to assess the movability of water, avoiding water that is too salty or putrid. Humans also tend to favour cold water to lukewarm; cold water is likely to cover fewer microbes. The test advertised in spring water or mineral water derives from the minerals dissolved in it, as pure H_2O is tasteless. As such, purity in spring and mineral water refers to purity from toxins, pollutants, and microbes.

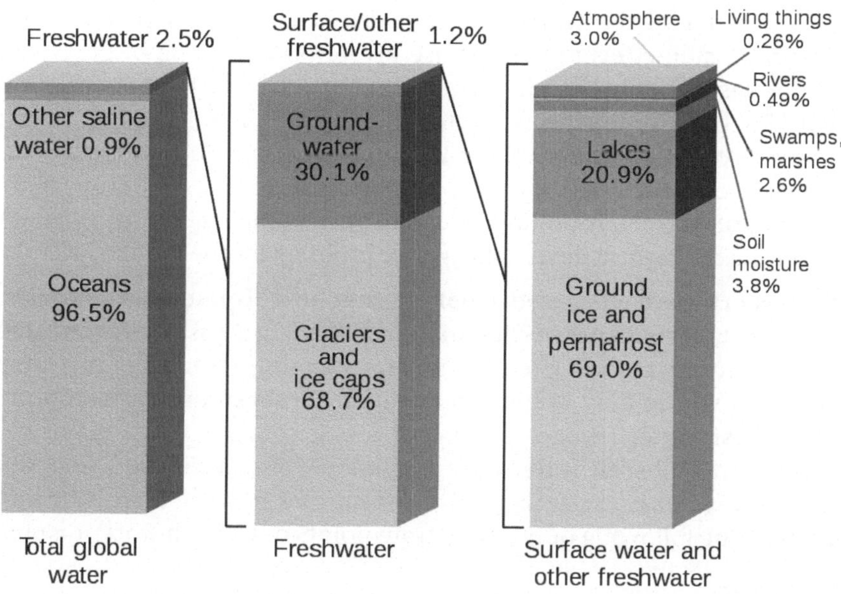

The Pie Chart above Shows that how much of Earth's Water is available for our Uses and in what Forms does it Exist.

In 2000, 5 per cent of the global population was projected to be water scarce, meaning less than 1000 m^3 of fresh water was obtainable per person per year; by 2025 this is projected to affect 31 per cent of the population. Many of these people are in countries with high population growth rates and their water problems are cumulative swiftly.

The water-scarce countries also have another water problem apart from scarcity–the water is often unsafe to drink. Every year, over 2 million people die from water-borne diseases. So having sufficient water is not enough; it must be uncontaminated water. Access to safe water varies with region, being lowest in Africa. The access in cities, about 94 per cent, is much higher than in rural areas, where it is only about 71 per cent. Globally, 18 per cent of the population lack access to safe drinking water – and that is over 1 billion people. Lack of safe water is due both to lack of investment in water supply systems and to inadequate maintenance of the systems. About a half of the water in supply systems in the developing world is lost to leakage, illegal abstractions and vandalism. In some countries, water is extremely subsidised for those connected to the system, while poorer people not associated rely on unsafe sources or expensive private sellers. Globally, 2.4 billion people lack access to satisfactory sanitation. In developing countries over 90 per cent of sewage is dumped unprocessed into waters where the water supplies can be polluted. Recognising these problems, the World Summit on Sustainable Development in Johannesburg in 2002 set water targets to address these matters

Even in parts of the world where there is little or no scarcity of water, environmental side-effects of water use are often becoming of great concern. This comprises not only the obvious pollution, particularly of sewage, nitrates and high TDS, but also the devastation of natural wetland homes by diversion of water elsewhere, falling water tables due to over-extraction, and the drowning of land by massive reservoirs. About half of the rivers and lakes in Europe and North America are still extremely polluted, despite developments in recent years. Water quality may be the biggest developing water problem for the industrialised world.

Water shortages have caused major international disputes in many parts of the world. Water management is predominantly difficult in areas where the catchment of a river crosses many countries. Egypt, for example, gets most of its water from the River Nile, whose flow originates mainly from seven upstream countries. In the Middle East, water resources are of strategic concern, and a major cause of political conflict. Full-scale water wars are unlikely, but tension between countries competing for water is escalating to the extent that in some areas war has been endangered. The following quotes illustrate points of view on water disputes:

Surface Water

India's average annual surface run-off made by rainfall and snowmelt is projected to be about 1869 billion cubic meter (BCM). However, it is projected that only about 690 BCM or 37 per cent of the surface water resources can actually be mobilised. This is because (i) over 90 per cent of the annual flow of the Himalayas Rivers occur over a four month period and (ii) potential to capture such resources is complicated by limited suitable storage reservoir sites.

Rainfall

The average annual rainfall in India is about 1170 mm. This is substantial difference in rain both provisionally and spatially. Most rain falls in the monsoon season (June-September), necessitating the creation of large storages for all-out consumption of the surface run-off. Within any given year, it is likely to have both situations of drought and of floods in the same region. Regional varieties are also extreme, ranging from a low value of 100 mm in Western Rajasthan to over 11,000 mm in Meghalaya in North-Eastern India. Possible changes in rainfall patterns in the coming decade, global warming and climate change and other predicted or observed long-term trends on water obtainability could affect India's water resources.

Groundwater

India's rechargeable annual groundwater potential has been measured at around 431 BCM in collective terms. On an all India basis it is projected that about 30 per cent of the groundwater possible has been appointed for irrigation and domestic use. The regional situation is very much dissimilar and large parts of India have already subjugated almost all of their dynamic recharge. Haryana and Punjab have exploited about 94 per cent of their groundwater resources. Areas with reducing groundwater tables are found in Rajasthan, Gujarat, most of western Uttar Pradesh and in all of the Deccan states. Occurrence of water availability at about 1000 cubic meters per capita per annum is a usually threshold for water indicating scarcity (UNDP). Investment to capture supplementary surface run-off will become increasingly more difficult and expensive in the future. Over time, both for surface and groundwater resources, a situation where resources were substantially underutilised and where substantial development potential existed, has malformed in little more than a generation to a situation of water scarcity and partial development options.

The water availability index comprises surface water only, yet groundwater is an significant constituent of water availability that factors significantly in the Indian economy. Groundwater is an significant source of drinking water and food security for India, supplying about 80 per cent of water for domestic use in rural areas and about 50 per cent of water for urban and industrial uses. Groundwater provides a very important percentage of water supply for irrigation, and contributed significantly to India's agricultural and overall economic development. With more than 17 million energised wells nationwide, groundwater now supplies more than 50 per cent of irrigated area.

Floods

Floods are caused by extremely intense or tremendously long rainfall events, by intense snowmelts, or a mixture of both. Floods are usually followed by inundation, when water spills beyond the river channel. Examples of recent plain floods are: the Storm Surge Flood in 1991, which killed 140 000 people, and calamitous floods in 1998 and 2007 in Bangladesh; the Midwest Flood of 1993 in the United States; two floods of a 100-yr flood size on the Rhine in 1993 and 1995; the flood in 1996 in South Korea; floods in 1996 and 1998 in China; the devastating flood in 1997 on the Oder in Poland and Germany; a flood in 2000 in Mozambique; a flood in August 2002 on the Elbe in Germany and the Czech Republic; and recent floods in 2005 and 2006 on the Danube. Based on data from 1950 to 1998, Berz (2001) evaluated the number of great flood disasters worldwide, which he defines as events in which international or inter-regional assistance is necessary. He concluded that their number has increased considerably over recent decades: six cases in the 1950s, seven in the 1960s, eight in the 1970s, 18 in the 1980s, and 26 in the 1990s. The same tendency has been observed in Europe. Between 1975 and 2001, 238 flood events were recorded in Europe, and the annual number of flood events has clearly increased (European Environment Agency (EEA) 2004).

Floods are recurring phenomena in India. Due to different climatic and rainfall designs in different regions, it has been the experience that, while some parts are suffering devastating floods, another part is suffering drought at the similar time. With the increase in population and development activity, there has been a tendency to subjugate the floodplains, which has resulted in damage of a more serious nature over the years. Often, because of the varying rainfall distribution, areas which are not conventionally prone to floods also experience severe inundation. Thus, floods are the single most recurrent disaster faced by the country.

Flooding is caused by the insufficient capacity within the banks of the rivers to contain the high flows brought down from the higher catchments due to heavy rainfall. Flooding is emphasised by erosion and silting of the river beds, resulting in a decrease of the carrying capacity of river channels; earthquakes and landslides leading to changes in river courses and obstructions to flow; harmonisation of floods in the main and tributary rivers; retardation due to tidal effects; infringement of floodplains; and haphazard and unintended growth of urban areas. Some parts of the country, mainly coastal areas of Andhra Pradesh, Odisha, Tamil Nadu and West Bengal, experience cyclones, which are often escorted by heavy rainfall leading to flooding.

Floods can be divided into 4 types: flash floods, due to heavy rainfall in a short period, *i.e.* a tropical cyclone causing very high rainfall intensity, over-bank floods, due to incessant heavy rainfall in the watershed area for a long period overcapacity the water course-the level of violence depends on the river size, quantity of rain and duration (normally, hydrological floods prolong more than flash floods); flood tides, which occur at the river mouth connected to the sea; and storm surges, which happen along the beach because of tropical cyclones in the sea.

Drought

A drought is an arid period with less or even no rainfall happening than is the usual case. It occurs whenever the groundwater level and river flow decreases and a water shortage for human, industry, service and agricultural uses occurs. It can be classified into 3 types: meteorological drought, evaporation from the soil and plants is more than the annual rainfall; hydrological drought, characterized by a meteorological phenomenal change with prolonged period with only a small quantity of average rainfall intensity-groundwater and river water levels remain low causing a lack of water during the dry season; and agriculture drought, the situation of water shortage for agriculture which may occasionally occur in the early rainy season and widely causes a destruction of agriculture.

Drought is a recurrent natural feature which results from the lack of precipitation over an extended period of time (*e.g.* a season or several years). It is a temporary deviation of rainfall and moisture conditions from the mean, thus differing from aridity and seasonal aridity. It is a creeping phenomenon and, unlike other hazards, can last for months and, in severe cases, years. Drought affects almost all climatic regions and more than one-half of the Earth is susceptible to droughts every year. Regions with advanced variability of rainfall and runoff are more vulnerable. Contingent on the likely impact, the phenomenon of drought can be categorized in several ways, such as meteorological, hydrological and agricultural. The three-dimensional extent of drought is much greater than for any other hazard and is not limited to basin or political boundaries. Long- lasting droughts lead to degradation of soil, plant and animal habitats and social disruption.

During a severe drought in 1917/1918, the Jhelum River in Kashmir dried up completely. Out of the 328 million ha geographical area of India, 107 million ha (nearly one-third), spread over administrative districts in several states, is affected by drought. It comprises about 39 per cent of cultivable land and about 29 per cent of our population. India has experienced 22 major droughts during the last 131 years. The 2002 drought, one of the severest in India, affected 56 per cent of its geographical area, the livelihoods of 300 million people and 150 million cattle in 18 states. The Government of India had to provide relief amounting to about US$ 4500 million.

Examples of some drought events are: severe droughts in the African Sahel since the late 1960s; a devastating drought in southern Africa from 1991 to 1992 that threatened some 18 million people with famine; the 1999–2000 drought in Ethiopia during which more than 10 million people suffered serious food shortages; droughts in northwest Europe from 1988 to 1992; several droughts in the UK in the early 1990s; the 1992 drought in Brandenburg, Germany, and in Poland; the 2000–2002 drought in the Amudarya; and the 1988 and 1999 droughts in the United States (Kundzewicz and Kaczmarek 2000, Demuth 2005). Also, the heat wave of 2003 in Europe affected France, Spain, Portugal, Germany, Italy, and other central and eastern European countries. The temperatures in the summer of 2003 in Germany were on average 3.4°C higher than they were from 1961 to 1990, corresponding to a recurrence period of 450 years (Demuth, 2005).

Sustainable Water Management

Main approaches for conservation of water are:

1. Reducing agricultural water expenditure by growing competence of irrigation. By the traditional method of irrigation plants absorb less than 50 per cent of the water applied to the soil, the rest is lost.
2. Reducing water wastage in industry by recycling the used water.
3. Reducing domestic water wastage by building wastewater treatment plants and recycling the treated water.
4. Rain water harvesting by employing practices to store rain water and recharge groundwater.
5. Afforestation and protection of watersheds to improve water economy.

Some important water management approaches to provide a sustainable supply of high quality water are:

1. Construction of dams and reservoirs to ensure year round supply of water and, in addition, controlling flood and producing electricity.
2. Desalinization of seawater and saline groundwater, making it fit for drinking and other purposes. Diversion of water bodies (*e.g.* through canal) to upsurge the natural supply of water to a particular area.

Any single substitute would not, by itself, be expected to provide the grade of efficiency wanted to deal with the harshness of floods and droughts. The potential actions are classified in terms of the joint consideration of structural and non-structural management of possible efficiency based on IWRM and CDRM as follows.

Structural Measures

The strategy components that can be taken to contribute to flood and drought management with structural events are namely: - provide local drainage, if essential using dikes to exclude overflow, - provision of local flood storage, natural depression and drainage capacity, - land use regulation, - augmenting the effectiveness of existing reservoirs and development of a new reservoir operating rule that gives the completest practical recognition of the objective, - improvement in channel drainage capacities, dikes and river cut off, - augmenting the flood water detention area, - drought prevention using supply management, *e.g.* new water resources large upland storage, medium/small-sized storage, coastal storage, basin transfer, *etc.*

Non-structural Measures

Options for non-structural management are as follows: - vital data collection, flood and drought risk mapping to identify areas where flood and drought resistance design standard, land use restriction, flood and drought forecasting and warning programs should be applied, - adoption of flood and drought warning and emergency relief programs together with the application of flood tolerant design and construction standards, - drought prevention using water management, *e.g.* conjunctive use, recycling of drainage water from domestic use for agriculture, efficiency savings, *etc.*, - drought prevention using demand management, *e.g.* application of water rights, fee/land charges, crop diversification/subsidies.

Dams and Reservoirs

A dam is a barricade that impounds water or streams. Dams usually serve the main purpose of retaining water, while other structures such as flood gates known as dikes are used to manage or prevent water flow into specific land regions. Hydro power and hydroelectricity are often used in consumption with dams to provide electricity for millions of consumers. It can also be used to collect water or for storage of water which can be evenly distributed between different locations. Dams and reservoirs have played a major role in moderating flood-peaks and providing safeguards for drought mitigation.

Masonry Dams

Arch dam: In the arch dam, stability is obtained by a mixture of arch and gravity action. If the upstream face is vertical the entire weight of the dam must be carried to the foundation by gravity while the distribution of the normal hydrostatic pressure between vertical cantilever and arch action will depend upon the stiffness of the dam in a vertical and horizontal direction. For this type of dam, firm dependable supports at the abatements are more important. The most desirable place for an arch dam is a narrow canyon with steep side walls composed of sound rock. The safety of an arc dam is dependent on the strength of the side wall abatements as well as the character of the rock should be carefully inspected.

Gravity Dams

In a gravity dam, stability is tenable by making it of such a size and shape that it will resist overturning, sliding and crushing at the toe. For this type of dam, impermeable foundations with high bearing strength are essential. Gravity dams can prove to be a better alternative to other types of dams. Since the fear of flood is a strong motivator in many regions gravity dams are being built in some examples where an arch dam would have been more economical.

Embankment Dams

These dams are made from compacted earth and have two main types; rock fill and earth fill dams.

Rock-fill dams: These dams are embankments of compacted free-draining granular earth with an impervious zone. The earth utilized often comprises a large percentage of large particles hence the term rock-fill. The impermeable zone may be on the upstream face and made up of masonry, concrete, plastic membrane, steel sheet piles, timber or other material. Rock fill dams are resistant to damage from earthquakes. However insufficient quality control during construction can lead to poor compaction and sand in the embankment which can lead to liquefaction of the rock-fill during an earthquake.

Earthen Dams (Earth-fill dams): Earth fill dams, also called earthen, rolled-earth or simply earth dams, are constructed as a simple embankment of well compressed earth. A homogenous rolled-earth dam is entirely built of one type of material but may contain a drain layer to collect seep water.

Dams Problems

- ☆ Disintegration and physical alteration of rivers.
- ☆ Serious influences on riverine ecosystems.
- ☆ Social significances of large dams due to displacement of people.
- ☆ Water logging and salinization of adjacent lands.

- Extricating animal populations, damaging their habitat and cutting off their migration routes.
- Fishing and travel by boat disrupted.

2.2.4 Mineral Resources

Almost all Earth materials are used by humans for something. We require metals for making machines, sands and gravels for making roads and buildings, sand for making computer chips, limestone and gypsum for making concrete, clays for making ceramics, gold, silver, copper and aluminium for making electric circuits, and diamonds and corundum (sapphire, ruby, emerald) for abrasives and jewellery. A mineral is a pure inorganic substance that occurs naturally in the earth's crust. More than two-thousand minerals have been identified and most of these are inorganic, which are formed by the various combination of elements. However, a small proportion of the earth's crust contains organic materials, consist of single elements such as gold, silver, diamond, and sulphur.

A mineral deposit is a attentiveness of naturally occurring solid, liquid, or gaseous material, in or on the Earth's crust in such form and amount that its extraction and its conversion into useful materials or items are profitable now or may be so in the future. Mineral resources are non-renewable and include metals (*e.g.* iron, copper, and aluminium), and non-metals (*e.g.* salt, gypsum, clay, sand, phosphates).

Minerals are valued natural resources being finite and non-renewable. They establish the vital raw materials for many basic industries and are a major resource for development. Management of mineral resources has, therefore, to be closely combined with the overall strategy of development; and exploitation of minerals is to be guided by long-term national goals and perspectives.

Use and Exploitation

The use of minerals depends upon its deposits. Some countries are rich in mineral deposits, while others have no deposits. The greatest use of minerals be contingent on its properties. For instance, Aluminium is light, strong and durable in nature, so it is used for aircraft, shipping, and car industries.

Minerals are used in almost all industries. Gold, silver, and platinum metal are used in the jewellery industry. Copper is used in the coin industry and for making pipes and wire. Silicon obtained from quartz is used in the computer industry.

In today's context of increasingly scarce mineral resources and growing demand for raw materials, Europe and France are setting up action plans to identify new resources, secure supplies and develop materials recycling.

Some more common metals are also of strategic importance. Copper, for instance, is used in a great many different products. More copper is processed than any other metal: it is a major component in electrical and electronic goods and used on a very large scale in construction and civil engineering.

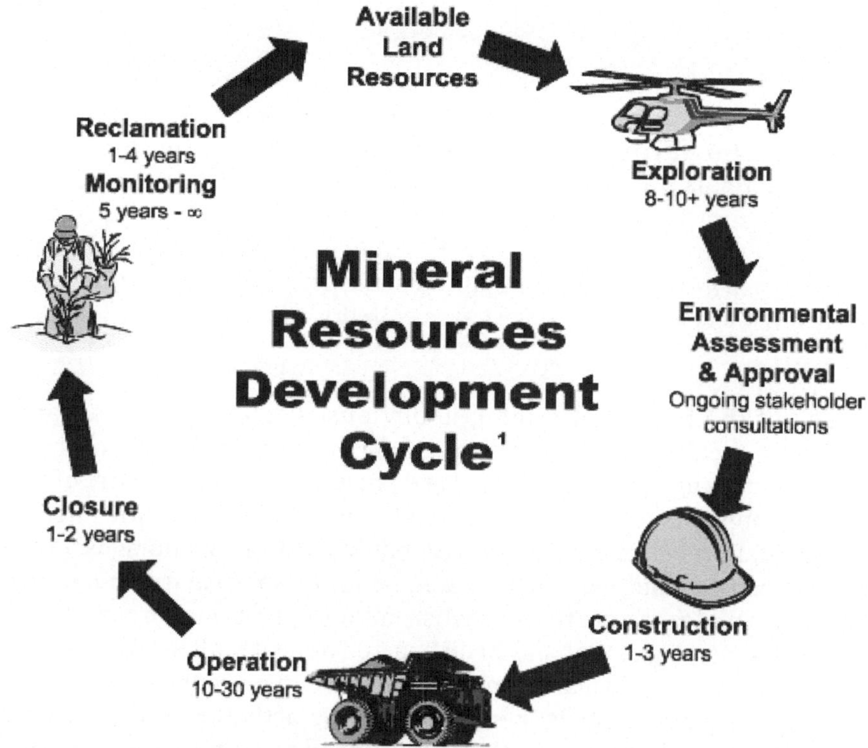

Because the processes that form ores operate on geologic time scales, the most economic mineral resources are essentially non-renewable. New deposits cannot be generated in human time-scales. But, as mentioned previously, as the reserves of materials become depleted it is possible to find other sources that are more costly to exploit. Furthermore, mineral resources are not evenly distributed. Some countries are mineral-rich; some are mineral-poor. This is a particular issue for strategic mineral resources. These strategic metals are those for which economical source do not exist in the U.S., must be imported from other potentially non-friendly nations, but are needed for highly specialized applications such as national security, defence, or aerospace applications. These metals include, Manganese, Cobalt, Platinum, and Chromium, all of which are stockpiled by the U.S. government in case supplies are cut off.

Environmental Effects of Extraction and Minerals' Utilization

Extraction and processing has great environmental impacts in terms of such things as air quality, surface water quality, ground-water quality, soils, vegetation, and aesthetics. Acid mine drainage is one example, Sulphide minerals newly exposed to Oxygen and water near the surface create sulphuric acid. Rainwater falling on the mine tailings becomes acidified and can create toxic conditions in the runoff. This can rally potentially dangerous heavy metals and kill organisms in the streams demanding the tailings.

Case Study

Aravilli mountains which covers about 10 per cent of geographical area is rich source of minerals wealth. This mountain range play important role in control of climate and act as mini water shed. On the request of environmentalist, Honourable Supreme Court has passed the order to stop these mines in Rajasthan. Marble mining near Rajsamant Lake has led to drying up of lake. Marble mining was stopped on December 2002.Recently, mining in Goa has attained the attention of the press and media and ultimately government has to take the decision to stop this mining.

Much of the impact of mining is obvious. The disruption of land then suitable for agricultural, urban or recreational use; the worsening of the instant environment through noise and airborne dust; and the creation of ore of the most unsafe environments for workers and potentially dangerous for the public are all environmental problems associated with mining. However, mining is a relatively short-term activity, and much can be done both to limit environmental damage during mining and to restore the land when mining operations are complete. Today, in many countries, legislation has been enacted at nearly all levels to ensure that extreme restrictions could make mining completely uneconomical. Inappropriately, the lack of adequate controls over some mining activities in the past has left numerous scars on the surface of the earth and led to confrontation among many members of the public toward new mining activities in their areas.

Luckily, many underground mines leave little evidence of their presence, even after mining operations have ceased. They are usually filled by infiltrating groundwater over time, but the rocks are usually strong enough to hold in spite of uncontrolled mine openings and passageways. Occasionally the old mines can be put to very good use. Old underground mines can be used as storage areas for grains, seeds, burial of nuclear wastes, and truck parking.

When an open pit mine closes, a large hole remains with no willingly available waste rock to fill it. The pit slopes are often too steep for estate. If the water table is high enough, the bottom of the pit may flood, creating an artificial lake. Therefore, very large open pit mines are difficult, if not impossible, to reclaim. Lesser open pits may be filled with waste rock. In some places, surface mines can be domestic to form small lakes and wetlands mat support fish, birds and other wildlife. Underground mines do not lead to such drastic disruptions of the surface as open pit and strip mining, but a new hazard known as subsidence can be encountered. Subsidence under towns and roads can leave homes derelict and transportation severely disrupted.

In addition to the influence that mining activities may have on the, landscape, the environment may be disrupted over a wider area by changes Ýn the delivery and chemistry of surface waters or groundwater. Water passing through the mines or dumps becomes acidified, later finding its way into rivers, streams or the local groundwater system. Many streams can be affected by uncontrolled mine works.

Well Drilling and Production

Drilling wells are used to explore and produce oil, gas, brine, geothermal fluids. Blow out and fire hazards can create severe pollution. Oil and brine spillage and seepage must be carefully controlled.

Effects of Used Resources on the Environment

The burning of fossil fuels in power stations, homes and automobile engines results in gases, particles and excess heat being emitted into the environment. The use of nuclear fuels generates toxic radioactive waste products requiring special disposal. Oil refining and metal/mineral production also generate wastes and pollutants.

Acid Pollution

The most important pollutant of the hydrosphere is acid in the form of acid rain and acid mine drainage. Acid mine drainage formed by dissolution of pyrite dissolves more pyrite, thus accentuating the effects.

Scorching Fossil Fuels

The burning of fossil fuels in automobiles, power plants and heating systems creates air pollution. The burning of solid waste and smelting also generates air pollution.

Removal of Nuclear Wastes

The mining and processing of uranium ores, the manufacture of nuclear fuels, the use of fuels in nuclear power stations, and nuclear weapons manufacture all generate waste products requiring disposal. The long-term disposal of radioactive wastes is still an unresolved problem.

Food Resources

It is projected that one or more parts of about 30,000 plant species can be bothered by people; however, interestingly only 15 plant and 8 animal species supply as much as 90 per cent of food eaten by us. In fact, only four crops (namely wheat, rice, corn and potato) together account for more of the world's total food production than all other crops combined. Two-third of the world human population endures on a diet of grains, mainly rice, wheat and corn. Perhaps this large fraction of population cannot afford meat products. With increase in income due to development, the grain ingesting generally increases because people consume more of meat, eggs and milk produced by grain eating livestock.

Human body needs food for various purposes. Food consumed by humans are of dissimilar types and a balanced diet is needed for all practical purposes, vitamins, proteins carbohydrates and minerals are primarily obtained from cereals, fruits, vegetables, pulses and spices, milk, butter, meat and eggs all of which obtained from different types of plants and animals. These are our main food resource. A large number of items are spent by human either in their natural states or after proper processing and cooking.

The food consumed by human is prejudiced by wide range of cultural and individual differences, mainly due to ecological as well as personal reasons. The source of much of the food consumed by man is terrestrial agricultural, which represents the most operated of all the non-urban ecosystems. There are two main types of agriculture (1) Crop agriculture in which plant production is harvested for use by man and (2) Animal agricultural where a crop from highly manipulated ecosystem is fed to domesticated animals.

It has been lead that even if the rapid development of conventional agriculture is continued protein deficiency will continue to exist. Consequently a search is in progress for supplementary sources of plant and animal proteins. Amongst these animal protein has been considered to be of more importance because of the similar amino acid composition of man. Besides, animal's flesh has the advantage that it more easily assailable. In case the animals are to be evaded, a food source which will yield plant protein in a digestible form or from which the majority of cellulose has been detached, fungi can be preferred.

These can be easily integrated and contains a good amount of protein. Other sources are fish and beef. The advantage of using mushroom is that it does not absorb much human or fossil energy in manufacture can be grown self-sufficiently of environment factors.

Young sea-birds, reptiles, amphibians, insects, molluscs, rodents and many other animals can be utilized as food. Fresh water and brackish water fish are other sources of animal proteins. It has been found that marine and fresh water fisheries provide best sources of animal protein.

Types of Food Supply

1. Crop Plants

Mostly produce grains about 76 per cent of the world's food.

E.g. rice, wheat and maize

2. Range Lands

It produces 17 per cent of world's food from trees and grazing animals.

E.g. fruits, milk and meat

3. Ocean

Fisheries – 7 per cent of world's food

World Food Problem

The food problem has been high on the international agenda for the past two years. The global food problem consists of the lack of food provision for the Earth's population. It establishes itself primarily in the poorest countries of the Third World and is currently aggravating as their populations grow. The total number of people suffering from a lack of food is over one billion people worldwide. The fact is that population growth out-paces agricultural production and the development of agricultural technologies. However, according to some approximations even the existing level of agrarian and cattle breeding methods would allow us to feed over ten billion people provided these methods use available resources rationally, and there is a just delivery system in place for the resulting products.

In the Earth's surface 79 per cent water out of total area. 21 per cent land (forest, desert, mountain and barren land). Less per cent cultivated land, at the same time population explosion is high therefore world food problem arises. Environmental degradation like soil erosion, water logging, water pollution, salinity affect agricultural land. Urbanisation affects agricultural land. Hence production of rice, wheat, corn and other vegetable is difficult.

The global food problem has the following contradictions:

Food production in terms of output, quality and efficiency is unevenly spread over different world regions;

On the whole, in terms of scale, food production worldwide formally meets the current world's requirements;

Food production is lacking in places where food shortages are felt the most;

There is no shortage of land worldwide theoretically suitable for food production to feed even between 14 and 33 billion people; at same time, fertile soil is being destroyed in cultivated land and industrial areas;

Over 1 billion people suffer from starvation and undernourishment worldwide and approximately 1.6 billion people from overeating and obesity.

Impact of Over-grazing and Modern Agriculture

Overgrazing represents an environmental hazard whereby wildlife or livestock excessively feeds on pasture. It is also the practice of grazing livestock on vegetation before it has recovered from a former grazing state, also known as intensive grazing. Otherwise stated, overgrazing takes place when vegetation or pasture is repeatedly removed from land and it is not given enough time to continue growing.

Soil Erosion

The continued crushing of frequent animals in an average forage land will act to accelerate the death of plants and vegetation cover. This is because the animals will graze even on the slightest shoots of new growth. The top fertile soil of the farmland is removed due to the excessive water supply. This leads to the loss of nutrient rich soil that disadvantaged the productivity. It also causes the global warming because the silt of water bodies induces the release of soil carbon from the particulate organic material.

Land Degradation

The acts of compaction and erosion as a result of overgrazing can cause marvellous land degradation. In drier areas, the experience is even worse as a large percentage of pasture and land cover is destroyed, causative to relentless progression of desertification. In fact, in some areas overgrazing has led to complete desertification. Overgrazing joint with overstocking has the most damaging consequences to the world's natural environment.

Loss of Valuable Species

The natural composition of plant population and their renewal capacity is meaningfully affected by overgrazing. The original pasture crops are composed of high quality pastures and herbs with great nutritional value. When animals intensively graze on such pastures, even the root stocks which cover the reserve food or renewal capacity get ruined. Once ruined, some other more adaptable species such as weeds and unpalatable plants take up their place. These subordinate plant species have less nutritional values and because they are highly adaptive, they substitute the native species thereby causing the loss of valuable plant species.

Food Shortage/Famine

As earlier stated, overgrazing is a main contributor to desertification because it changes arable or pasture land into unproductive piece land. The subsequent soil is thus not suited for growing food since it loses its essential nutrients. The loss of land productivity directly results in the loss of food available for consumption. This intensifies food supply reduction and if population growth is still registered, it causes starvation and economic challenges.

Impact of Modern Agriculture

Modern agriculture is an evolving approach to agricultural inventions and farming practices that help farmers increase efficiency and reduce the number of natural resources like water, land, and energy essential to meet the world's food, fuel, and fibre needs. The agribusiness, intensive farming, organic farming, and maintainable agriculture are other names of modern agriculture.

Modern agriculture makes use of hybrid seeds of single crop variety, technologically advanced equipment, fertilizers, pesticides and water to produce great quantities of single crop.

Problems using Fertilizers

Micro-nutrient imbalance: Chemical fertilizers have nitrogen, phosphorus and potassium (N, P and K) which are essential macro-nutrients. Excessive use of fertilizers cause micro-nutrient imbalance. For example, excessive fertilizer use in Punjab and Haryana has caused deficiency of the micro-nutrient Zinc in the soils, which is affecting productivity of the soil.

Nitrate pollution: Excess Nitrogenous fertilizers applied in fields leach deep into the soil contaminating the groundwater. If the concentration of nitrate in drinking water exceeds 25 mg/L it leads to a fatal condition in new-born babies. This condition is termed "Blue Baby Syndrome"

Eutrophication: The application of excess fertilizers in fields leads to wash off of the nutrient loaded water into nearby lakes causing over-nourishment. This is called "Eutrophication". Eutrophication causes the lakes to be attacked by "algal blooms". Algal blooms use nutrients rapidly and grow fast. Their life is short, they die and pollute water thereby affecting aquatic life in the lake.

Problems in using Pesticides

In order to improve crop yield, pesticides are used indiscriminately in agriculture. Pesticides are of two types:

1. First generation pesticides that use Sulphur, Arsenic, Lead or Mercury to kill pests
2. Second generation pesticides such as Dichloro Diphenyl Trichloroethane (DDT) used to kill pests. These pesticides are organic in nature. Although these pesticides protect our crops from severe losses due to pests, they have several side-effects as listed below:
    - ☆ Death of non-target organisms: Several insecticides kill not only the target species but also several beneficial not target organisms
    - ☆ Pesticide resistance: Some pests that survive the pesticide generate highly resistant generations that are immune to all kinds of pesticides. These pests are called "superpests"
    - ☆ Bio-magnification: Most pesticides are non-biodegradable and accumulis ate in the food chain. This is called bio-accumulation or bio-magnification. These pesticides in a bio-magnified form are harmful to human beings.
    - ☆ Risk of cancer: Pesticide enhances the risk of cancer in two ways (i) It acts as a carcinogen and (ii) It indirectly suppresses the immune system.

Water Logging

Over irrigation of croplands by farmers for good growth of their crop usually leads to water logging. Inadequate drainage caused excess water to accrue underground and slowly forms a incessant column with the water table. Under water-logged conditions, pore-spaces in the soil get fully soaked with water and the soil- air gets depleted. The water table increases while the roots of plants do not get passable air for respiration, Mechanical strength of the soil declines, the crop plants get lodged and crop yield falls. In Punjab and Haryana, extensive areas have become water-logged due to passable canal water supply or tube-well water.

Case Study

Salinity and water logging in Punjab, Haryana and Rajasthan:

The first alarming report of salt-affected wasteland formation due to irrigation practices came from Haryana in 1858.

Several villages in Panipat, and Delhi lying in Western Yamuna Canal were suffering from salinity problems.

The floods of 1947, 1950, 1952, 1954-55 in Punjab resulted in aggravated water logging with serious drainage problems.

Introduction to canal irrigation in 3.3 m ha in Haryana resulted in raise in water table followed by water-logging and salinity in many irrigated areas as a result of fall in crop productivity.

Rajasthan too has suffered badly in this regard following the biggest irrigation project "Indira Gandhi Canal Project".

Preventing excessive irrigation, sub-surface drainage technology and bio-drainage with trees like Eucalyptus are some of the remedial events to prevent water-logging.

Causes of Water Logging

- ☆ Excessive water supply to the croplands
- ☆ Heavy rain
- ☆ Poor drainage

Measures to Prevent Water Logging

- ☆ Avoid and prevent excessive irrigation
- ☆ Sub-surface drainage technology
- ☆ Bio-drainage by trees like Eucalyptus
- ☆ Salinity

At current day, one third of the total cultivable land area of the world is affected by salts. Saline soils are characterized by the accumulation of soluble salts like sodium chloride, sodium sulphate, calcium chloride, magnesium chloride *etc.* in the soil profile. Their electrical conductivity is more than 4 ds/m. So, soils have carbonates and bicarbonates of sodium, the pH usually exceed 8.0 and the exchangeable sodium percentage (ESP) is more than 15 per cent.

The salinity of the soil is one of the reasons of low productivity just because of the improper management of farm drainage. In this situation, the roots of plants do not get enough air to respiration then it leads to low crop yield as well as low mechanical strength.

Remedy

1. The most common method for getting rid of salts is to flush them out by applying more good quality water to such soils.
2. Another method is laying underground network of perforated drainage pipes for flushing out the salts slowly.

2.2.5 Energy Resources

Energy is the capacity to do work and is required for life processes. An energy resource is something that can produce heat, power life, move objects, or produce electricity.

Matter that stores energy is called a fuel. Human energy consumption has grown progressively throughout human history. Early humans had modest energy requirements, mostly food and fuel for fires to cook and keep warm. In today's society, humans consume as much as 110 times as much energy per person as early humans. Most of the energy we use today come from fossil fuels (stored solar energy). But fossils fuels have a disadvantage in that they are non-renewable on a human time scale, and because other potentially harmful effects on the environment. In any event, the exploitation of all energy sources (with the possible exception of direct solar energy used for heating), ultimately rely on materials on planet Earth.

Energy resources are the occasions an area offers to generate electricity based on its natural conditions and circumstances. Some of these energy resources are obvious; an area might contain coal, oil, wood, or gas. But others, like renewable resources such as wind, solar, hydroelectric, and wave power are not so understandable– they're based on the natural weather patterns and features of an area. Energy sources can be classified into two types: non-renewable and renewable. Non-renewable resources, such as fossil fuels and nuclear material, are removed from the earth and can be depleted. These resources have been the most used type of energy in the modern era.

Renewable resources, such as wind, water, solar, and geothermal, come from sources that regenerate as fast as they are consumed and are continuously available. Some, such as biofuel produced from food crops and other plants, are refilled every growing season. In the early part of the twenty-first century, renewable sources have become more popular as non-renewable sources have begun to be depleted.

Solar Energy

Solar energy can be used directly or indirectly for human welfare. Direct solar energy is the radiant energy, whereas indirect solar energy is energy obtained from materials that have previously incorporated the sun's radiant energy. Solar energy can be used for direct heating, and alternatively the heat converted into electricity (thermal electric generation). Photovoltaic cells convert direct solar energy into electricity. A backup system is required to store and generate electricity when solar power is not operative at night or during cloudy days. Among various energy resources where solar energy is utilized indirectly, biomass energy is the most important one. Biomass energy is obtained from those materials whose origin can be traced to photosynthesis such as live plant material and dried residues, fresh water and marine algae, agricultural and forest residues (*e.g.* straw, husks, corn cobs, bark, sawdust, roots, animal wastes) *etc.* Biomass also includes biodegradable organic wastes from industries like sugar mills, breweries, *etc.* At least half of the world's population relies upon biomass as their main source of energy. In India, fuel wood is still a major source of energy for domestic purposes in rural areas. Biomass fuel, which can be a solid, liquid or gas, is burned to release its energy. Solid biomass includes wood, charcoal, animal dung and peat. Biomass can be converted to liquid fuels especially methanol and ethanol which can be used in internal combustion. Biomass, particularly animal waste, can also be converted into biogas in biogas digesters by using the process of microbial decomposition. Biogas is a clean fuel whose combustion produces fewer pollutants than other combustible energy resources. It is composed of a mixture of gases and can be stored and transported easily. Production of biomass for energy requires sufficient area of land and water. Energy plantations of plant species, showing high calorific value and growth rate, are raised in selected areas to produce biomass.

The Rising Demand for Energy

Energy is derived from non-renewable (conventional) and renewable (non-conventional) resources and the former are in the process of depletion. These are fossil fuels – oil, coal and natural gas. It took millions of years to build up these resources. Renewable resources are solar energy, wind energy, water energy and biomass.

Approximately 80 per cent of the world's energy is produced by fossil fuels. However, in France, the French Atomic Energy Commission established nuclear reactors which produce enough energy to meet 70 per cent of country's requirement.

World demand for oil (according to UN reports) rose from 436 million tonnes in 1960 to 2189 million tonnes in 1970 and to 3200 million tonnes in 1999. The corresponding figures for coal are 1043, 1635 and 2146 and for natural gas the figures are 187, 1022 and 2301. The demand will continue to grow. Of the developing countries, China has the highest per capita consumption of energy. For India, per capita consumption is lower than that of China.

The use of solar energy is through photovoltaic cells. The photovoltaic news reported that world's photovoltaic production climbed from 0.1 megawatt to 200 megawatts in 1999. The biomass resources are various types of cultivated or uncultivated vegetation. Wood forms the chief resource and is the primary fuel for the people in Africa and Asia. Excessive use of wood has led to depletion of forests.

Renewable and Non-renewable Resources

There are nine major areas of energy resources. They fall into two categories: non-renewable and renewable. Non-renewable energy resources, like coal, nuclear, oil, and natural gas, are available in limited supplies. This is usually due to the long time it takes for them to be replenished. Renewable resources are replenished naturally and over relatively short periods of time. The five major renewable energy resources are solar, wind, water (hydro), biomass, and geothermal.

Renewable Energy

Renewable energy, often mentioned to as clean energy, comes from natural sources or processes that are constantly replenished. For example, sunlight or wind keep shining and blowing, even if their availability depends on time and weather.

While renewable energy is often thought of as a new technology, harnessing nature's power has long been used for heating, transportation, lighting, and more. Wind has powered boats to sail the seas and windmills to grind grain. The sun has provided warmth during the day and helped kindle fires to last into the evening. But over the past 500 years or so, humans increasingly turned to cheaper, dirtier energy sources such as coal and frocked gas.

Wind power is limited to areas with high consistent winds, and so is limited to very specific areas. The wind mills are not aesthetically pleasing to look at, make a lot of noise and kill large numbers of birds, all problems that would need to be overcome to expand this resource.

As for hydroelectric resources, they will not likely increase, since most rivers are already dammed and there are few places left where new hydroelectric facilities could be built.

Geothermal energy is limited to areas of known thermal activity (mainly recently active volcanic areas). It is a great local resource, but will never play a major role as an energy resource.

Solar energy is a huge source, but requires other resources (Li, Rare Earth Elements) to exploit. Many of these problems might be overcome with new research and the development of new technologies.

Hydrogen Fuel Cells are another promising resources with plenty of supply, but needs further research and technological development.

Future energy resources have huge environmental, political and economic implications that could change the world order. Still, the geologic aspects of energy resources will play a large role.

Non-Renewable Energy

Non-renewable, or "dirty," energy includes fossil fuels such as oil, gas, and coal. Non-renewable sources of energy are only available in limited amounts and take a long time to replenish. When we pump gas at the station, we're using a finite resource refined from crude oil that's been around since prehistoric times.

Non-renewable energy sources are also typically found in specific parts of the world, making them more plentiful in some nations than others. By contrast, every country has access to sunshine and wind. Prioritizing non-renewable energy can also improve national security by reducing a country's reliance on exports from fossil fuel–rich nations.

First we look at the reserves of various non-renewable energy resources. Note that Uranium (for nuclear energy) and Coal appear to be most plentiful, while Tar sands and oil shale are currently not economical. The current known oil reserves will likely run out sometime between 2050 and 2150.

Currently we are consuming oil at a rate 3 times that of the discovery of new resources. Even in terms of 4,000 years of human history, the oil age will be very short lasting only 150 to 200 years.

Coal reserves could last for about 300 years if we can cope with the associated pollution. Natural Gas is cleaner and can probably last for another 200 years. Nuclear seems like a good bet in terms of available resources, but can it be made cheap, clean, and safe? Will the recent problems with nuclear reactors during the March 11, 2011 earthquake have an effect on the future of nuclear energy?

Tar Sands and Oil Shale will require research to find more efficient way to extract, the resource, but will likely be necessary to replace oil in the short term.

Alternate Sources of Energy

There are 5 fundamental sources of energy:

- Nuclear fusion in the Sun (solar energy)
- Gravity generated by the Earth and Moon.
- Nuclear fission reactions.
- Energy in the interior of the Earth.
- Energy stored in chemical bonds.
- Solar Energy

Solar Energy arrives from the Sun by electromagnetic radiation. It can be used directly for heat and converted to electricity for other uses. It is a nearly unlimited source, it is renewable, and largely, non-polluting.

Gravity Generated by the Earth and Moon

Gravitational pull of the Moon on the Earth causes tides. Tidal flow can be harnessed to drive turbines. This is also a nearly unlimited source of energy and is largely non-polluting.

Combining both solar energy and gravity provides other useful sources of energy. Solar radiation heats air and evaporates water.

Gravity causes cooler air to sink and condense water vapour. Gravity then pulls condensed water back to Earth, where it flows downhill. The circulation of the atmosphere by the process is what we call the wind. Energy can be extracted from the wind using windmills. Water flowing downhill has a result of gravity can also be harnessed for energy to drive turbines and generate electricity. This is called hydroelectric energy. This sources of energy are mostly renewable, but only locally, and are generally non-polluting.

Nuclear Fission Reactions

Radioactive Uranium is concentrated and made into fuel rods that generate large amounts of heat as a result of radioactive decay. This heat is used to turn water into steam. Expansion of the steam can then be used to drive a turbine and generate electricity. Once proposed as a cheap, clean, and safe way to generate energy, Nuclear power has come under some disfavour. Costs of making sure nuclear power plants are clean and safe and the problem of disposing of radioactive wastes, which are unsafe, as well as questions about the safety of the plants under human care, have contributed to this disfavour.

Energy in the Interior of the Earth

Decay of radioactive elements has produced heat throughout Earth history. It is this heat that causes the temperature to increase with depth in the Earth and is responsible for melting of mantle rocks to form magmas. Magmas can carry the heat

Case Study- Energy schemes in the rural developing world

Abellon Clean Energy, India

Pollution is rampant in Gujarat, one of India's most industrialised states. Factories rely mainly on lignite as fuel and the state's agricultural base suffers from high salinity and erratic rainfall. Farmers traditionally burn their crop waste to clear the land, thus further polluting the air.

Abellon, based in Ahmedabad, India, with offices in the U.S., Italy and Ghana, got its start in 2008 when the founders saw the opportunity to tackle both of these problems by replacing the coal and lignite used in factories with a fuel made from the farmers' crop waste.

Abellon now pays 8,500 local farmers a small income for crop residues such as cotton stalks and cumin stems. Along with sawdust from nearby saw-mills, these residues are made into pellets and sold to local industries.

This year the two pellet plants currently in operation produced 65,000 tonnes of pellets, mostly for large industrial customers. The company, which reported US$3.2 million in income in 2010 aims to open two additional pellet plants in Gujarat and to treble sales in India over the next five years.

It also plans to expand into international markets. In addition to its biomass work, Abellon is working in solar energy, setting up pilot solar power projects combining multiple solar technologies.

upward into the crust. Groundwater circulating in the vicinity of igneous intrusions carries the heat back toward the surface. If this hot water can be tapped, it can be used directly to heat homes, or if trapped at great depth under pressure it can be turned into steam which will expand and drive a turbine to generate electricity.

Energy Stored in Chemical Bonds

Energy stored in chemical bonds drives chemical reactions. When the reactions take place this energy is either released or absorbed. If it is absorbed, it is stored in the chemical bond for later use. If it is released, it can produce useful heat energy.

Biomass Energy is another example. It involves burning (a chemical reaction) of wood, or other organic by-products. Such organic material is produced by photosynthesis, a chemical process which derives energy from the Sun and stores that energy until the material is burned.

Fossil Fuels - Biomass energy that is buried within the Earth where it is stored until humans extract and burn it to release the energy. Among these sources are petroleum (Oil and natural gas), oil shale, tar sands, and coal. All of which will be one of the primary topics of our discussion here.

2.2.6 Land Resources

Land is a finite resource is subject to competing pressures from urbanisation, infrastructure, increased food, feed, fibres and fuel manufacture and the provision of key ecosystem services. But it's also a shrinking resource. This is a world-wide problem. Demands for areas to settle grow food and biomass is rising around the world, and climate change is likely to impact on land demand, obtainability and degradation. The Rio+20 conferences documented that land and soil degradation are a global problem, and proposed to strive for "a land degradation neutral world in the context of sustainable development".

Land Resources Issues

1. Human settlements, agriculture, commerce, industry and tourism development have historically been the major competing uses for limited land resources in many small island developing States. As human needs and population grow, the pressure on land and other natural resources continue to increase.
2. Many factors and causal relationships influence the management of land resources. Increased demand for monetary income has led to greater production of cash crops for export and to inappropriate tourism development. For agriculture, this has meant increased areas under cultivation and more mechanized production systems. Land pressures in some small island developing States have been further worsened by the intensification of animal production, particularly high-input production chains that are dependent on concentrate feed. Unsustainable agricultural practices have contributed to deforestation; changes in cropping pattern,

with consequent losses of biodiversity across the landscape; loss of soil fertility; and agrochemical pollution of soils, freshwater and coastal resources downstream.

3. National decisions and the capacity to manage land resources sustainably can also be constrained by other factors, including a lack of institutional capacity to properly negotiate rights to exploit natural resources by foreign companies. Poorly designed projects financed by donors have sometimes been based on purely economic considerations, without taking due account of local environmental conditions and national priorities, as a result of inappropriate forestry and tourism policies.

Land Degradation

Land degradation is caused by multiple forces, counting extreme weather conditions particularly drought, and human activities that pollute or degrade the quality of soils and land utility negatively affecting food production, livelihoods, and the production and provision of other ecosystem goods and services. Land degradation is one of the world's most pressing environmental problems and it will worsen without rapid remedial action. Globally, about 25 per cent of the total land area has been degraded. When land is degraded, soil carbon and nitrous oxide is released into the atmosphere, making land degradation one of the most important sponsors to climate change. Scientists recently warned that 24 billion tons of fertile soil was being lost per year, largely due to unsustainable agriculture practices. If this trend continues, 95 per cent of the Earth's land areas could become degraded by 2050.

Globally, 3.2 billion people are affected by land degradation, especially rural communities, smallholder farmers, and the very poor. The world population is projected to increase by about 35 per cent to 9.7 billion in 2050, with rising demands for agricultural products including food, feed, fibre, and fuel. However, pressure on the global land resource is increasing due to other factors as well, such as agricultural production systems made less resilient by the loss of biodiversity, and natural factors such as climate variability and extreme weather events. Climate change exacerbates variations in yields and income from agriculture, threatening the resilience of agro-ecosystems and stability of food production systems.

The problems are particularly severe in the driest parts the planet. Dry land landscapes cover approximately 40 per cent of the world's land area and support two billion people. The vast majority of people who depend on dry lands live in developing countries, where women and children are most vulnerable to the impacts of land degradation and drought.

Soil Erosion

Soil is the earth's fragile skin that anchors all life on Earth. It is included of countless species that create a dynamic and complex ecosystem and is among the most precious resources to humans. Increased demand for agriculture commodities generates incentives to convert forests and grasslands to farm fields and pastures. The transition to agriculture from natural vegetation often cannot hold onto the soil and many of these plants, such as coffee, cotton, palm oil, soybean and wheat, can actually increase soil erosion beyond the soil's ability to maintain itself.

The possessions of soil erosion go beyond the loss of fertile land. It has led to augmented pollution and sedimentation in streams and rivers, clogging these waterways and causing declines in fish and other species. And tainted lands are also often less able to hold onto water, which can worsen flooding. Sustainable land use can help to reduce the impacts of agriculture and livestock, stopping soil degradation and erosion and the loss of valuable land to desertification.

The health of soil is a primary concern to farmers and the global community whose livelihoods depend on well managed agriculture that starts with the dirt beneath our feet. While there are many challenges to maintaining healthy soil, there are also solutions and a dedicated group of people, including WWF, who work to innovate and uphold the fragile skin from which biodiversity springs. Erosion degrades land, which means it can support fewer plants that can take in climate-warming carbon dioxide. Soils themselves could potentially sequester enough greenhouse gases in a year to equal about 5 per cent of all annual human-made GHG emissions. Better land management can help keep soils intact so they can grow more carbon-sucking vegetation. This is already happening in China, where the Grain-for-Green project in the Yellow River basin conserved soil and water and abridged carbon emissions.

Desertification

Desertification is land degradation in arid, semi-arid, and dry sub-humid areas, together known as dry lands, resulting from many factors, including human activities and climatic variations. The range and intensity of desertification have increased in some dry land areas over the past numerous decades. Desertification does not occur in linear, easily map able patterns. Deserts advance erratically, forming patches on their borders. Areas far from natural deserts can degrade quickly to barren soil, rock, or sand through poor land management. The attendance of a nearby desert has no direct relationship to desertification. Unfortunately, an area undergoing desertification is brought to public attention only after the process is well underway. Often slight or no data are available to indicate the previous state of the ecosystem or the rate of degradation. Scientists still question whether desertification, as a process of global change, is enduring or how and when it can be halted or reversed.

Effects of Desertification

Farming becomes next to impossible. If an area becomes a desert, then it's almost impossible to grow substantial crops there without special technologies. This can cost a lot of money to try and do, so many farmers will have to sell their land and leave the desert areas.

- **Hunger:** Without farms in these areas, the food that those farms produce will become much scarcer, and the people who live in those local areas will be a lot more likely to try and deal with hunger problems. Animals will also go hungry, which will cause even more of a food shortage.
- **Flooding:** Without the plant life in an area, flooding is a lot more eminent. Not all deserts are dry; those that are wet could experience a lot of flooding because there is nothing to stop the water from gathering and going all over the place. Flooding can also negatively affect the water supply, which we will discuss next.
- **Poor Water Quality**: If an area becomes a desert, the water quality is going to become a lot worse than it would have been otherwise. This is because the plant life plays a significant role in keeping the water clean and clear; without its presence, it becomes a lot more difficult for you to be able to do that.
- **Overpopulation:** When areas start to become desert, animals and people will go to other areas where they can actually thrive. This causes crowding and overpopulation, which will, in the long run, end up continuing the cycle of desertification that started this whole thing anyway.
- **Poverty:** All of the issues that we've talked about above (related to the problem of desertification) can lead to poverty if it is not kept in check. Without food and water, it becomes harder for people to thrive, and they take a lot of time to try and get the things that they need.

Role of an Individual and Communities in Conservation of Natural Resources

The proper use of natural resources for long-lasing human welfare is known as conservation. The term natural resources includes all land, waters, vegetation,minerals and wildlife useful to the society in the maintenance of civilization. The wise and judicious use of natural resources without wasting them and the efforts of replacement like planting tree whenever possible are called conservation. The term conservation came into use in the late 19th century and referred to the management, mainly for economic reasons, of such valuable natural resources as fish, topsoil, pasture-land and minerals and also to the preservation of forests, wilderness and watershed areas.

Need for Conservation

- Deforestation caused the loss of energy resources.
- Use of natural resources is increasing but the amount of these resources by decreasing.
- Relational and international capacities conserving the resources are not properly organized, must have some common conservation strategy.
- To maintain the essential ecological processes *i.e.* food chain recycling of mineral resources *etc.*

- To ensure the availability and sustainability of resources which assumes the survival of all the species is a healthy and easy manner.
- To preserve the diversity at the specific habitat levels.
- To maintain the essential life support system – soil, water, air, pond, plants and animals *etc.*

Role of an Individual

Natural resources like forests, water, soil, food, minerals and energy resources play an important role in the economy and development of a nation. Humans can play significant role in conservation of natural resources. A petite effort by individuals can help to jam these resources which are a gift of nature to the mankind. Brief description of role of individual to conserve different types of natural resources is given below:

Conservation of Energy

- Switch off lights, fans and other appliances when not in use.
- Use solar heater for cooking your food on sunny days, which will cut down your LPG expenses.
- Grow trees near the houses and get a cool breeze and shade. This will cut off your electricity charges on A/C and coolers.
- Use always pressure cooker.

Conservation of Water

- Use minimum water for all domestic purposes.
- Check the water leaks in pipes and repair them properly.
- Reuse the soapy water, after washing clothes for washing courtyard, carpets *etc.*
- Use drip irrigation.
- Rain water harvesting system should be installed in all the houses.
- Sewage treatment plant may be installed in all industries and institution.
- Continuous running of water taps should be avoided.
- Watering of plants should be done in the evening.

Conservation of Food Resources

- Cook required amount of food.
- Don't waste the food, give it to someone before spoiling.
- Don't store large amount of food grains and protect them from damaging insects.

Conservation of Forest

- ☆ Use non timber product.
- ☆ Plant more trees.
- ☆ Grassing must be controlled
- ☆ Minimise the use of paper and fuel.
- ☆ Avoid the construction of dam, road in the forest areas.

Promote Sustainable Agriculture

- ☆ Diversify the existing cropping pattern for sustainability of agriculture
- ☆ Cultivate need based crop
- ☆ Maintain soil fertility
- ☆ Make optimum use of fertilizers, pesticides and other chemicals for production and processing of agriculture products
- ☆ Save grains in storage to minimise the losses
- ☆ Improve indigenous breeds of mulch animals for sustainable dairy production systems.

Role of Communities

- ☆ Equitable use of resources for sustainable life style:
- ☆ Sustainable development: Development of healthy environment without damaging natural resources.
- ☆ Unsustainable development; Degradation of the environment due to over utilisation of natural resources.
- ☆ Life style in more developed countries: 22 per cent of world population, 88 per cent of its natural resources and 85 per cent of total global income.
- ☆ Consumption is more and pollution is more.
- ☆ Life style in less developed countries: 78 per cent of world population, 12 per cent of its natural resources and 15 per cent of total global income.
- ☆ Consumption is less and pollution is less.

Causes of Unsustainability: Main Cause – difference between MDCs and LDCs

Sustainable life style: MDCs should have to reduce the utilisation of natural resources that should have to be diverted to LDCs. This will reduce the gap between MDCs and LDCs, leads to sustainable development of the entire world.

Considering preservation and conservation of environment, the United States Environmental preservation is viewed or seen as the setting aside of earthly resources for preventing damage normally caused by contact with humans or by certain human activities, such as logging, mining, hunting, and fishing, only to replace them with new human activities such as tourism and recreation.

Furthermore regulations and laws may be enacted for the preservation of natural resources. Being earth friendly is very essential as this will save our planet at the time making a better place to live in for us, for future generations. The responsibility lies more on the human population because they have got the thinking power and the wisdom to judge good and had man should realize that he is not alone in this world. There are others to use the available resources. Hence responsibility should be for all human being for an equitable use of natural resources for sustainable use of natural resources for sustainable life styles of all in this mother earth.

Equitable Use of Resources for Sustainable Lifestyles

Scarcity of resources is the burning problem of modern technology. The twenty-first century will see growing human needs for resources since many parts of the world are using natural resources at a rate faster than the natural processes can replenish it.

In last 50 years, the consumption of resource in the society has increased many folds. There is a big gap in the consumers lifestyle between developed and developing countries. Urbanisation has changed the life style of middle class population in developing countries creating more stress on the use of natural resources. It has been estimated that More Developed Countries (MDC) of the world constitute only 22 per cent of world's population but they use 88 per cent of natural resources. These countries use 73 per cent of energy resources and command 85 per cent of income and in turn they contribute very big proportion of pollution. On the other hand less developed countries (LDCs) have moderate industrial growth and constitute 78 per cent of world's population and use only 12 per cent of natural resources, 27 per cent of energy and have only 15 per cent of global income.

There is a huge gap between rich and poor. In this age of development the rich have gone richer and the poor is becoming poorer. This has led to unsustainable growth. There is an increasing global concern about the management of natural resources. The solution to this problem is to have more equitable distribution of resources and income. Two major causes of unsustainability are over population in poor countries and over consumption of resources by rich countries. A global consensus has to be reached for balanced distribution of natural resources.

For equitable use of natural resources more developed countries/rich people have to lower down their level of consumption to bare minimum so that these resources can be shared by poor people to satisfy their needs. Time has come to think that it is need of the hour that rich and poor should make equitable use of resources for sustainable development of mankind.

It is vital to sustain the natural resources. We should conserve natural resources so that it may yield sustainable benefit to the present generation while maintaining its potential to meet the needs of the future generation. There are three specific objectives to conserve living resources:

1. To ensure that any utilisation of the ecosystem is sustainable.
2. To preserve biodiversity and
3. To maintain essential ecological processes.

Reduction of the unsustainable and unequal use of resources, and control of our population growth are essential for the survival of our nation and indeed of human kind everywhere.

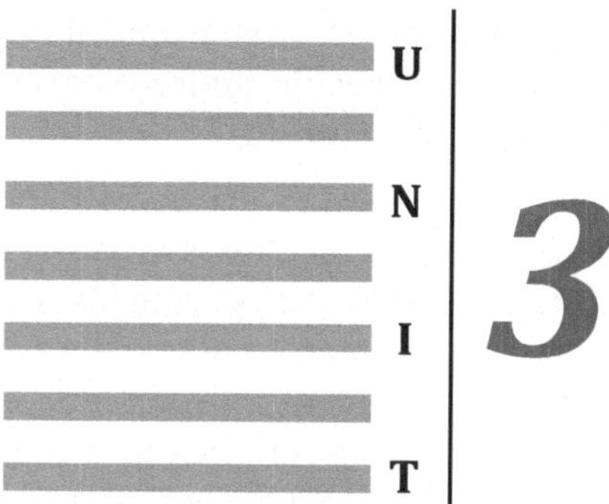

Ecosystems

3.1 Concept of Ecosystem

In biology, an ecosystem is a community of creatures and their physical environment. The notion of an ecosystem distinguishes the many ways that an organism interacts with and depends on various parts of its environment. The ecosystem idea generalizes the "food chain" and "food web" concepts, letting for more relations than just consumption. For example, plants provide not just food for animals but also shelter, shade, moisture, *etc.*

An ecosystem, a term very often used in biology, is a communal of plants and animals interacting with each other in a given area, and also with their non-living environments. The non-living environments include weather, earth, sun, soil, climate and atmosphere. The ecosystem relates to the way that all these different organisms live in close proximity to each other and how they interact with each other. For example, in an ecosystem where there are both rabbits and foxes, these two creatures are in a relationship where the fox eats the rabbit in order to survive. This association has a knock on effect with the other creatures and plants that live in the same or similar areas. For instance, the more rabbits that foxes eat, the more the plants may start to thrive because there are fewer rabbits to eat them.

According to Wikipedia

"An ecosystem is a community of living organisms in conjunction with the non-living components of their environment (things like air, water and mineral soil), interacting as a system. These biotic and abiotic components are regarded as linked together through nutrient cycles and energy flows. As ecosystems are defined by the network of interactions among organisms, and between organisms and their environment, they can be of any size but usually encompass specific, limited spaces (although some scientists say that the entire planet is an ecosystem)."

The concept of ecosystem was first put forth by A.G. Tansley (1935). Ecosystem is the major ecological unit. It has both structure and functions. The structure is related to species diversity. The more complex is the structure the greater is the diversity of the species in the ecosystem. The functions of ecosystem are related to the flow of energy and cycling of materials through structural components of the ecosystem. According to Woodbury (1954), ecosystem is a complex in which habitat, plants and animals are considered as one interesting unit, the materials and energy of one passing in and out of the others.

The study of ecosystems mainly consists of the study of certain processes that link the living, or biotic, components to the non-living, or abiotic, components. The two main processes that ecosystem scientists study are Energy transformations and biogeochemical cycling. As we learned earlier, ecology generally is defined as the interactions of organisms with one another and with the environment in which they occur. We can study ecology at the level of the individual, the population, the community, and the ecosystem.

3.1.1 Examples of an Ecosystem

Ecosystems are unified and stable systems that provides for the advantageous relationship between the living and non-living things in the environment. An ecosystem is consequently a natural self-sufficient unit and is defined by the natural functions of both living and non-living things which includes instances such as lakes, deserts, forests, ponds, gardens, oceans and grasslands.

1. *Lakes*: Since the lake includes both biotic and abiotic things interconnected with physical and chemical interactions, it qualifies as an ecosystem. It is an example of a stationary ecosystem as the enabling environment is relatively still.
2. *Deserts*: The desert is considered as an ecosystem because it has an ecology characterized by both biotic and abiotic functions and processes in arid areas. It includes both plant, animal, and bacterial populations and their interaction with desert climate, air and soil. In desert ecosystems, both plants and animals are specially adapted to withstand the harsh prevailing conditions.
3. *Forests*: In the forest, there are many animal and plant species interacting with non-living things. For example, precipitation and sun's energy is taken up by the plants and used for growth and then consumed by primary consumers (the herbivores). The primary consumers are later preyed on by secondary consumers particularly the carnivores and the system continue through simple and complex food web and food chain respectively.

4. *Ponds*: A pond refers to a small lake or a pool of water. Just like the lake, it qualifies as an ecosystem because it contains both living and non-living things that depend on each other for survival. For instance, there are various living microorganisms in ponds that live in the ponds and are supported by the prevailing environmental condition of the pond.
5. *Gardens*: Gardens encourage natural processes and various functions in the local areas. They help in buffering the detrimental impacts of strong winds, regulating water drainage, and providing shelter and food for indigenous wildlife.
6. *Oceans*: The oceans have diverse range of natural components that enable its continuity and the constant exchange of materials between living and non-living things. Aquatic animals such as fish obtain essential nutrients such as phosphorous, nitrogen, oxygen and food from within the water. The same applies to the plants living under water. The ocean ecosystems, also known as marine ecosystems, are regarded as the most extensive of the planets ecosystems.
7. *Grasslands*: Grasslands are ecosystems in temperate regions that do not get so much rain as compared to other regions such as the tropical rainforests.

3.2 Structure of Ecosystem

The structure of an ecosystem is fundamentally a description of an environment's organisms and physical features. This includes the quantity and delivery of energy in the environment. Moreover, the construction of ecosystem provides information on various climatic conditions prevailing in the environment. It is described under two major headings:

- Abiotic components
- Biotic components

Abiotic Components

Abiotic components of an ecosystem include basic inorganic elements and compounds such as water, air, soil, phosphates, calcium carbonates, and various organic compounds, which are usually by-products of organic activities.

Abiotic components also include physical factors and elements such as solar radiation, wind currents, and moisture. It is also worth noting that the sun's radiant energy is the only significant source of energy for an ecosystem. The abiotic component is further divided into two that is; the climatic factors and edaphic factors.

- *Climatic factors:* Climatic factors include elements namely rain, light, temperature, humidity, wind and air.

- *Edaphic factors*: These are elements such as minerals, soil, topography and pH just to mention a few.

Biotic Components

The biotic components encompass all living organisms present in an ecosystem, and they fall under any of the following three categories:

- Producers or Autotrophic components
- Consumers
- Decomposers and transformers

3.2.1 Producers, Consumers and Decomposers

Producers: Producers, also known as autotrophs, are the green plants with chlorophyll which gives them the ability to use solar energy to manufacture their own food through a process termed as photosynthesis.

Consumers: Consumers are on the second level. Since they lack chlorophyll, they depend on producers for food. Also known as heterotrophs, consumers are additionally grouped into primary consumers, secondary consumers, tertiary consumers, and quaternary consumers.

- **Primary consumers:** Also referred to as first order consumers or herbivores, the animal species under this category primarily feeds on plants or primary producers. Examples include zebras, goats, cattle, rabbits, grasshoppers, deer *etc.*
- **Secondary consumers:** species under this category are also known as second order consumers or primary carnivores. They mainly feed on herbivores and examples here are rats, foxes *etc.*
- **Tertiary consumers:** Tertiary consumers are the third order consumers (large carnivores) that feed on the secondary consumers. Examples are the wolves, owls, snakes.
- **Quaternary consumers:** These are the last order consumers and they are considered the largest carnivores. They feed on primary, secondary and tertiary consumers and are not eaten by any other animal. They are animals with little or without natural enemies. Examples are the tigers, cheaters, lions, hawks, polar bears, alligators, sharks *etc.*
- **Reducers or Decomposers:** Reducers also known as detrivores (such as worms, dung flies, vultures and crabs) are responsible for breaking down dead organic matter. Decomposers, also known as saprotrophs, are the fungi and bacteria which finish the process by further breaking down the dead remains of plants and animals into simple nutrients through decomposition/biodegradation. The nutrients (by-products of the dead organic materials) are then used by producers resulting in cyclic exchange of materials in the ecosystem.

3.2.2 Types of Ecosystem

Mainly there are two types of ecosystems. Namely terrestrial ecosystem and aquatic ecosystem.

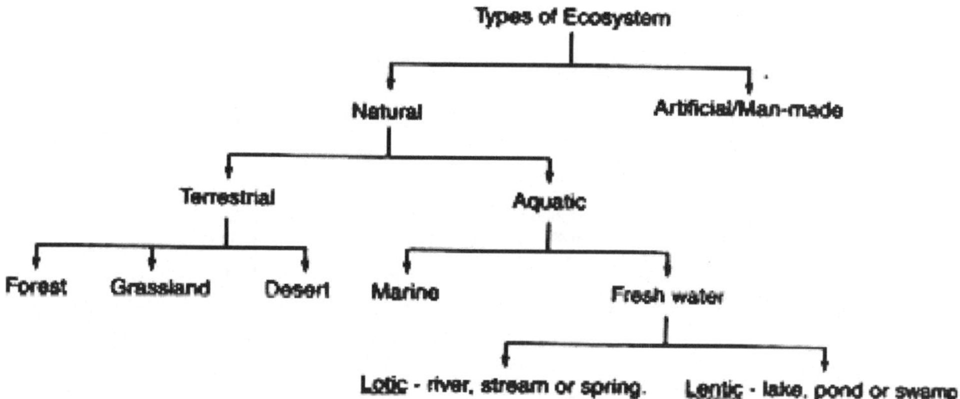

1. Terrestrial Ecosystem

This is the ecosystem which subsists on land. It can be further divided into the following types,

- **The Forest Ecosystems**: In forest ecosystems, a massive number of organisms can live in fairly small space. These are ecosystems in which you find a lot of flora. It suffices to say that the density of living organisms in forest ecosystems is pretty high.

  A slight change in forest ecosystem can affect the entire balance, effectively killing the whole ecosystem. These ecosystems also feature a wide range of fauna. They can further be divided into Tropical evergreen forest, Tropical deciduous forest, temperate evergreen forest, temperate deciduous forest, and Taiga.

- **The Desert Ecosystem:** Desert ecosystems fall within regions that obtain an annual rainfall not more than 25mm. The Earth is made of about 17 per cent desert ecosystem. These ecosystems are characterized by intense sunlight, extremely high temperature, and low availability of water. Flora and fauna are not only rare but also poorly developed. The plants feature stems and leaves that are modified in a bid to conserve as much water as possible. For instance, some common desert plants have succulent stems to store water, such as the spiny-leafed cactus. Animals in this ecosystem are also adapted to various conditions that prevail in desert environments. The animals include reptiles, camels, birds, and insects.

- **The Grassland Ecosystem**: Grassland ecosystems can be found in both the temperate and tropical regions across the globe, but with slight variations. These ecosystems mainly comprise grasses with a few trees and shrubs. Grasses are the main vegetation here, along with legumes, which belong to the composite family.

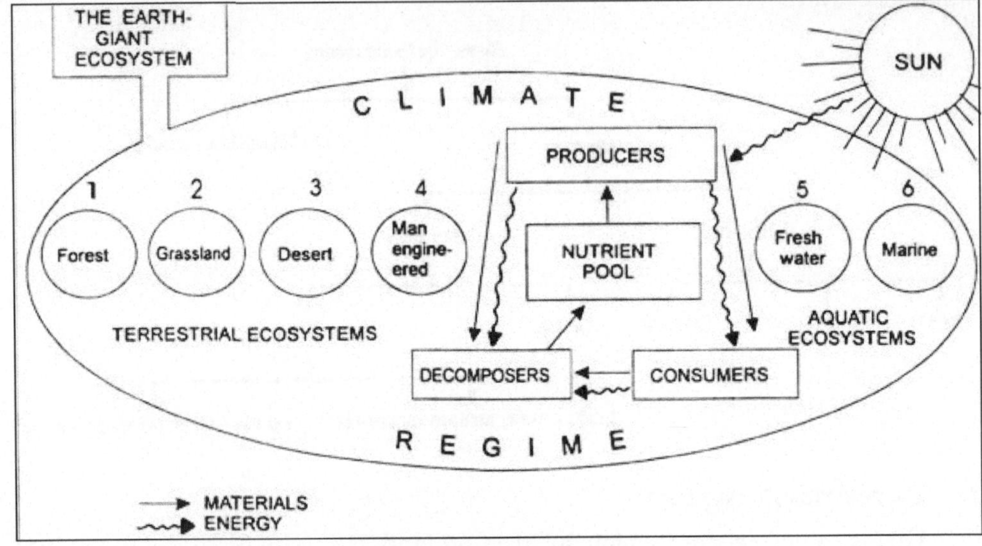

Basic Types of Ecosystem.

Grassland ecosystem is also home to a variety of grazing animals, herbivores, and insectivores. This ecosystem is further divided into savannah and prairies ecosystems.

- **The Mountain Ecosystem**: Mountain land offers a dispersed and diverse range of habitats where various plants and animals can be found. The higher altitudes are characterized by harsh environmental conditions, which can only support the lives of treeless alpine vegetation. The animals found in this habitat have thick fur coats, which protect them from cold. At the lower slopes of the mountain environment, we mostly have coniferous forests.

2. Aquatic Ecosystem

This is the ecosystem which exists in water. It can be further divided into,

- **Freshwater ecosystem** (Pong or lake or river ecosystem): It is worth noting that freshwater ecosystems account for only 1.8 per cent of the Earth's surface total, which makes it the smallest of the three main types of ecosystems. This ecosystem encompasses relatively small fish, amphibians, a variety of insects, as well as plants.
- **Marine ecosystem (Ocean ecosystem):** Covering about 71 per cent of Earth's surface and containing about 97 per cent of all the Earth's

water, ocean ecosystem is certainly the largest ecosystem. Water in this ecosystem is characterized by high amounts of dissolved salts and minerals. There is a vast variety of organisms that live in oceanic ecosystems. These organisms include the corals, the brown algae, echinoderms, cephalopods, sharks, and din flagellates.

3.3 Function of Ecosystem

An ecosystem is a practical and life supporting environmental system. The environmental system contains of biotic and abiotic components. Biotic components include living organisms and abiotic components includes in organic matter and energy. An ecosystem is a distinct structural, functional and life sustaining environmental system. The environmental system covers of biotic and abiotic mechanisms in a habitat. Biotic component of the ecosystem comprises the living organisms; plants, animals and microbes while the abiotic component comprises inorganic matter and energy

Abiotic components provide the medium for the synthesis and continuation of organic components (protoplasm). The synthesis and perpetuation processes involve energy exchange and this energy comes from the sun in the form of light or solar energy.

In an ecosystem there are three functional components.

1. Inorganic constituents
2. Organism
3. Energy input

These three components interact with each other to form an environmental system. The chief producers convert inorganic constituents into organic components by photosynthesis using the energy from the solar radiations. The herbivores make use of the energy from the producers and they themselves serve as a food for the carnivores. Animals of different types accrue organic matter in their body which is taken as food. They are known as secondary producers. The dead organic matters of plants and animals are decomposed by bacteria and fungi which break the complex molecules and liberate inorganic components. These are known as decomposers. During this procedure some amount of energy is released in the form of heat. The ecosystem of different habitats are interrelated with one another.

The functions of the ecosystem are as follows:

- ☆ It regulates the essential ecological processes, supports life systems and renders the stability.
- ☆ It is also responsible for the cycling of nutrients between biotic and abiotic components.
- ☆ It maintains a balance among the various trophic levels in the ecosystem.

- It cycles the minerals through the biosphere.
- The abiotic components help in the synthesis of organic components that involves the exchange of energy.

3.4 Energy Flow in the Ecosystem

The flow of energy in ecosystems is tremendously significant to the thriving of life on Earth. Nearly all of the energy in Earth's ecosystems invents within the Sun. Once this solar energy reaches Earth, it is dispersed among ecosystems in a tremendously complex way. A simple way to examine this delivery is through a food chain or food web.

All organisms, dead or alive, have potential for energy transmission in an ecosystem. For example, a leaf is eaten by a caterpillar, which is eaten by a small bird, which is eaten by a hawk. If the leaf was left uneaten, it would fall to the ground and be decomposed by smaller organisms. Therefore little matter is really wasted in ecosystems.

However, energy is a dissimilar story. Due to the second law of thermodynamics, not all energy can be made full use of. Throughout the food chain the energy must be transformed into useful work, which always yields wasted energy as heat.

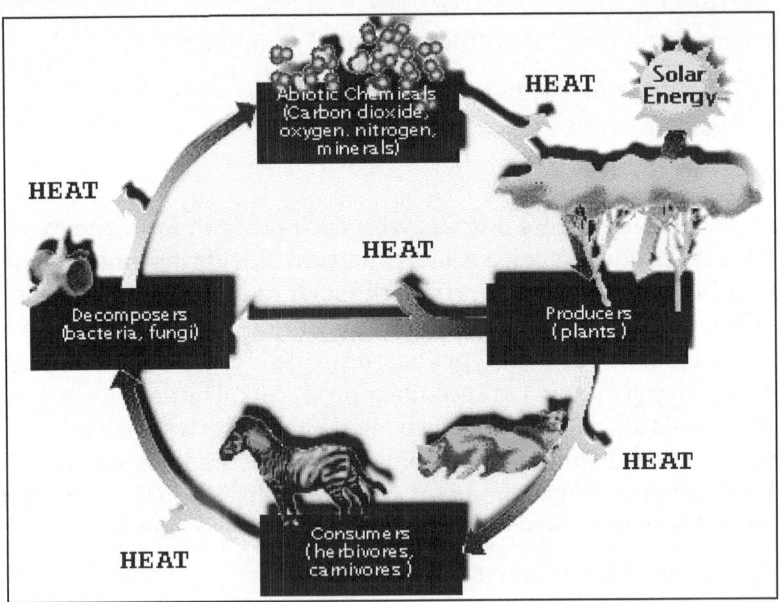

A food chain catalogues the movement of this energy and nutrients from one organism to another.

Energy enters the biological system as light energy, or photons, is transformed into chemical energy in organic molecules by cellular processes including photosynthesis and respiration, and ultimately is converted to heat energy. This energy is dissipated, meaning it is lost to the system as heat; once it is lost it cannot be recycled. Without the continued input of solar energy, biological systems would quickly shut down. Thus the Earth is an open system with respect to energy.

Elements such as carbon, nitrogen, or phosphorus enter living organisms in a diversity of ways. Plants obtain elements from the neighbouring atmosphere, water, or soils. Animals may also attain rudiments directly from the physical environment, but typically they obtain these mainly as an importance of consuming other organisms. These materials are malformed biochemically within the bodies of organisms, but sooner or later, due to excretion or decomposition, they are returned to an inorganic state (that is, inorganic material such as carbon, nitrogen, and phosphorus, instead of those elements being bound up in organic matter). Often bacteria complete this process, through the process called decomposition or mineralization (see next lecture on microbes).

During decomposition these materials are not demolished or lost, so the Earth is a closed system with respect to elements (with the exception of a meteorite entering the system now and then.). The elements are pedalled boundlessly between their biotic and abiotic states within ecosystems. Those elements whose supply inclines to limit biological activity are called nutrients.

3.4.1 The Transformation of Energy

The transformations of energy in an ecosystem instigate first with the input of energy from the sun. Energy from the sun is captured by the process of photosynthesis. Carbon dioxide is joined with hydrogen (derived from the splitting of water molecules) to harvest carbohydrates (the shorthand notation is "CHO"). Energy is stored in the high energy bonds of adenosine triphosphate, or ATP (see lecture on photosynthesis).

The prophet Isaiah said "all flesh is grass", earning him the name of first ecologist, because almost all energy available to organisms originates in plants. Because it is the first step in the production of energy for living things, it is called primary production (click here for a primer on photosynthesis). Herbivores obtain their energy by consuming plants or plant products, carnivores eat herbivores, and detritivores consume the droppings and carcasses of us all.

This figure represents a simple food chain, in which energy from the sun, seized by plant photosynthesis, flows from trophic level to trophic level via the food chain. A trophic level is composed of organisms that mark a living in the similar way that is they are all primary producers (plants), primary consumers (herbivores) or secondary consumers (carnivores). Dead tissue and waste crops are manufactured at all levels. Rummagers, detritivores, and decomposers together account for the use of all such "waste" – consumers of carcasses and fallen leaves may be other animals, such as crows and beetles, but eventually it is the microbes that finish the job of decomposition. Not astonishingly, the quantity of the main production varies a great deal from place to place, due to differences in the quantity of solar radiation and the obtainability of nutrients and water.

Usually when we think of food chains we visualize green plants, herbivores, and so on. These are mentioned to as grazer food chains, because living plants are directly spent. In many conditions the main energy input is not green plants but dead organic matter. These are called detritus food chains.

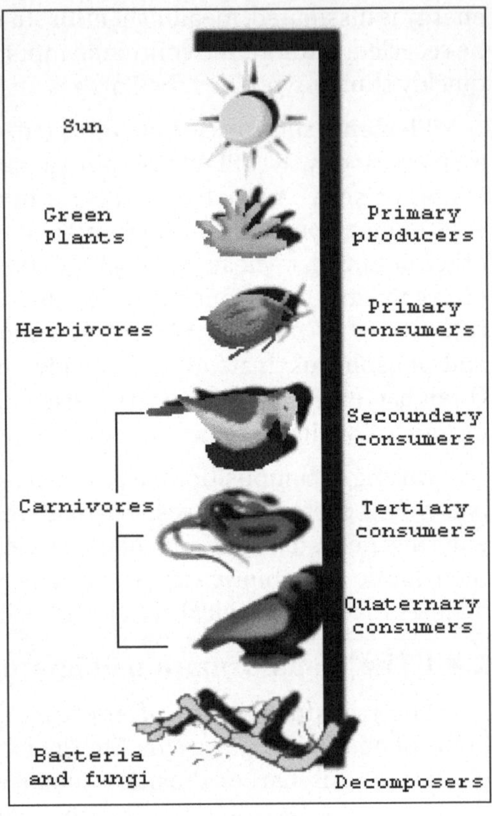

Finally, although we have been talking about food chains, in reality the group of biological systems is much more complicated than can be signified by a simple "chain". There are many food links and chains in an ecosystem, and we refer to all of these linkages as a food web. Food webs can be very complex, where it appears that "everything is connected to everything else" (this is a major take-home point of this lecture), and it is important to understand what are the most significant linkages in any particular food web. Ecosystem scientists use several different tools, which can be described generally under the term "biogeochemistry".

3.5 Food Chains, Food Webs and Ecological Pyramids

3.5.1 Food Chains

The sun is the ultimate source of energy on earth. It delivers the energy vital for all plant life. The plants exploit this energy for the course of photosynthesis, which is used to synthesise their food. During this biological process, light energy is transformed into chemical energy and is approved on through successive levels. The flow of energy from a producer, to a customer and ultimately, to a top predator or a detritivore is called the food chain. Dead and decaying matter, along with organic

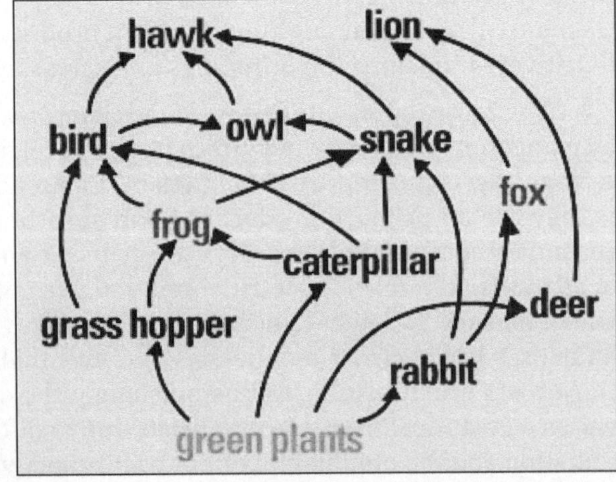

debris, is wrecked down into its constituents by scavengers. The reducers then engross these constituents. After gaining the energy, the reducers liberate molecules to the environment, which can be used again by the producers.

In the ecosystem, green plants alone are capable to trap in solar energy and change it into chemical energy. The chemical energy is protected up in the numerous organic compounds, such as carbohydrates, fats and proteins, present in the green plants. Since virtually all other living organisms depend upon green plants for their energy, the efficiency of plants in any given area in capturing solar energy sets the upper limit to long-term energy flow and biological activity in the community. The food contrived by the green plants is utilized by themselves and also by herbivores. Animals feed recurrently. Herbivores fall prey to some carnivorous animals. In this way one form of life supports the other form. Thus, food from one trophic level reaches to the other trophic level and in this way a chain is recognised. This is known as food chain.

A food chain may be distinct as the transmission of energy and nutrients through a succession of organisms through repeated process of eating and being eaten. In food chain initial link is a green plant or producer which produces chemical energy available to consumers. For example, marsh grass is consumed by grasshopper, the grasshopper is consumed by a bird and that bird is consumed by hawk.

3.5.1.1 Types of Food Chains

Food chains are of three types:

1. **Grazing food chain**: The grazing food chain starts from green plants and from autotrophs it goes to herbivores (primary consumers) to primary carnivores (secondary consumers) and then to secondary carnivores (tertiary consumers) and so on. The gross production of a green plant in an ecosystem may meet three fates–it may be oxidized in respiration, it may be eaten by herbivorous animals and after the death and decay of producers it may be utilized by decomposers and converters and finally released into the environment. In herbivores the integrated food can be stored as carbohydrates, proteins and fats, and distorted into much more multifaceted organic molecules.
2. **Parasitic food chain**: It goes from large organisms to smaller ones without absolute killing as in the case of predator.
3. **Saprophytic or detritus food chain**: The dead organic remains including metabolic wastes and exudates derived from browsing food chain are usually termed detritus. The energy contained in detritus is not lost in ecosystem as a whole, rather it serves as a source of energy for a group of organisms called detritivores that are distinct from the grazing food chain. The food chain so moulded is called detritus food chain.

3.5.1.2 Importance of Food Chains

Food chains are important because they show the intricate relationships in ecosystems.

They can disclose how each organism be contingent on someone else for survival. Food chains also display what happens when a problem occurs and a producer or consumer is lost.

Entire communities can collapse. Food chains can help scientists learn more about ecosystems and how to help them stay balanced.

Depending on the food chain you are examining, the same organism can be considered to be at more than one trophic level.

For example, seals could be considered apex predators at the highest trophic level in certain environments where they eat fish that are primary or secondary consumers.

However, in other communities where seals become prey for sharks, they could be considered to be at a lower trophic level. These relationships are easier to see in food webs and harder to notice in food chains or pyramids.

3.5.1.3 Examples of Food Chains

You can find fascinating examples of food chains in environments ranging from forests to lakes. For example, meerkats can be a top predator in one food chain by eating insects and worms. However, in other food chains, predators like eagles can eat the meerkats.

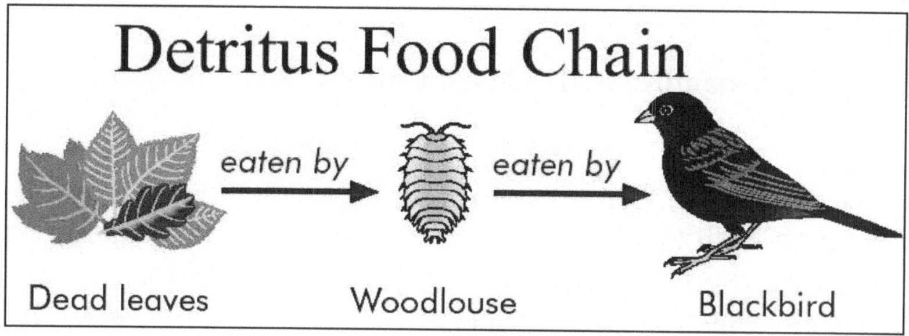

An example of a humble food chain starts with grass, which is a producer. The next level is the grasshopper or chief consumer and herbivore that eats the grass. Then, the secondary consumer is the frog that eats the grasshopper. Finally, the tertiary consumer is the hawk that eats the frog.

Another instance of a food chain starts with a tree that has delicious leaves. Insects are the primary consumers that eat the leaves. Then, woodpeckers are the secondary consumers that eat the insects. Finally, a feral cat acts as the tertiary consumer and eats the woodpeckers.

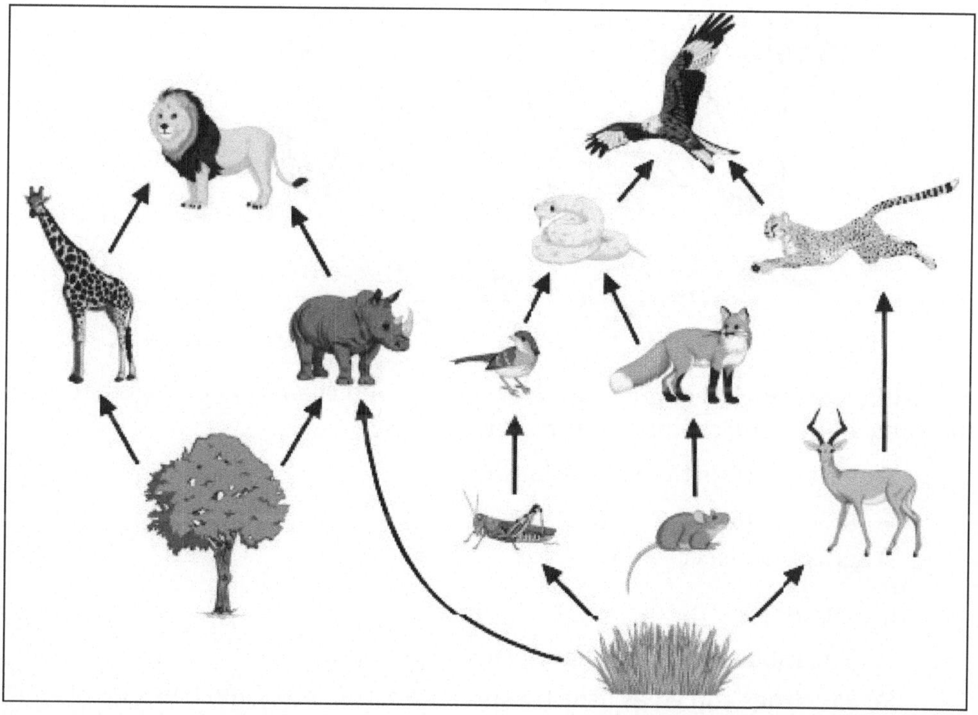

3.5.2 Food Webs

Food web is a significant ecological concept. Basically, food web signifies feeding relationships within a community (Smith and Smith 2009). It also suggests the transmission of food energy from its source in plants through herbivores to carnivores (Krebs, 2009). Usually, food webs consist of a number of food chains meshed together. Each food chain is a expressive diagram including a series of arrows, each pointing from one species to another, signifying the flow of food energy from one feeding group of organisms to another.

Food webs show how numerous food chains are related. It is a more multifaceted depiction of how the plants and animals in an ecosystem relate. A food web may instigate with prairie grass, which would be eaten by insects, mice or rabbits, which would be eaten by different predators. More species are included in a food web, which uses a series of arrows to describe relationships. When all the food chains in an ecosystem are joined up together, they form a food web. Here is an example of a food web:

Although it looks complex, it is just several food chains joined together. Here are some of the food chains in this food web:

- ☆ grass → insect → vole → hawk
- ☆ grass → insect → frog → fox
- ☆ grass → insect → vole → fox

Notice that:

- ☆ The rabbits and slugs have just one predator
- ☆ The frogs and voles have two predators (the foxes and hawks)
- ☆ The insects have three predators (frogs, voles and thrushes)

This leads to some interesting effects if the population of a particular organism in the food web decreases. Some animals can just eat more of another organism if food is in short supply, while others may starve and die. This in turn can affect the populations of other organisms in the food web.

There are two types of food chains: the grazing food chain, beginning with autotrophs, and the detrital food chain, beginning with dead organic matter (Smith and Smith, 2009). In a grazing food chain, energy and nutrients move from plants to the herbivores consuming them, and to the carnivores or omnivores preying upon the herbivores. In a detrital food chain, dead organic matter of plants and animals is broken down by decomposers, *e.g.*, bacteria and fungi, and moves to detritivores and then carnivores. Food chains and webs comprise different types of consumers. A producer and its seeds or fruit are always at the lowest level, followed by primary consumers, secondary consumers and tertiary consumers. Trees and grass are producers. Examples of primary consumers, which eat producers, are mice and insects. Secondary consumers eat primary consumers. Examples are snakes and toads. Tertiary consumers, such as red tail hawks or other raptors, eat secondary consumers.

Many food chains exist in an ecosystem, but as a matter of fact these food chains are not independent. In ecosystem, one organism does not depend wholly on another. The resources are shared specially at the beginning of the chain. The marsh plants are eaten by variety of insects, birds, mammals and fishes and some of the animals are eaten by several predators.

3.5.3 Ecological Pyramids

Ecological pyramid came into presence due to well-known scientist such as Charles Elton, G. Evelyn Hutchinson, and Raymond Lindeman. It has many names like Eltonian pyramid after Charles Elton, energy pyramid, trophic pyramid, and food pyramid. It is a explicit demonstration of the association between organisms

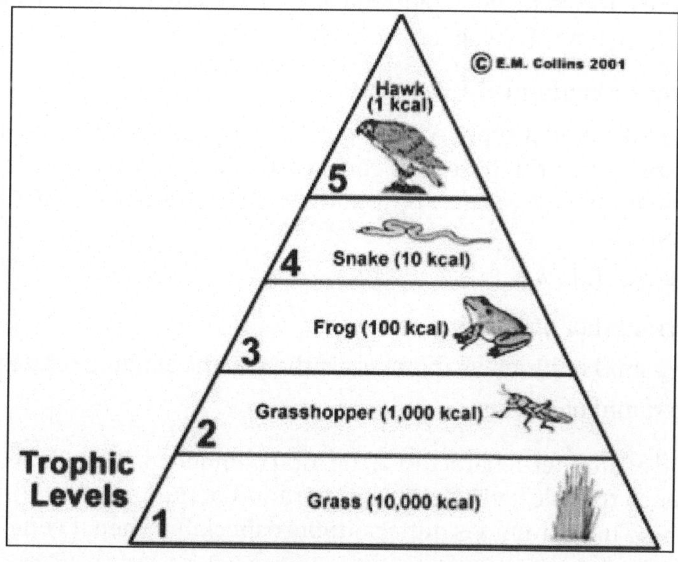

at various trophic levels in a food chain. The foundation of an ecological pyramid is the biomass, energy, and number. Just as the name proposes ecological pyramids are in the form of a pyramid. The notion was originally presented by Charles Elton, the forerunner British Ecologist.

The lowest of an ecological pyramid is the broadest and is engaged the producers, which form the first trophic level. Producers are at the bottommost level. Just as in a food chain, the producers are consumed by the primary consumers, in an ecological pyramid; the next level is occupied by the primary consumers. The next level of the pyramid is employed by the secondary consumers and the last, by the tertiary consumers.

Ecological pyramid also demonstrates the flow of energy or biomass at each trophic level in a particular ecosystem. Biomass relates to the amount of living or organic matter in an organism. Biomass pyramids are shaped that way to show that biomass is largest at the base, and lessening in amount as it goes through the apex. The two types of biomass pyramids are the upright and the inverted. The upright pyramid is originated in most ecosystems. It results when the combined weight of producers is larger than the combined weight of consumers. An upturned type results when the combined weight of producers is smaller than the combined weight of consumers.

Example: In this figure, the grass in the food pyramid is the Producer.

The grasshopper in the food pyramid is a Primary Consumer.

The frog in the food pyramid is a Secondary Consumer, because it eats the Primary Consumer (the grasshopper).

The snake in the food pyramid is the Tertiary Consumer, because it eats the Secondary Consumer (the frog).

Lastly, the hawk in the food pyramid is a Quaternary Consumer, because it eats the Tertiary Consumer (the snake).

3.5.3.1 Types of Ecological Pyramids

There are dissimilar types of ecological pyramids. In each ecological pyramid, producer level forms the base and consecutive levels make up the peak. Three types of pyramidal relatives may be found among the organisms at different levels in the ecosystem.

These are as follows:

1. Pyramid of numbers,
2. Pyramid of biomass (biomass is the weight of living organisms), and
3. Pyramid of energy

Pyramid of numbers characterise the total numbers of individuals (population) present in each trophic level. The term "pyramid of numbers" was first coined by Elton in 1972. This pyramid is quite suitable especially when it comes to counting the number of organisms. Counting is a simple task and it can be done over the years to determine the changes in a specific ecosystem. Nevertheless, some organisms are hard to count, particularly when it comes to young forms. This type of pyramid is best obtainable by taking an example of Lake Ecosystem. In this type of pyramid the base trophic level is engaged by producer elements–algae, diatoms and other hydrophytes which are most plentiful. At the second trophic level come the herbivores or zooplanktons which are lesser in number than producers.

Pyramid of numbers is divided into two different forms depending on the number of organisms: upright and inverted.

In the Upright pyramid of numbers, the numbers of organisms mostly decrease from bottom to top. It typically happens in pond and grassland ecosystems where plants occupy the base of the pyramid. The next levels of the pyramid include the consumers. An inverted pyramid is actually the opposite of the upright pyramid. It can closely be experiential in tree ecosystem, where trees are the producers and insects are the consumers.

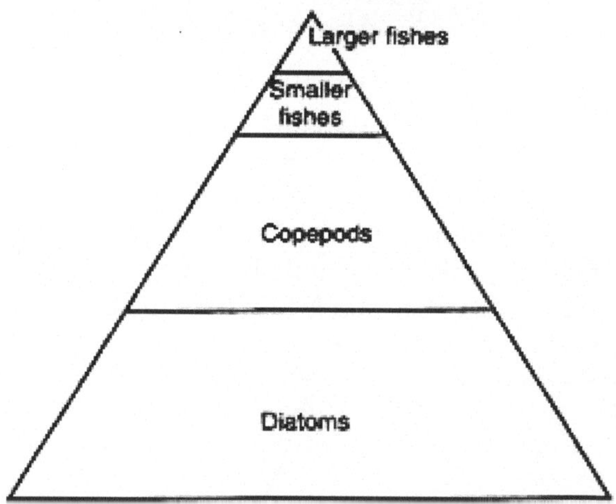

Pyramid of Biomass designates the total mass of organisms in a particular trophic level. A pyramid of numbers shows the association between the amount of food obtainable and how much energy is being passed at each trophic. Biomass spent by animals mostly is either converted to living tissue, converted to energy or remain as undigested products. The pyramid is typically larger at the bottom but as it goes up it reduces in size and becomes smaller. There is always a discount of biomass with an increase in trophic level. Around 10 per cent to 20 per cent of the biomass is passed from one trophic level to the other.

An example of an upright pyramid is the terrestrial ecosystem. It has a great base mainly consists of primary consumers, the smaller trophic levels are located at the top. This pyramid can be used to solve the particular issue in the pyramid of numbers because it demonstrates the meticulous representation of the amount of energy present in each trophic level. When there is a reduction of biomass with a rise in trophic levels, it indicates wastage and consumption of biomass at every transmission level.

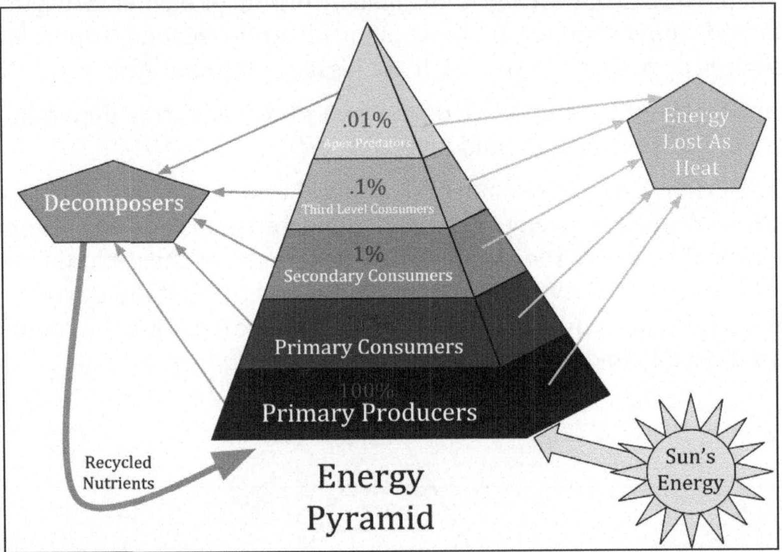

It also has some restrictions, since it's actually impossible to measure the mass of every single individual. Only an example is taken, this leads to mistakes. Also, the precise time of the year when data is taken should be put into consideration, since breeding seasons in diverse organisms differ.

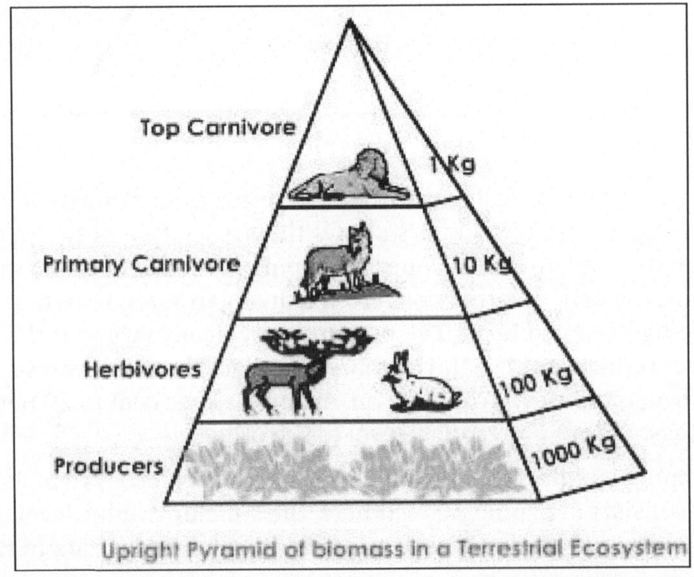

Upright Pyramid of biomass in a Terrestrial Ecosystem

Pyramid of energy is an erect pyramid that exemplifies the flow of energy from producers to consumers. Additionally, it indicates the actual role played by various organisms in energy transfer. Energy pyramids indicate how much energy is required in the next trophic level as it flows upwards. This portrays not only the amount of total energy used by the organisms at each trophic level of food chain but more significant, the actual role of numerous creatures in transfer of energy. At the producer level the total energy will be much greater than the energy at the consecutive higher trophic level.

In the energy flow process, two things become obvious. Initially there is one way along which energy moves *i.e.* unidirectional flow of energy. Energy comes in the system from outside source *i.e.* sun. The energy took by autotrophs does not go back to the sun, the energy that passes from autotrophs to herbivores does not revert back and as it moves progressively through the various trophic levels, it is no longer obtainable to the previous levels.

3.6 Ecological Succession

A distinguishing feature of biological groups is that their structure and composition changes according to certain changes in environmental conditions. Some of these variations occur in a more foreseeable and orderly fashion. The phenomenon through which these vicissitudes occur in ecological communities is Ecological succession.

This is a significant aspect of the study of ecology and procedures the core of ecological science. This ecological sequence can be triggered by some form of commotion or even due to the formation of new habitats in the ecosystem.

The preliminary point of ecological succession can be traced back to millions and millions of year back, where living organisms never happened. Slowly, as ecological changes happened, over time, new organisms developed, and the diversity of the planet started showing changes.

Every species has a set of conservational conditions under which it will produce and reproduce most optimally. In a given ecosystem, and under that ecosystem's set of environmental circumstances, those species that can grow the most efficiently and produce the most viable offspring will become the most abundant organisms. As long as the ecosystem's set of environmental conditions rests constant, those species optimally adapted to those conditions will show. The "engine" of succession, the cause of ecosystem change, is the impact of established species have upon their own environments. A consequence of living is the occasionally understated and sometimes overt alteration of one's own environment. The original environment may have been optimum for the first species of plant or animal, but the newly altered environment is often optimal for some other species of plant or animal. Under the changed conditions of the environment, the previously leading species may fail and another species may become ruling.

Ecological succession may also occur when the conditions of an environment suddenly and drastically change. A forest fires, wind storms, and human activities like agriculture all greatly alter the conditions of an environment. These massive forces may also destroy species and thus alter the dynamics of the ecological community triggering a scramble for dominance among the species still present.

3.6.1 Types of Ecological Succession

Ecological succession is mainly of two types- Primary succession and secondary succession. The primary succession is a slow process that initiates in areas where there are no living organisms. Secondary succession instigates in areas that were once occupied but destroyed due to environmental disturbances. This is usually faster, as most of the other issues are already present.

3.7 Forest Ecosystem

3.7.1 Introduction

A forest ecosystem is a public of organisms that lives within a forest. A forest is typically defined as a large group of trees. A natural woodland area making it a appropriate place for the existence of biotic and abiotic components, is usually labelled as a forest ecosystem. A forest ecosystem consists of various plants, animals, and other micro-organisms, making it a normal habitat for them.

The forest ecosystem plays an indispensable role in the environment. It helps to equilibrium the climate of the planet. One of the main roles that forests play is that it provides us oxygen to breathe. Trees help to maintain the balance of carbon dioxide in the atmosphere. Besides this, forests also help to avert soil erosion, rainfall, purifying the air, *etc.* A forest ecosystem defines the community of plants, animals, microbes and all other creatures in interaction with the chemical and physical features of their environment: Exactly, a terrestrial environment conquered by trees growing in a closed canopy – a forest, in other words.

The organisms complicated in a forest ecosystem meaning are symbiotic on one another for survival and can be broadly classified according to their ecological role as producers, consumers and decomposers. To designate forest ecosystem dynamics, you'll use a well-known example of such an ecosystem as your model: the Amazon rainforest of South America. In forest ecosystems, a massive number of organisms can live in fairly small space. These are ecosystems in which you find a lot of flora. It suffices to say that the density of living organisms in forest ecosystems is pretty high.

A minor change in forest ecosystem can affect the whole balance, efficiently killing the whole ecosystem. These ecosystems also feature a wide range of fauna. They can additional be divided into Tropical evergreen forest, Tropical deciduous forest, temperate evergreen forest, temperate deciduous forest, and Taig.

3.7.2 Types of Forest Ecosystems

Forest ecosystem has been confidential into three major types – tropical forest ecosystem, temperate forest ecosystem and boreal ecosystem. The types of forest ecosystems have been divided based on latitude and different characteristics. The different types of forest ecosystems are as follows:

Tropical Forest Ecosystem

Tropical forests, also known as tropical rainforest, receives almost 100 inches of rain every year. The tropical forest has a wide variety of species among all other types of forest ecosystems. Tropical forests are usually found in latitude between 23.5 degrees North and 23.5 degrees South. The temperature recorded in tropical forests is between 68 degrees and 77 degrees Fahrenheit. Heavy rainfall in the tropical forest leads to poor quality of soil due to a lack of nutrients. High-level rainfall throughout the year causes leaching of soil nutrients.

The vegetation of the tropical rainforest mostly comprises broad-leafed trees that are very tall about 82-115 feet height. Due to dense canopy, the sun finds it quite tough to reach the forest floor. The tropical forest is a home for millions of animals that includes a massive variety of birds, mammals, amphibians, reptiles, *etc.* Here you can find almost half the species of animals that lived on Earth.

The tropical forest ecosystem is again divided into different categories and types. These are:

- ☆ *Evergreen forest*: The evergreen forest receives heavy year-round rainfall with no dry season.

- *Seasonal forest*: Seasonal forest host evergreen flora with a short-time dry season.
- *Dry forest*: As the name suggests, dry forests experience a long-period of the dry season.
- *Montane forest*: Montane forests received most of the precipitation in the form of fog, hence also known as cloud forests.
- *Sub-tropical forest*: The sub-tropical forest is found in North and South regions of tropical forests. The flora of this region is adapted to survive in the adverse conditions of summer drought.
- *Tropical and subtropical coniferous forest*: This type of forest experiences dry and hot climatic conditions with conifers that are adapted to resist variable weather.

Temperate Forest Ecosystem

Temperate forests are typically found in North America, Eurasia, Japan, *etc.* Temperate forest obtains less rainfall as compared to tropical forests approximately 30-60 inches every year.

Unlike tropical forests, temperate forests experience all the four seasons with difference in temperature. The winters in the temperate forest quite often experience temperature below freezing point, and in summers, the temperature becomes very high with a high level of humidity.

The soil of temperate forest is rich in organic matter that allows a huge variety of vegetation to grow in the temperate forest. The temperate forest provides natural habitat to many animals such as squirrels, deer, black bears, raccoons, coyotes, various birds like warblers, owls, woodpeckers, hawks, *etc.*

The temperate forest ecosystem is further divided into two parts – Temperate Deciduous Forest and Temperate Coniferous Forest.

- **Temperate Deciduous Forest:** Temperate Deciduous forests experience four specific seasons. It receives annual rainfall between 30-60 inches. In winter, the temperate deciduous forest experience snowfall. The soil of this forest is quite fertile. As a result, you will find a wide variety of vegetation in the temperate deciduous forest, such as ferns, wildflowers, mosses, oak, birch, maple, *etc.*
- **Temperate Coniferous Forest:** The temperate coniferous forest ecosystem is usually found in coastal areas. Coniferous forests receive heavy rainfall throughout the year, approximately 50-200 inches. The floor of coniferous forests is mostly covered with a thick layer of decomposed matter. Temperate coniferous forest is covered with evergreen tall conifers. The most common plant species found in coniferous forests include cedar, Douglas fir, spruce, maple, cypress, pine, redwood, ferns, mosses, *etc.*

Boreal Forest Ecosystem

The boreal forest is also known as Taiga forests are usually found in Siberia, Northern Asia, Canada, and Scandinavia. One of the chief characteristics of the boreal forest is that it involvements short summers and very long winter seasons. Boreal forests receive approximately 15-40 inches precipitation every year (mostly receives in the form of snowfall). The trees found in boreal forests are the evergreen type, such as pine, fir, spruce, *etc.* The boreal forest has a dense canopy that hardly allows the sun to reach the forest surface. This is why the vegetation is quite less in the forest floor.

The animals found in the boreal forest ecosystem are typically enclosed with thick fur to protect them from a long period of cold winters. Some instances of animals that lived in boreal forests are – elk, caribou, lynxes, wolverines, deer, snowshoe hare, moose, wolves, *etc.* Boreal forests are home to numerous of the plants and animals typically thought of as forest-dwellers, such as foxes, moose, reindeer, bears, squirrels and wolves. Plants like fungi, mosses and lichen do well in the boreal climate, because they are tough sufficient to endure what can often be extremely cold, icy winters.

Savannah Forest Ecosystem

Savannah ecosystem is generally found in South America, Australia, and Africa. Savannah forests are quite vulnerable to forest fires; on the other hand, it has characterized by the ability to re-grow much faster.

The sceneries of the Savannah forest ecosystem are covered with large areas of green lands, bushes and clusters of feeble trees. Savannah and woodland ecosystems have a vulnerability to fires and the ability to rejuvenate and re-grow. Prevalent in South America, Africa and Australia, savannahs and woodlands are categorised by vast areas of grasslands, bush thickets and clusters of sparse trees with flattened crowns.

Because of pollution and deforestation, many types of forests around the world are in need of extra protection. It's important to recall how significant a role forests play in the planet's ecosystem, and to do what you can in order to help the ecologists fighting to protect the plants and animals that live within the world's forests.

3.7.3 Characteristic Features of Forest Ecosystems

Forest ecosystems are so rich and diverse, and they have so many thrilling and charming features. Below, you will find a discussion of some of the key features of these ecosystems.

Forest Canopy

The forest canopy is an important feature of a forest ecosystem. It relates to the top portion of a community of trees or plant crowns. A forest canopy serves as the border between the atmosphere and the land. The canopy is also the upper habitat for other biological organisms in a forest ecosystem. It is frequently composed of large trees. The construction of forest canopy is not the similar in every forest ecosystem because it depends on the availability of nutrients, tree arrangement and differences in biological species. More than half of the plant species are found in a forest ecosystem, so the biodiversity is greatest in the forest canopy. Most organisms are able to survive in forest awning because it is directly exposed to sunlight and rainwater.

Forest Floor

The forest floor is the most separate feature of a forest ecosystem. It is calm of fallen leaves, stems, twigs, branches and bark on the surface of the soil. A forest floor also covers organic and inorganic substances. Many living organisms, such as the fungi, bacteria and other microorganisms, inhabit the forest floor. It is rich in nutrients and mineral contents. The forest floor has a important role in the transfer of nutrients in the life cycle of the forest ecosystem. Most of the carbon and energy from the forest ecosystem is added to the forest floor over time. The majority of nutrients of the forest ecosystem comes from the forest floor due to the decomposition of organic substances.

- ☆ Attractive to insects: Many insects live in tree bark, leaf mulch or flowers and as such they find forest ecosystems very attractive places to make their homes.

Forest Soil

The soil is a feature of a forest ecosystem that is affected by the changes in climate, geology, amount of rainfall and vegetation. The soil of temperate forests is more fertile because trees' leaves drop to the ground every fall. This litter donates to the layers of organic material found in forest soil. The old leaves become a source of food for bacteria and fungi. These creatures facilitate the breaking down of the leaves and other organic material. Decomposition enriches the forest soil as it provides more nutrients to the living trees and plants in the forest ecosystem. However, the soil in tropical rain forests has poor quality because of the torrential rains. The continuous rain erodes and dissolves soil nutrients before the trees can advantage from them.

- ☆ **Seasonality:** In countries that have seasonal climates, forest ecosystems will change with the seasons.
- ☆ **Deciduous or evergreen:** A forest may be deciduous (*i.e.* it sheds its leaves in winter) or evergreen (*i.e.* its leaves stay green and intact all the time), or it may be a mix of both deciduous and evergreen trees.
- ☆ **Different levels:** Some forest ecosystems such as rain forests, feature several distinct levels – such as the forest floor, the lower canopy, the upper canopy and the tree tops.
- ☆ **Attractive to birds:** Many bird species nest in tree tops and this makes forest ecosystems attractive to birds.

3.7.4 Structure and Function of Forest Ecosystems

A forest ecosystem includes the living organisms of the forest, and it spreads precipitously upward into the atmospheric layer pervasive forest canopies and downward to the lowest soil layers affected by roots and biotic processes. Ecosystem analysis is a mix of biogeochemistry, ecophysiology, and micrometeorology that highlights "the circulation, transformation, and accumulation of energy and matter through the medium of living things and their activities" (Evans, 1956). For instance, rather than absorbed on the growth of individual trees, the ecosystem ecologist often expresses forest growth as net primary production in units of kilograms per hectare per year. Ecosystem ecology is less worried with species diversity than with the contribution that any composite of species makes to the water, carbon, energy, and nutrient transmission on the landscape.

The operative of forest ecosystems is categorised by energy and nutrient flows and cycling, biomass production and the decomposition of dead organic matter. Plants need nutrients and carbohydrates for biomass production and for other vital functions. Carbohydrates are produced by photosynthesis from carbon dioxide, water and nutrients. Autotrophic plants can directly utilise solar energy. Most bacteria, fungi and animals are heterotrophs that use plants as their direct or indirect source of energy. Organisms and their environment are continuously connected via energy and nutrient flows. An ecosystem can be distributed into structural and functional levels of organisation. Changes in environmental factors affect the organisms at all levels of organisation ranging from the cell to the ecosystem, and finally the biosphere. Changes in structure and functioning are rapid on the cellular level, but slower at higher levels of organisation. Structural changes are reflected in the functioning of the same organisation level. Functional changes do not, however, require changes in structure. Both functional and structural changes are related to the developmental stage of a tree stand and the variation in environmental factors.

I. Biotic Components

The various biotic components, representatives from the three functional groups, of a forest ecosystem are:

1) Producer Organisms

In a forest, the producers are mainly trees.

Trees are of different kinds depending upon the type of forest developed in that climate.

Apart from trees, climbers, epiphytes, shrubs and ground vegetation.

Dominant species of trees in major types of forest ecosystems are: *Tectona grandis*, Acer, Betula, Picea, Pine, Cedrus.

2) Consumers

In a forest, consumers are of three main types;

- ✰ Primary Consumers

  These are Herbivores which feed directly on producers.

  e.g.: Ants, Beetles, Bugs, spiders *etc.* feeding on tree leaves.

  Larger animals such as Elephants, Deer, giraffe *etc.* grazing on shoots and/or fruits of trees.

- ✰ Secondary Consumers

  These are carnivores and feed on primary consumers.

  E.g.: Birds, Lizards, Frogs, Snakes and Foxes.

- ✰ Tertiary Consumers

  These are secondary carnivores and feed on secondary consumers.

  These include top carnivores like Lion, Tiger.

3) Decomposers

These include wide variety of saprotrophic micro- organism like;

- ✰ Bacteria (Bacillus Sp., Clostridium sp., pseudomonas.
- ✰ Fungi (Aspergillus sp., Ganoderma sp., Fusarium.
- ✰ Actinomycetes (Streptomyces).

They attract the dead or decayed bodies of organisms and thus decomposition takes place.

Therefore, nutrients are released for reuse.

II. Abiotic Components

These include basic inorganic and organic compounds present in the soil and atmosphere.

In addition dead organic debris is also found littered in forests.

3.8 Grassland Ecosystem

3.8.1 Introduction

The word grassland is slightly self-explanatory. A grassland is a wide open grassy space. A grassland may also cover low shrubs and other plants, but its main feature is that it is a place where plenty of grass grows. Ecological systems (ecosystems) consist of all the living organisms in an area and their physical environment (soil, water, air).

Ecosystems are prejudiced over time by the local climate, the parent material under the plants, variations in the local landscape, disturbances such as fire and floods, and by the organisms that live in them. Grassland ecosystems in British Columbia generally occur in areas where the climate is hot and dry in summer and cool to cold and dry in winter. The parent material is often composed of fine sediments, and grasslands are most often in valley or plateau landscapes. The organisms that live in them comprise plants and animals that have adapted to the dry climatic conditions in a variety of ways.

Differences in elevation, climate, soils, aspect, and their location in relation to mountain ranges have caused in many differences in the grassland ecosystems of British Columbia. The mosaics of ecosystems found in our grasslands, including wetlands, riparian areas, aspen stands and rocky cliffs, allow for a rich diversity of species.

A grassland ecosystem is a public of creatures living together within a grassy space. These creatures can include various types of grasses, insects, and animals, *etc.* Grassland ecosystems can be found throughout the world, for example:

In the tropics, near to the equator.

In the temperate zones of the earth, between the equator and the Polar Regions.

Grassland Ecosystem is an area where the vegetation is conquered by grasses and other herbaceous (non-woody) plants. It is also called provisional landscape because grassland ecosystems are dominated by the grass with few or no trees in the area where there is not enough for a forest and too much of a forest. In India, they are found mainly high Himalayas. The rest of India's grasslands are mainly composed of the Steppes and Savana. Steppe formations occupy big areas of sandy and saline soils.

3.8.2 Types of Grassland Ecosystems

A grassland consists of large rolling fields of grasses, flowers and herbs. Grasslands ecosystems emerge due to low levels of irregular precipitation that is only considerable enough to support smaller plants. Grass survives in these arid conditions because of its deep and highly intricate root system that enables it to access dampness hidden deep in the soil. There are two main types of grasslands – tropical and temperate – with several subcategories within each type.

Tropical Grasslands

Tropical grasslands are warm all year round with established rainy and dry seasons. During the rainy season, tropical grasslands obtain between 50 and 130 centimetres of rain. Most notable of the tropical grasslands is the African savannah, which has an infrequent tree and is home to many of the world's most remarkable species, such as elephants, giraffes, lions and zebras. Savannah grass is typically quite short, making for excellent grazing and hunting grounds.

Temperate Grasslands

Temperate grasslands also have two seasons, growing and dormant.

During the dormant season, no grass or crops grow because it's too cold. These grasslands make for excellent farming because of their deep and nutrient-rich soils. Temperate grasslands obtain between 25 and 75 centimetres of rain a year. The most distinguished temperate grasslands are the North American prairies, which are home to a wide variety of wildlife, including pronghorn antelope, mice, jack rabbits, foxes, snakes and coyotes. Grass here can grow quite tall, reaching up to seven meters in height. Trees are restricted to where there is a large attentiveness of water such as a river or lake. Because of their profound rich soil, more than half of temperate prairie grasslands have been converted to farmland that provides a large chunk of the world's food.

The components of the Grassland Ecosystem are discussed below:

Biotic Components

The biotic components of a grassland ecosystem are the living organisms that occur in the system. These organisms can be classified as producers, consumers, or decomposers.

Producers are able to capture the sun's energy through photosynthesis and engross nutrients from the soil, storing them for future use by themselves and by other organisms. Grasses, shrubs, trees, mosses, lichens, and cyanobacteria are some of the several producers found in a grassland ecosystem. When these plants die they deliver energy for a host of insects, fungi and bacteria that live in and on the soil and feed on plant debris. Grasses are an significant source of food for large grazing animals such as California Bighorn Sheep, Mule Deer and Elk, and for much lesser animals such as marmots, Pocket Gophers, and mice.

Consumers are creatures that do not have the ability to capture the energy produced by the sun, but consume plant and/or animal physical to gain their energy for growth and activity. Consumers are further divided into three types based on their ability to digest plant and animal material:

- Herbivores eat only plants–for example, the elk that graze the grasslands of the Columbia Valley, or an insect nibbling on the leaf of a sticky geranium.
- Omnivores, such as the black bear, eat both plants and animals.
- Carnivores eat only animals–for instance, the red-tailed hawk or western rattlesnake.

Decomposers include the insects, fungi, algae and bacteria both on the ground and in the soil that help to break down the organic layer to deliver nutrients for growing plants. There are many millions of these organisms in each square metre of grassland. Soil has many biotic functions in a grasslands ecosystem. It delivers the material in which plants grow, holds moisture for plants to absorb, is the "recycling bin" for plant and animal matter, and provides an important habitat for soil organisms. Soil is a energetic link between the biotic and abiotic parts of a grassland ecosystem.

Abiotic Components

The abiotic components of a grassland ecosystem are the non-living features of the ecosystem that the living organisms depend on. Each abiotic component influences the number and variety of plants that grow in an ecosystem, which in turn has an effect on the diversity of animals that live there. The four main abiotic components are: climate, parent material and soil, topography, and natural disturbances.

- ☆ **CLIMATE:** Climate includes the rainfall, temperature and wind patterns that occur in an area, and is the most significant abiotic component of a grassland ecosystem. Temperature, in tandem with precipitation, controls whether grasslands, forests, or some combination of these two, form. The amount and distribution of the rainfall an area receives in a year influences the types and productivity of grassland plants. The climate in our grassland ecosystems is usually hot and dry in the spring and summer growing season, and cool or cold in winter dormant season. Precipitation in the winter falls typically as snow rather than rain. During the hottest months of the year (the height of summer) more water evaporates from parts of the grasslands than falls as rain, creating a moisture deficit.
- ☆ **PARENT MATERIAL AND SOIL:** Parent material is the biological material that lies on top of the bedrock and is the foundation on which soil has developed. Much of the parent material underlying BC's grasslands was deposited as the last ice sheets melted away. The actual composition of the material at any specific location depends on how and where it was deposited in relation to the ice. In the Rocky Mountain Trench, for example, some material was deposited under a moving glacier, while on the Chilcotin plateau some was deposited under a stationary ice sheet; in many places throughout the grasslands material was carried and deposited by water on, in, or under the ice.
- ☆ **SOIL PROFILE**: Grassland soils developed on the glacial till (C) deposited by the ice as it melted 12,000 to 10,000 years ago. These soils have a deep organic-rich layer (A) that results from the breakdown of the roots and plant material each year. The organic layer increases in depth with increases in elevation and moisture.
- ☆ **TOPOGRAPHY:** Topography is the variety of shapes found on the landscape determined by slopes, promotion and aspects. The topography of grassland ecosystems is a varied landscape of gently rolling hills and prairies, rock outcrops, cliffs, gullies, and low lying areas. Diverse topography is what gives incredible variety to grassland ecosystems.
- ☆ **NATURAL DISTRUBANCE:** Natural disturbances change grasslands in many ways, adding to the diversity of these ecosystems. Some types of disturbance, such as annual flooding of riparian areas along rivers and streams, can be predicted while others, such as a fire after a lightning storm, happen unexpectedly.

3.8.3 Characteristic Features of Grassland Ecosystem

Different types of grasslands share similar features. Savannahs have dispersed trees and prevail in certain parts of Africa, South America, Australia and Asia. Temperate grasslands are largely devoid of trees, receive less rainfall than savannahs and bear broader temperature extremes.

The two types of temperate grasslands are steppes and prairies. Prairies have smaller grasses, and prairies have taller grasses due to advanced precipitation. You can find both types of temperate grasslands throughout North America and Europe.

Grassland biomes are ecosystems in which the main vegetation type comprises of various grasses rather than trees or large shrubs. A grassland ecosystem may be divided into numerous types, including savannahs, temperate grasslands, tall grass prairies, steppes, alpine tundra, and flooded grasslands, among others.

The kind of grassland that occurs in a particular region is based on the latitude, terrain, local climate, precipitation, and wildfire regime. The faunal groups and grassland plants reinforced by these different types of grasslands vary on the basis of characteristics of the grasslands themselves, as well as by geographic location.

Temperate grasslands lack the scattered trees that are typical of savannahs. Temperate grasslands have distinct hot and cold seasons and a moderate amount of precipitation in the late spring and early summer. Taller grasses control in the temperate grasslands that receive more rainfall, whereas shorter grasses can be found in drier regions.

Temperate grasslands have deeper, richer soils than savannahs and support a dissimilar suite of animal species, including gazelles, zebras, rhinoceroses, horses, lions, wolves, deer, jackrabbits, foxes, skunks and prairie dogs. The exact grassland plants and animal compositions of temperate grasslands are driven by their geographic location, local climatic regime, and soil type.

Flooded, montane, tundra, and desert grasslands are additional distinct grassland biomes. Flooded grasslands or flooded savannahs are wetland habitats dominated by grasses, for example, the Florida Everglades. Montane or alpine grasslands may be tropical, subtropical or temperate and occur in cool temperatures at high elevations, such as the steppes of the Tibetan plateaus.

The highest-elevation montane grasslands are called the alpine tundra. Grasslands that grow in deserts, which are distinct as regions receiving less than 50 centimetres of rain per year, are known as desert grasslands.

Precipitation in the Grasslands: Grasslands make up 25 per cent of the Earth's land surface and dominate in regions with partial rainfall, which prevents forest growth. This is the result of nearby mountain ranges that cause rain shadows over adjacent open range lands. Typically, grasslands have not only limited but also unpredictable rainfall, and droughts are common. Where rainfall is even less, deserts will form. Savannahs, on average, receive roughly 76 to 101 centimetres (30 to 40 inches) of rain per year, but steppes only average 25 to 51 centimetres

(10 to 20 inches) per year. Prairies tend to be intermediate between savannahs and steppes with 51 to 89 centimetres (20 to 35 inches) per year.

Temperature in the Grasslands: Temperatures vary much more in temperate grasslands than they do in savannahs. Savannahs are in warm climates with average annual temperatures that only vary between 21 and 26 degrees Celsius (70 and 78 degrees Fahrenheit). They characteristically have only two seasons, a wet and a dry season. Temperate grasslands have hot summers where temperature can surpass 38 degrees Celsius (100 degrees Fahrenheit) and cold winters that can drop below negative 40 degrees Celsius (negative 40 degrees Fahrenheit).

Fire in the Grasslands: Fires are an significant grassland characteristic. Regular fires endorse the growth of native grasses but limit the growth of trees. Native grasses have deeper root systems that can survive fires, but aggressive plants tend to have shallower roots and succumb to fires. Development has curtailed the number and extent of grassland fires, and the lack of seasonal fires threatens the health of the world's grasslands. Only 5 per cent of the world's grasslands are being protected and maintained, and they remain the most endangered biome in the world.

Flora and Fauna: Savannahs are home to some of the main mammals on the planet such as elephants, giraffes, rhinos, lions and zebras. Temperate grasslands are also home to large mammals, chiefly bison and horses, medium-sized mammals like deer, antelope and coyotes, as well as small mammals such as mice and jack rabbits. The type of grasses that grow depend upon the amount of rainfall. Shorter steppe grasses often contain of buffalo grass, and savannah grasses will contain taller grasses like bluestem and rye.

3.8.4 Structure and Function of Grassland Ecosystem

According to the Grasslands Conservation Council of British Columbia, grasslands function as a habitat for biotic mechanisms or living organisms "classified as producers, consumers or decomposers." For example, trees and plants produce nutrients through the process of photosynthesis, whereas consumers such as animals eat plants and other animals to get energy.

I. Biotic Components

1) *Producer Organisms*

In grassland, producers are mainly grasses; though, a few herbs and shrubs also contribute to primary production of biomass.

Some of the most common species of grasses are:

Brachiaria sp., *Cynodon* sp., *Desmodium* sp., *Digitaria* sp.

2) *Consumers*

In a grassland, consumers are of three main types;

a) Primary Consumers

The primary consumers are herbivores feeding directly on grasses. These are grazing animals such as

Cows, Buffaloes, Sheep, Goats, Deer, Rabbits *etc.*

Besides them, numerous species of insects, termites, *etc.* are also present.

b) Secondary Consumers

These are carnivores that feed on primary consumers (Herbivores)

These include;-Frogs, Snakes, Lizards, Birds, Foxes, Jackals *etc.*

c) Tertiary Consumers

These include hawks *etc.* which feed on secondary consumers.

3) Decomposers

These include wide variety of saprotrophic micro- organism like: Bacteria; Fungi; Actinomycetes. They attract the dead or decayed bodies of organisms and thus decomposition takes place. Therefore, nutrients are released for reuse by producers.

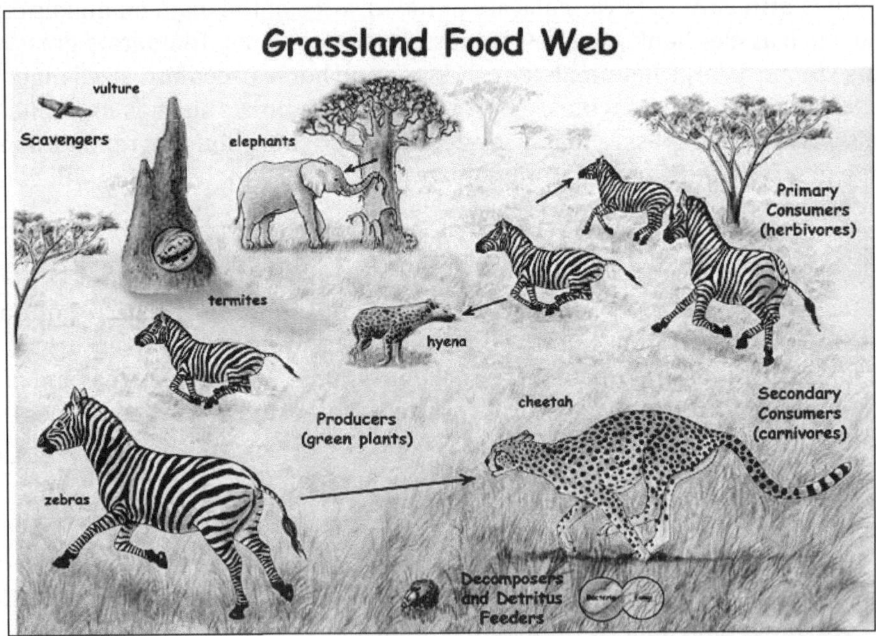

II. Abiotic Components

These include basic inorganic and organic compounds present in the soil and aerial environment. The essential elements like C, H, N, O, P, S *etc.* are supplied by water, nitrogen, nitrates, sulphates, phosphates present in soil and atmosphere.

Function of Rangeland Grasslands

Grasslands composed of grasses, such as wheat, provide grazing for livestock, as well as insect herbivores such as grasshoppers, locusts, and Mormon crickets.

Grasslands Provide Food

"Grasslands have been the seedbeds for the ancestors of major cereal crops, including wheat, rice, rye, barley, sorghum, and millet," states the World Resources Institute (WRI).

Grasslands Are Breeding Areas

Twenty-three Endemic Bird Areas include grasslands, which are important breeding grounds for thousands of bird species. The Andes in Peru, Central Chile, and southern Patagonia "rank highest for biological importance," states the WRI.

3.9 Desert Ecosystem

3.9.1 Introduction

Deserts are categorised by extreme environmental conditions. They get at most 50 centimetres (cm), or 20 inches, a year of precipitation; more normally, they're fortunate to get half of that. Most of them are found at low latitudes, that is, earlier to the equator than to the poles. The massive Sahara, probably the most famous desert on Earth and its third largest, lies just north of the equator in Africa. While they are far less densely populated than other biomes owed to how dry they are and being ill hospitable overall, most deserts do feature a range of vegetation as well as both vertebrate and invertebrate animal life. The word desert comes from the Latin word 'desertus'. 'Desertus' means waste, or somewhat that has been left. From this we get English words such as desert and deserted.

An ecosystem is a system of organisms that live together as a community. So, putting these words together, we can say that a desert ecosystem is a community of organisms that live together in an environment that seems to be deserted wasteland.

Desert ecosystem is the driest ecosystem of the earth and this is the reason it has less vegetation and less diversity of life. It is one of the parts of the terrestrial ecosystem. The plants and animals of the desert ecosystem have mastered the art of survival in harsh conditions. A desert ecosystem is basically devoid of any rainfall or precipitation.

While deserts are dry, they can be found all around the world. While we may think of a desert as a hot, dry piece of land, it can be cold as well. Irrespective of the region, any desert is usually cold at night and receives very little rainfall. However, they do produce plants, which have adapted to such living conditions. Although a desert is harsh and uninviting to humans, it's a natural part of the global environment. Some places will always get more precipitation, which means other will be left to survive without the presence of regular rainfall.

Inaptly, some of the desert animals are being compromised because of human involvement. Expanded housing options mean that people are starting to move into desert climates and import food and other resources from lush areas. We can live almost anywhere now as long as we're connected, which is threatening the biodiversity of desert wildlife.

While these ecosystems are significant, they're far more resilient than other climates in the world. Animals and plants have adapted to the most extreme conditions on the planet, but we need to curb human involvement if we want to continue observing these fascinating and marvellous creatures.

Several things make up a desert ecosystem. Among those are:

- Structure
- Characteristics
- Animals

The ecosystem is dependent upon the type of desert; temperate deserts, also referred to as cold deserts, or hot or subtropical deserts. Hot deserts and cold deserts have different kinds of ecosystems. However, despite being very different, the two kinds of deserts have a few similarities.

3.9.2 Types of Desert Ecosystem

When we hear the word desert, we usually think of a very hot, sandy environment. But, this is just one type of desert ecosystem. Depending on the natural environment and geographical location there are different types of deserts present of the earth. They are:

Hot Deserts

Hot deserts are the desert area which is situated closer to the equator line. The Sahara Desert is one of the examples of the hot desert. The temperature around these types of the desert is highest all around the globe. The existence of any plant and animals is very stimulating sure to a shortage of food and water. Still, many plants and animals have adapted themselves to survive in such hot weather.

The seasons are sincere to quite hot year-round, and owing to the low humidity of these environments, the temperature swing from the hottest time of day to the coldest time of day may be extreme – over 45 C (about 80 F) in some regions. This is mostly because the surface receives twice as much solar radiation during the day as would the surface in comparable but more humid environment, and loses twice as much heat at night.

Rainfall is characteristically very sparse in hot-and-dry deserts, and evaporation rates routinely outpace rainfall rates. Falling rain has even been eminent to evaporate before attainment the ground. What little rain these deserts get occurs in short, brief and sometimes intense bursts, though monsoons and leftovers of tropical systems that drift into some deserts can provide abundant moisture at times. The Atacama Desert in Chile on the western coast of South America, known as the world's driest place, receives an average of 1.5 cm per year of rain – barely a half-inch.

Plants in hot-and-dry deserts are mostly low shrubs and short, woody trees. Animals include small nocturnal carnivores, with comparatively high populations of burrowers and kangaroo rats. Insects, arachnids, reptiles and birds are also shared. The animals hide from the sun and then come out to forage at dusk or at night, when the desert is coolest.

Cold Deserts

Hot deserts usually exist at low altitudes. Desertification can exist at high altitudes too, however – and when this happens, the desert will be cold. A good example is the deserted rocky peaks of a mountain. A cold desert may be sandy or rocky, but it will be a harsh environment where organisms have modified in weird and wonderful ways so that they can survive. Cold winter deserts, also called semiarid deserts, are characterized by moderately long, dry summers and winters that include brief intervals of rain. This pattern is like that of hot-and-dry deserts, but the overall temperatures are somewhat cooler. U.S. examples include the sagebrush zones of Utah, Montana and the Great Basin. They also include northern, but subarctic, portions of North America, Newfoundland, Greenland, Russia, Europe and northern Asia.

Summer temperatures in these deserts usually average between 21-27 C (70-80 F). It normally does not rise above 38 C (100 F), and evening temperatures are cool, at around 10 C (50 F). Annual rainfall can be as low as only 2 to 4 cm (about 0.8 to 1.5 inches).

Soil can variety from sandy and fine-textured to loose rock fragments, gravel or sand. There is no subsurface water in these environments. As for vegetation, cacti (the plural of "cactus") are found here. The spines of cacti and other plants in cold winter deserts provide protection in a difficult natural setting. The multitude of spines offer enough shade for the surface of these plants to reduce water losses through transpiration. Many plants have glossy leaves, which lets them reflect more light energy. Semiarid desert plants comprise Creosote bush, bur sage, white thorn, cat claw, mesquite, brittle bushes, lyceums and jujube.

As for animals, insects and jack rabbits are seen during the day, staying in shadow as much as possible. Many animals seek protection in burrows underground, where they are insulated from the hot, dry air. These include kangaroo rats, rabbits, skunks, some insects, birds and reptiles.

Ice or Polar Deserts

Ice deserts are another type of cold desert. Here, in its place of a sandy or rocky wasteland, we have a seemingly uninhabited region that is calm of ice. Ice deserts can be found towards the north and south poles of the planet, though they may also be located high up on mountain peaks. Polar deserts or cold deserts are curiosities, like almost everything about the Earth's poles. Compared to other desert biomes, they receive a veritable flood of precipitation, especially in the winter months. Mean annual precipitation is about 15 to 26 cm (6 to 10 in). Winter in the Arctic Polar Desert – which spans 5.4 million square miles in parts of Alaska, Canada, Greenland, Iceland, Norway, Sweden, Finland and Russia – occurs between mid-December and mid-March, while that in the 5.5-million-square-mile Antarctic Desert that spans the continent after which is was named falls between mid-June and mid-September.

Polar desert plants are widely scattered over the vast lands in which they grow. Plant heights can reach 122 cm (about 4 feet) in some areas. The main plants are deciduous, meaning that they have leaves they shed seasonally with most of these having spiny leaves. Fungi and dwarf shrubs are also common.

Temperate Deserts

Although both types of deserts have similar features, their climates are very diverse. Neither is very friendly to human or plant life, and animals have to adapt to survive in both temperate and subtropical desert ecosystems.

As we covered briefly above, temperate deserts are dry tundra, where freezing temperatures force animals to store extra fat to stay warm. The animals in this area make the most out of what warmth they can get by fascinating sunlight through their fur.

Coastal Deserts

They are found in regions that are generally cool to moderately warm. Portions of the aforementioned Atacama Desert in Chile represent the coastal desert biome. Here, cool winters alternative with relatively long and warm summers. Temperatures are reasonable compared to the two desert biomes already discussed. Average summer temperatures range from 13-24 C (55-75 F); winter temperatures are 5 C (41 F) or colder. The maximum annual temperature is close to 35 C (95 F), and the minimum is about -4 C (25 F).

Rainfall, while sparse, exceeds that of hot-and-dry and cold-winter deserts, averaging around 8 to 13 cm (3 to 5 inches) per year. The soil in these deserts is high in salt and other nutrients. Some of the plants have wide root systems, unlike the flora in the aforementioned desert types. These plants are almost botanical analogues of camels in that they can store very large quantities of water for future use when it is available. These plants include the salt bush, buckwheat bush, black bush, rice grass, little leaf horse brush, black sage and chrysothamnus.

Coastal-desert animals boast special adaptations for dealing with heat and lack of water. For example, some toad species seal themselves in burrows with sticky, gel-like secretions and remain inactive for eight or nine months until a heavy rain washes them out. Amphibians that include larval stages of development feature accelerated life cycles, bettering their chances of reaching maturity before the rainwater evaporates. Some insects lay eggs that are able to stay dormant in adverse conditions, maturing only when their environment is more suitable for hatching; fairy shrimps do the same. Coastal-desert mammals include coyotes and badgers; birds include the famed great horned owl, golden eagle and the bald eagle. Lizards and snakes are the chief reptilian representatives.

3.9.3 Characteristic Features of Desert Ecosystem

Here are some basic Desert ecosystem characteristics:

- **Dryness**: It is the common characteristic of all the deserts on the earth. Aridity simply implies the deficiency of moisture's or dryness. Desert experience very less rainfall and thus result in aridity.
- **Fewer rainfall/drizzle**: Less precipitation is one of the major features of deserts and also the reason behind the dryness. The rainfall in deserts is seasonal and occurs only for a limited duration. The annual rainfall that a desert receives every year is just 25-30 centimetres.
- **Thrilling temperature**: Desert ecosystems experience extreme temperatures during day and night. The days are very hot and the nights can be extremely cold. It is the sole characteristic of all the desert ecosystems either hot or cold all lacks moisture.
- **Velocity of wind**: It tends to be very high in a desert ecosystem. This is the reason deserts experience sandstorms/dust storms of high intensity resulting in the formation of huge sand dunes.
- **Scarcity of water**: Due to less rainfall, there is a shortage of water in a desert ecosystem. Due to the scarcity of water deserts have to face the situation of drought half of the year.
- **Mugginess**: The humidity level in a desert ecosystem is very low in the daytime and relatively high at night.
- **The feature of the soil**: In deserts is very low to grow vegetation. It is dry, rocky, thin, sandy, mainly grey in colour and has no organic contents like nitrogen, phosphorus *etc.* which are essential for the growth of plants.
- **Biodiversity in a desert ecosystem**: However, the survival in a desert ecosystem is very hard but despite the fact, deserts are home to various plants and animals. The plants and animals have adapted to survive in the harsh and extreme conditions of the desert.
- **The population density**: Density is very low in deserts and nearby areas as there is there a scarcity of water, food and climatic conditioner are too harsh.
- **Plants grow very gradually**: In this ecosystem that includes basically vegetation's with spines like cacti that can easily survive in such conditions.

3.9.4 Structure and Function of Desert Ecosystem

I. Biotic components

1) *Producer Organisms*

In a desert, producers are mainly shrubs/bushes; some grasses and a few trees.

Dominant plant species include: Succulents (water - retaining plants adapted to arid climate or soil conditions) and hardy grasses.

Besides some lower plants such as lichens and xerophytes mosses are also present.

2 *Consumer Organisms*

These include animals such as insects, reptiles which are capable of living in xeric conditions

Besides some nocturnal rodents, birds and some mammalians like camel *etc.* are also found.

3) *Decomposers*

Due to poor vegetation with very low amount of dead organic matter, decomposers are poor in desert ecosystem. The common decomposers are some bacteria and fungi, most of which are thermophillic.

II. Abiotic Components

Due to high temperature and very low rainfall, the organic substances are poorly present in the soil.

Deserts as Climate Pointers

Desert is one biome type that researchers and conservationists do not want to expand all the while ensuring that the deserts we currently have do not disappear. The main reason for this is that desert is of low agricultural quality, low biodiversity, and an extreme environment. Climate change presents a risk of desertification of present marginal landscapes. We know this occurs from paleoenvironmental data (24). Many of our deserts contain the remnants of ancient lakes, designated by former water stations and conserved biotic and abiotic materials indicative of previous presence of water. As deserts are the most extreme environments on the planet, they are disposed to to the most change. Radiation is intense - both in terms of how much deserts obtain and how much they reflect back into the environment. The expansion of desert can directly impact global average temperatures additional, increasing water evaporation in adjacent areas. Even though the plant and animal species that exist in hot deserts are well-adapted to those environments, we know from studies that such organisms are treading a fine line over environmental tolerance; some are even at their limits, according to the IPCC (25).

Desert Resources

As well as biodiversity and their cultural implication, deserts are also useful frugally due to their unique geological profiles or their environmental formation procedures. The desert biome is one of the most significant. There are 15 mineral deposit kinds on our planet and 13 of them are found in deserts. This makes the desert a significant place for mineral resources and for local and global economy. The aim this is the case is how water leaches through the ground or through evaporation so swiftly that mineral deposits are left behind. We discussed the problem with irrigation in Mesopotamia in the archaeology section where increased salt deposits made agriculture harder. This is a similar but natural process and it leads to the creation of large metal credits in particular. Typical mineral resources found in desert regions (both hot and cold) include salt and borates, and gypsum. Borates occurred in high density in the Great Basin Desert here in the US. From there, borate was taken to Death Valley to the emerging railroads and all over the country. This special kind of salt is used to manufacture glass, enamel and other ceramics, and in the medicinal and agrichemical industries. All of it was mined from the dried beds of paleolakes that once existed in the region. It is estimated that the value to the US economy of borates alone has topped $1 billion. The US deserts are also home to copper.

3.10 Aquatic Ecosystem

3.10.1 Introduction

An aquatic ecosystem is an ecosystem that is water based. The word 'aquatic' comes from the Latin word for water. An ecosystem is a separate community of organisms in an exact environment.

So, we can say that an aquatic ecosystem is a community of organisms that live together, interact, and to a degree be contingent on each other in a water based environment. An Ecosystem which contains the water is the main source of the habitat is known as the Aquatic Ecosystem. An aquatic ecosystem is essentially a group of interrelating organisms dependent on one additional and their water environment for nutrients (*e.g.*, nitrogen and phosphorus) and shelter.

While, familiar examples are ponds, lakes and rivers, but aquatic ecosystems also comprise areas such as floodplains and wetlands, which are flooded with water for all or only some parts of the year.

- ✰ Fresh Water Ecosystem: The Salt Contains the Fresh Water Bodies is Very Low always than the 5ppt.
- ✰ Marine Ecosystem: The Water Bodies Contains the Salt Concentration Equal or Above the Sea Level.
- ✰ Brackish Water Ecosystem: These Water Bodies have Salt in the Content between 5 to 35 PPT.

Aquatic ecosystems provision a wide range of organisms, including microorganisms, invertebrates, insects, plants, and fish. Some hydrologists work in understanding the trophic systems within aquatic ecosystems and their fitness as a function of environmental conditions such as water temperature and turbidity. Aquatic biodiversity is a main concern in water conservation and restoration projects, as well as water resource management.

An aquatic ecosystem is a group of interacting organisms dependent on one another and their water environment for nutrients (*e.g.*, nitrogen and phosphorus) and shelter. An aquatic ecosystem is an ecosystem that is based in water, whether it is a pond, lake, river or an ocean. It involves living aquatic organisms (*e.g.*: fish, planktons, annelids, *etc.*), which constitute as the biotic factors and their relationship with their environment, which together can be referred to as the abiotic factor. Composing more than 70 per cent of the earth's surface, aquatic ecosystems are not only the leading features of earth but are also very diverse in species and complexity of interaction among their physical, chemical and biological components.

An aquatic ecosystem comprises freshwater habitats like lakes, ponds, rivers, oceans and streams, wetlands, swamp, *etc.* and marine habitats include oceans, intertidal zone, reefs, seabed and so on. The aquatic ecosystem is the home for water-dependent living species including animals, plants, and microbes.

3.10.2 Types of Aquatic Ecosystem

Generally, a body of water can be classified as being freshwater, marine, or estuarine. A freshwater body of water has fewer dissolved compounds, or salts, present, while a marine body of water has numerous salts dissolved in it, henceforth the term 'salt water'. The average salinity of salt water is around 35 parts per thousand. Estuarine areas are those that experience a flux of both fresh and salt water, contingent on the tides and water currents. For example, the area where the Mississippi River meets the Gulf of Mexico is considered estuarine because there is a constant mix of fresh and salt water.

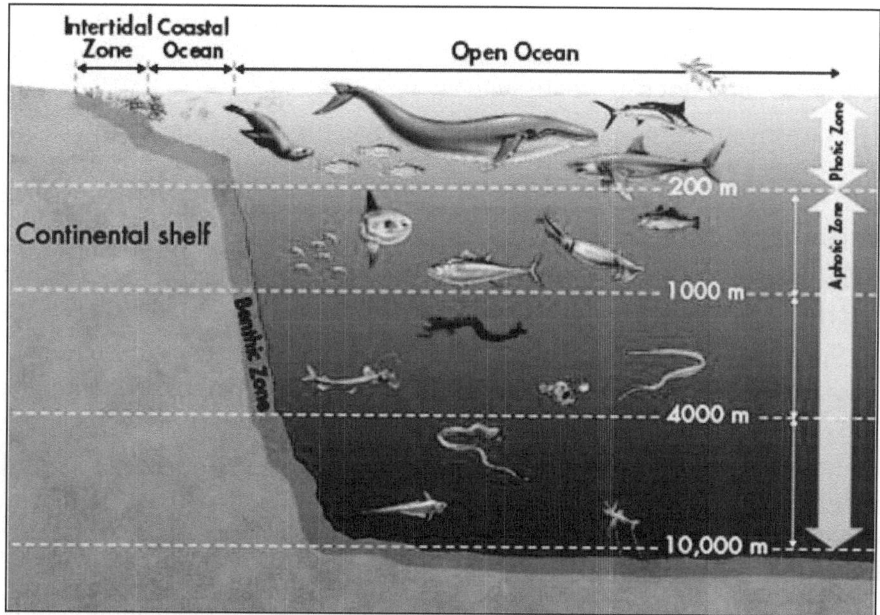

We can classify aquatic systems even further if we look at patterns of water movement. Lentic water bodies have very slow-moving or stagnant water. These comprise lakes and ponds. Lotic water bodies have faster-moving water, like rivers and streams. You can remember the difference by thinking about how lotic waters flow, the 'o' in each is pronounced the same. Finally, we have wetlands, which are precisely what they sound like. The soil of a wetland is saturated or flooded with water at least part of the time. Wetlands often occur in the zones linking land and large water bodies.

Aquatic ecosystems are any water-based environment in which plants and animals interact with the chemical and physical features of the aquatic environment. Aquatic ecosystems are generally divided into two types –the marine ecosystem and the freshwater ecosystem. The largest water ecosystem is the marine ecosystem, covering over 70 per cent of the earth's surface. Oceans, estuaries, coral reefs and coastal ecosystems are the various kinds of marine ecosystems. Freshwater ecosystems cover less than 1 per cent of the earth and are subdivided into lotic, lentic and wetlands.

As with other types of water, the aquatic ecosystem also has numerous types or several kinds. This type of water ecosystem depends on the location of the waters. Regarding water ecosystems which contain of several types, here are the kinds of water ecosystems:

River Ecosystem or Freshwater Ecosystem

The first type of water ecosystem is the river ecosystem. The ecosystem we are going to talk about is also often referred to as a freshwater ecosystem. This river ecosystem or bargain ecosystem is an ecosystem that is in a river atmosphere. It is mentioned to as a freshwater ecosystem because the water around this ecosystem has a tasteless and not salty taste such as seawater.

This freshwater ecosystem or river ecosystem turns out to have many features. The characteristics controlled by freshwater ecosystems are a differentiator from other ecosystems. Some of the characteristics possessed by river or freshwater ecosystems include:

Have a low salinity or salt level. The salt content in the river or freshwater ecosystem is very low, even lower than the cytoplasm.

Temperature is not extreme. These freshwater or river ecosystems have variations or levels of temperature changes that are not too extreme. The point is that between day and night, there is no difference in temperature that is too striking.

Limited sunlight can enter. In freshwater or river ecosystems, there is sunlight but in limited quantities, even less. This is because sunlight can only penetrate up to a few meters in the water. As a result, only certain areas or parts can enjoy the sunlight.

Prejudiced by climate and weather. Although the water environment only has a slight influence from conditions that exist in nature, this water environment still has influence. Influence is mainly due to the weather and also the climate. Although not too much affected by weather and climate such as terrestrial ecosystems, this water ecosystem is still influenced by both of these.

Those are the features of a freshwater ecosystem or river ecosystem. This freshwater ecosystem turns out to be divided into 2 types based on the motion of the water. Both types of freshwater ecosystems are limber ecosystems and lotic ecosystems. Limber ecosystems are ecosystems where the water is calm or not flowing, and the lotic ecosystem is an ecosystem where the water moves. The main types of freshwater ecosystem includes pond ecosystem, Lake Ecosystem and river ecosystem.

Pond Ecosystems

A pond is discernibly a closed, self-contained environment which houses a community of organisms. Ponds are usually freshwater ecosystems, however they can also be made up of brackish (salty or briny) water. Many diverse plants, fish and animals can live in these types of ecosystems. Frogs, newts, water weeds and water lilies are all examples of pond creatures. In addition, various types of fish can live in a pond. Ponds can be natural or human made ecosystems; if human made, it is not unusual for goldfish or ornamental carp (such as koi carp) to live in a pond ecosystem. In accumulation, certain birds and insects may visit the pond ecosystem with orderliness. For example, we might see dragonflies or herons around the pond. It may be up for debate whether these visitors are truly part of the ecosystem as they may also visit other ecosystems. But, it is certain that they have an impact on the ecosystem – and that it has an impact on them.

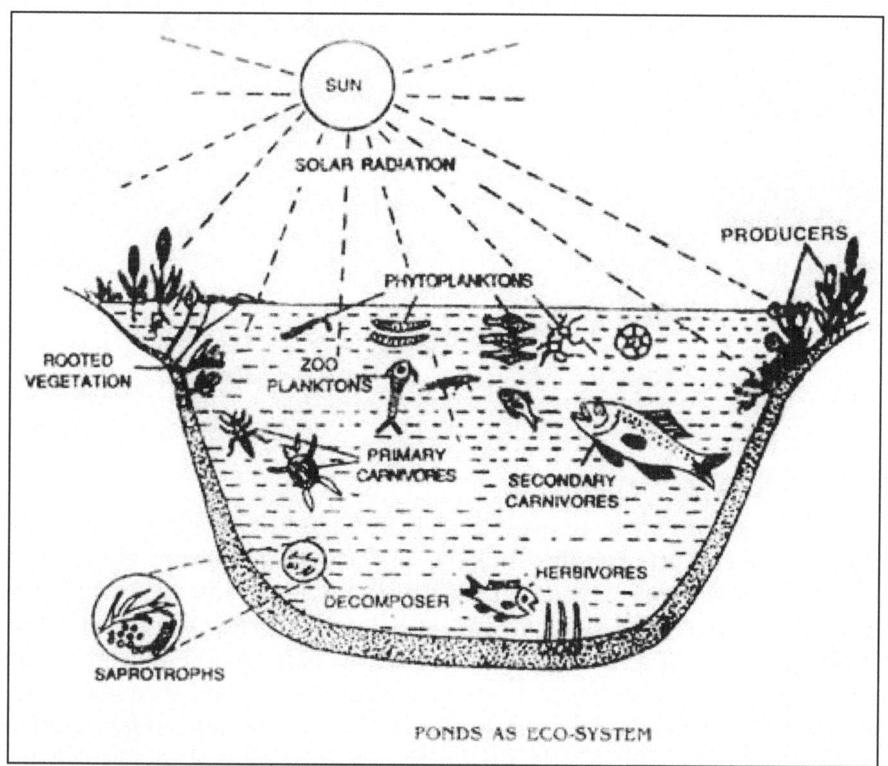

PONDS AS ECO-SYSTEM

Lake Ecosystems

Because they tend to be physically enclosed by the earth, rock or mountains round them, freshwater lakes are also recognisable as a distinct habitat that is occupied by a distinct community of organisms. In a freshwater lake ecosystem, we can find all kinds of diverse organisms, including crustaceans (such as shrimp and crayfish), fish (like carp, trout and pike) and many birds, reptiles and amphibians. Freshwater lakes can be home to some beautiful plant life, such as tall purple irises, and the flora and fauna that abound within them may also change with the seasons. Some animals may only use lakes for looking after their offspring in, such as frogs that may leave frog-spawn in a lake before leaving to inhabit other ecosystems.

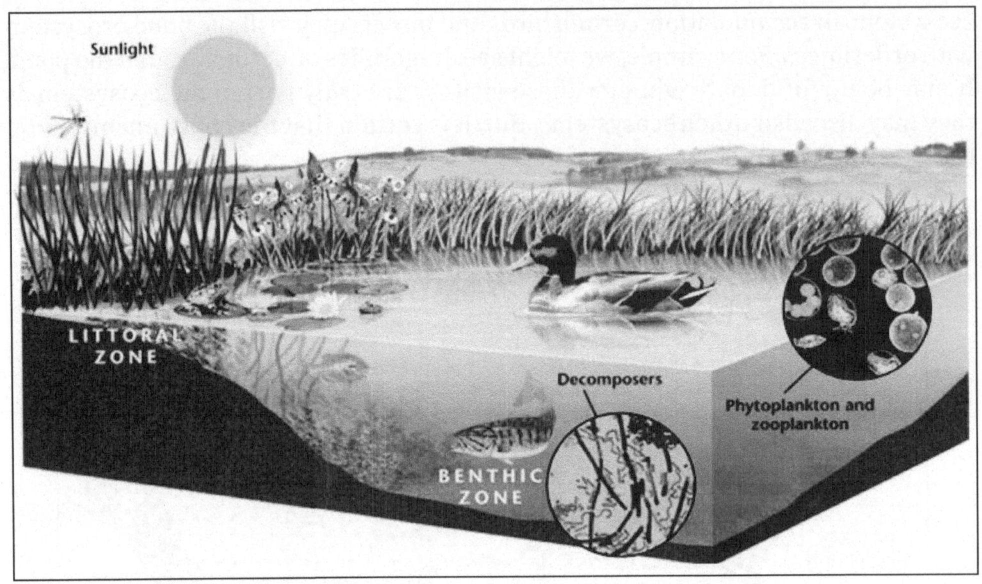

Freshwater River Ecosystems

River ecosystems are slightly different to ponds and lakes because whereas the concluding two ecosystems offer still (static) water, river water is always flowing. That means that these river ecosystems are the homes of animals and plants that are best modified to living in flowing water. Salmon are a key instance, as they use the flowing motion of a river to help them with their annual migration. And, in general, organisms that prefer to travel – whether to seek food or to seek a partner – are often to be found in freshwater river ecosystems because the motion of the river suits their style of life (whilst they, in their turn, have evolved to suit a flowing environment). Rivers tend to flow into the sea, and in this way river ecosystems and marine ecosystems meet each other. It may well, therefore, be up for debate to what degree river ecosystems are closed systems. But it is certainly clear that these are distinct types of fresh water ecosystems.

Marine Ecosystem

The second type of aquatic ecosystem is the marine ecosystem. As the name implies, marine ecosystems are a type of ecosystem that is in the territorial waters of the sea. As with freshwater ecosystems, the marine ecosystem also has its own features. The features controlled by marine ecosystems are special properties and are only possessed by the seawater ecosystem. Some of the properties possessed by seawater ecosystems include:

Has a high salt content. Unlike the freshwater ecosystem, this seawater ecosystem actually has a high salt content. The salt gratified of this seawater ecosystem is 0.3 per cent. The amount of salt with this amount is similar to the density of protoplasm.

There is life in all its parts. This marine ecosystem is a type of ecosystem that has life in every part of it. All depths in the sea have life, except on the very seabed. A deep seafloor is a place that is not suitable for life.

Interconnected ecosystems. This seawater ecosystem is a type of ecosystem that is interconnected with each other. This is one of them due to the long and wide sea area. Therefore, this ecosystem is very possible to mix with one another because of the circulation of seawater.

There is a waste of energy. The marine ecosystem is a type of ecosystem that has a relatively long food chain. Therefore, there is a lot of energy waste along the food chain.

Those are the features of marine ecosystems. Similar to freshwater ecosystems or river ecosystems, these marine ecosystems in each depth have their own characteristics. The surface area has more production activities compared to the inside or the base. This is because on the surface of the sun's light, it is easier to enter.

Ocean Ecosystems

Our planet earth is gifted with the five major oceans, namely Pacific, Indian, Arctic, and the Atlantic Ocean. Among all these five oceans, the Pacific and the Atlantic are the main and deepest ocean. These oceans serve as a home to more than five lakh aquatic species. Few beings of these ecosystems include shellfish, shark, tube worms, crab small and large ocean fishes, turtles, crustaceans, blue whale, reptiles, marine mammals, seabirds, plankton, corals and other ocean plants. Oceans are home to different species of life. The waters of the Arctic and Southern Oceans are very cold, yet filled with life. The major population of krill (small, shrimp-like marine creatures) lies under the ice of the Southern Ocean.

Estuaries Ecosystem

Estuaries are places where rivers meet the sea and may be distinct as areas where salt water is weak with fresh water. River mouths, coastal bays, tidal marshes and water bodies behind barrier beaches are some examples of estuaries. They are biologically productive as they have a special kind of water circulation that traps plant nutrients and stimulates primary production.

An estuary is a semi enclosed with a coastal body of water with one or more rivers or streams flowing into it.

It has free connection with an open sea.

The complete salinity range from 0-35 PPT is seen from the head to the mouth of an estuary.

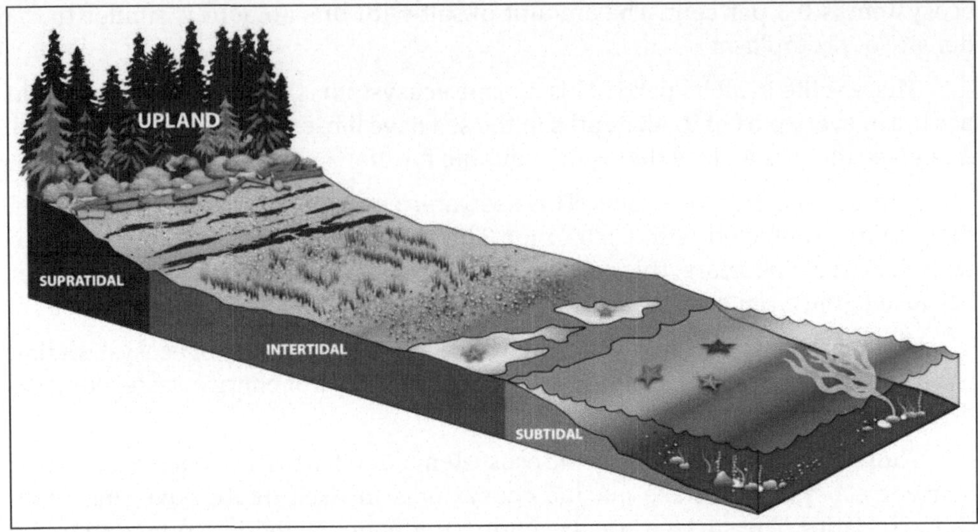

Estuaries are typically classifies by their geomorphological features or by water circulation patterns and can be referred by many different names such as bays, harbours, lagoons, inlets, *etc.*

Estuaries are usually biologically highly productive zones.

3.10.3 Characteristic Features of Aquatic Ecosystem

Ecosystems are one natural or environmental individual in which there are biotic and abiotic components. Therefore, in nature, there is not only one type of ecosystem but there are many types of ecosystems, considering that the environment on Earth is not just waters. Broadly speaking, the ecosystems on Earth are divided into two types, namely terrestrial ecosystems and aquatic ecosystems or water ecosystems. Because there is more than one type of ecosystem, these ecosystems have their own characteristics.

Aquatic ecosystems have several special features that will differentiate them from other types of ecosystems. Besides its environment which is dominated by waters, there are still other characteristics that are known to be overcome by this aquatic ecosystem. Regarding the features possessed by a water ecosystem, here are the features of the water ecosystem:

The environment is dominated by waters – Water ecosystems are ecosystems whose habitat is dominated by water. Nevertheless, there are still parts of this ecosystem that are not in the form of waters but with a very small number or percentage. An example is the surface of the water, although this part is still included in the water ecosystem but has mixed with air.

Inhabited by creatures that live in water or amphibians – This water ecosystem is inhabited by creatures that live in water, as well as in two realms (water and land). So from that, in this water ecosystem, we will meet animals and plants like that.

Limited sunlight – Unlike terrestrial ecosystems that are exposed too much sunlight, in this aquatic ecosystem, we will find partial or less sunlight because this sunlight can only infiltrate a few meters in the water. So, this ecosystem only gets limited sunlight.

Temperature changes are not too extreme – In contrast to land ecosystems that have extreme temperature differences between day and night, these aquatic ecosystems do not have extreme differences between the two conditions. Terrestrial ecosystems have extreme temperatures because they get a lot of sunlight during the day and are exposed to lots of wind so that this causes a very noticeable temperature difference while this water ecosystem only gets limited sunlight and is not too much exposed to the wind. So, the temperature difference is not too extreme.

There are consumers and producers – Just like terrestrial ecosystems, these aquatic ecosystems also have creatures that act as consumers and also producers. If on land acting as a producer are plants, then in the water which acts as a producer is phytoplankton. While those who act as consumers are animals and also water predators.

Those are the features of a water ecosystem. Water ecosystems do have some similarities and also differences with land ecosystems. This is a form of diversity in an arrival on Earth as a planet that has planetary features that can be inhabited by living things. The health of aquatic ecosystems is crucial to the health of the planet as a whole. Our earth is not called the blue planet for nothing: the seas with their fish, weeds, invertebrates and mammals and the rivers, lakes, streams, swamps and ponds of this world are all precious repositories of biodiversity. The seas help to control the world's temperature, too, and to lock carbon away from the atmosphere. Though we should all try and cut down on fish as a food source, there is no repudiating that fish and other aquatic organisms are irreplaceable links in the food chain for many terrestrial animals (*i.e.* animals that live on earth) as well.

3.10.4 Structure and Function of Aquatic Ecosystem

Aquatic ecosystems perform numerous valuable environmental functions. They recycle nutrients, purify water, recharge groundwater, augment and maintain stream flow, and deliver habitat for a wide variety of flora and fauna and recreation for people. A fast population surge accompanied by unplanned developing works has led to the pollution of surface waters due to residential, agricultural, commercial and industrial wastes/effluents and decline in the number of water bodies. Increased demands for drainage of wetlands have been accommodated by channelization, resulting in further loss of stream habitat, which has led to aquatic organisms becoming extinct or endangered in increasing numbers and to the damage of many beneficial uses of water, including drinking, swimming and fishing. Various anthropogenic activities have altered the physical, chemical and biological processes within aquatic ecosystems. A combined and accelerated effort toward environmental restoration and preservation is needed to stop further degradation of these fragile ecosystems. Failure to restore these ecosystems will result in sharply increased environmental costs later, in the death of species or ecosystem types, and in permanent ecological damage.

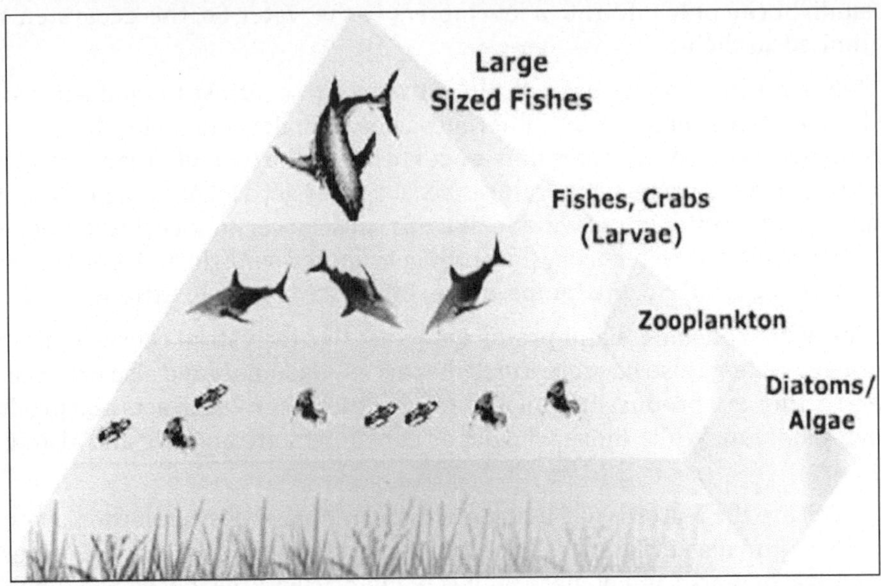

Aquatic Ecosystem Restoration means to reinstate degraded ecosystems to a level that can be enduringly continued through protection and conservation. More convincingly, restoration mitigates, effects or remediates tainted ecosystems to stages that have a higher order of ecological sustainability. Restoration can occur obviously, but usually involves reductions of stresses such as nutrient or contaminant loads. Valuations of tainted aquatic ecosystems allow decisions to be made as to what to control in order to remediate effects, or how much can be trusted on nature to clean itself. To achieve ecosystem stability or sustainability, decisions on what to do, including *in situ* options such as bioremediation or bio-manipulation, as well as the development of ecosystem indicators of progress towards restoration have to be taken. Restoration towards a less tainted, but not unavoidably pristine ecosystem, requires decisions as to how far the ecosystem has to be reinstated. To arrive at such deductions and to monitor progress towards them requires the development of indicators of ecosystem health, stability and sustainability. If these criteria are met, the ecosystem can be declared as remediated to acceptable conditions, perhaps even reinstated. The level of protection or control or regulation required to reach this state is dependent on the recovery or restoration of the ecosystem and the state that needs to be maintained. Restoration means the degraded aquatic ecosystems are remediated to some level of stability or sustainability involving minimization of stresses, *in situ* treatments, and probably conservation of components of the total aquatic ecosystem that have not yet been degraded.

Biotic component, which is a component containing of living creatures that have characteristics as living things (can move, eat and drink, develop prickly, and are sensitive to stimuli). Examples of biotic components in aquatic ecosystems include fish, phytoplankton, and zooplankton, plants that live in water, amphibians, and other aquatic animals.

An abiotic component, a component that is the conflicting of the biotic component. That is a component that contains of objects that do not live but their being has an influence on the survival of living things in the environment. Abiotic components that exist in aquatic ecosystems include sunlight, rock, air or oxygen, and temperature.

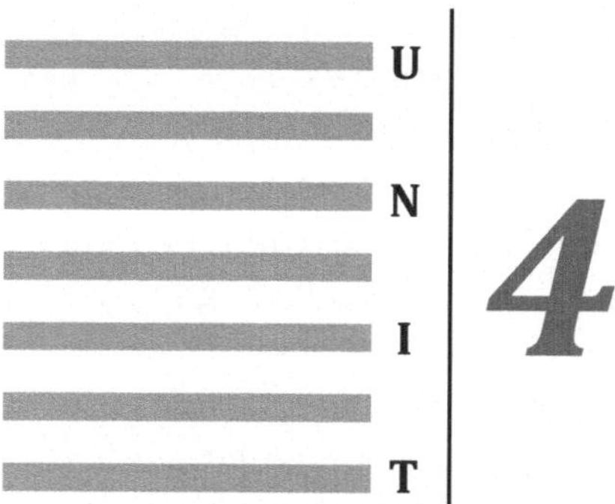

Biodiversity and its Conservation

4.1 Introduction

Biodiversity is the variety of living species on Earth – plants, animals and microorganisms – and the ecosystems they form. An ecosystem is the name given to all living species that live together in a stable community, interacting with one another and their physical environment. Biodiversity includes the diversity within species and between different species within terrestrial, freshwater and marine ecosystems. Ecosystems need a stable and diverse number of species to thrive.

Definition

Biodiversity refers to the numerous life forms that exist on earth, including animals, plants, micro-organisms, and the entire ecosystem they live in. Biodiversity is in the form of biological resources including genes, species, organisms, and ecosystem. Thus, the four main levels of biodiversity are species, genetic, ecosystem, and global biodiversity. Because of these features, biodiversity plays a fundamental role in upholding the aesthetic value of the environment, the integrity of the natural environment, and promotes the overall well-being of all plant and animal life. This calls for the need for biodiversity conservation for the survival of all living things and their natural habitat.

What is Biodiversity?

The term biodiversity was coined as a reduction of biological diversity by E.O. Wilson in 1985. Biodiversity may be defined as the variety and variability of living organisms and the ecological complexes in which they exist. In other words, biodiversity is the incidence of different types of ecosystems, different species of organisms with the whole range of their variants and genes adapted to different climates, environments along with their interactions and processes.

The sum total of all the variety of living organisms makes Biodiversity. The term Biodiversity was first coined by E.O. Wilson in the year 1985. It comprises all the species on the Earth, *i.e.* different plants, animals, organisms and their genes, water ecosystems, terrestrial ecosystem, marine ecosystem, *etc.* It is significant for our being as it provides important building blocks for numerous goods and services and delivers healthy environment to lead happy and healthy life.

Biodiversity plays a precarious role in sustaining human populations across the globe. We depend on it for continued food growth, for clean air and water and for medicine and shelter. It is no surprise then that ecosystem degradation threatens our most basic need – a healthy environment to live and thrive in. This is particularly true of biodiversity hotspots, which house some of the main diversity of species in the world and provide significant life-support services to the people who live in and around them.

In adding, biodiversity is embedded in human cultures and is important for our sympathetic of the world. Biodiversity enriches our lives – it has economic, cultural, recreational, religious and aesthetic importance across the world. We have famous it in art, music and literature throughout history. More than 190 countries acknowledge its importance to human populations through a show of support for the Convention on Biological Society.

Following image clarifies the biodiversity in an area. It comprises of plants and animals at one place. Plants like food and cash crops, intercropping, kitchen crop, trees, *etc.* and animals like farm animals, birds and insects.

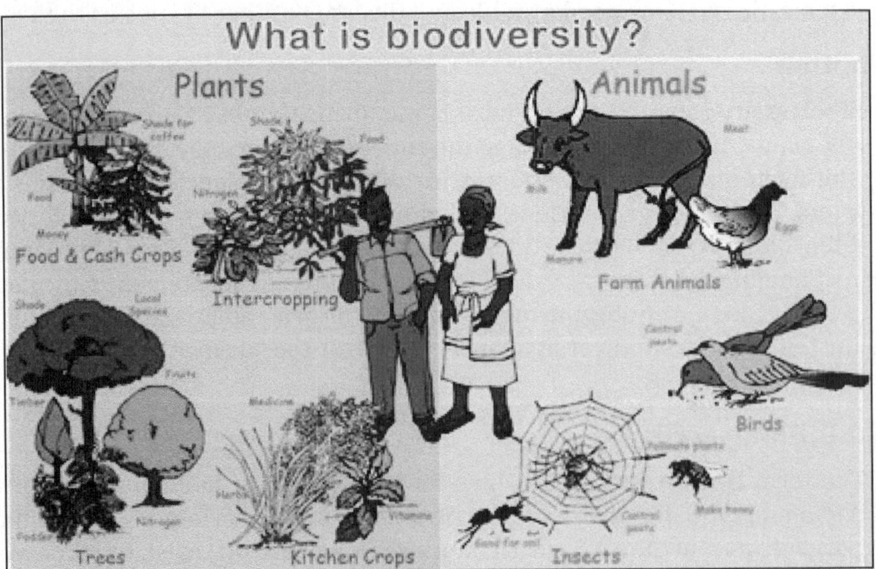

Types of Biodiversity Conservation

Going by the raw definition, biodiversity is the entirety of genes, species, and ecosystems of a given region. Edward Wilson coined this term for the first time. Conservation of biodiversity can be studied in different segments. They are:

- ☆ Genetic diversity
- ☆ Species diversity
- ☆ Ecological/Ecosystem diversity
- ☆ Global Biodiversity

The discussion on each type of diversity is given below:

1. Genetic Diversity

It describes the variation in the number and types of genes as well as chromosomes present in different species. The greatness of variation in genes of a species increases with upsurge in size and environmental parameters of the habitat. It includes difference in the type and number of genes; and chromosomes present in dissimilar species. The magnitude of diversity increases with the rise in environmental parameter and size of the habitat.

Following image shows the genetic diversity in Agriculture:

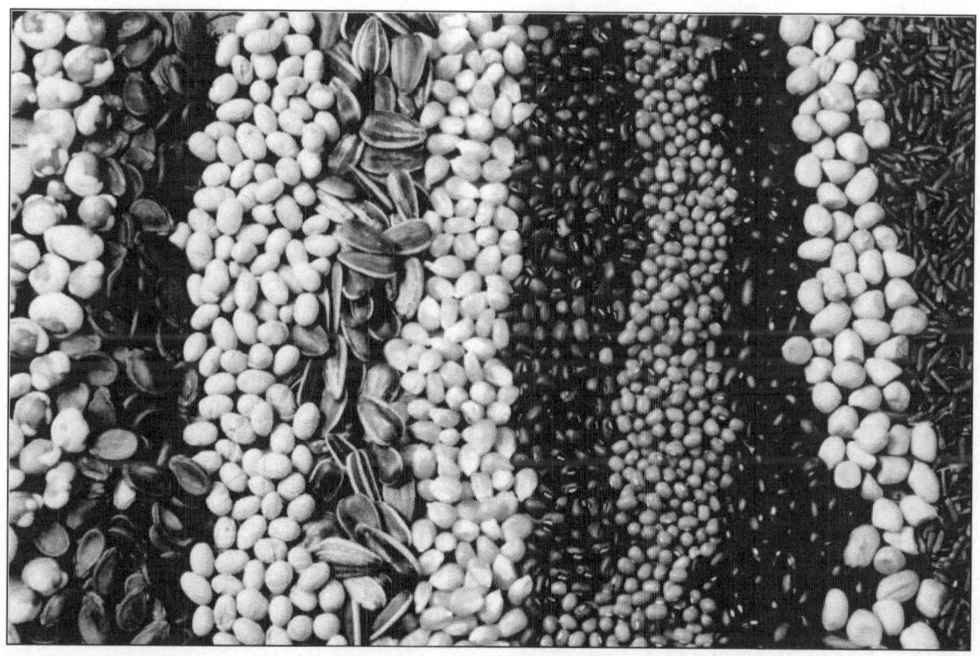

The genetic variation arises by gene and chromosome mutation in individuals and in sexually reproducing organisms and it is spread in the population by recombination of genetic materials during cell division after sexual reproduction. Genetic diversity encompasses all the variety of different genetic info contained in animals, microorganisms and the plants. A single species of organisms with diverse genetic diversity portray more adaptability and survival mechanism in contradiction of adverse environmental conditions likened to organisms of a single species with the same genetic make-up. Organisms of a single species that have greater genetic diversity tend to have more sustenance against environmental disturbances. Genetically uniform species are usually prone to disease harsh environments.

Genetic diversity has the following importance:

- It helps in evolution or speciation of new species.
- Genetic diversity is important for productivity and development of agriculture.
- It is important in adaptation to change as per the change in surroundings.

2. Species Diversity

It describes the variety in the number and richness of the spices with in a region. The species richness may be defined as the number of species per unit area. The richness of a species tells about the extent of biodiversity of a site and provides a means for comparing different sites.

Species diversity refers to the variety of organisms in the environment. They can be further sub-divided into.

Species richness: It is the total number of species in a given region or locality. It is measured using thc Menhink and Mangalet indices.

Special abundance: This is the relative numbers of organisms among the species (the number of individuals per species). When species have equal abundance, their variation becomes high, which also makes the diversity to be high. A low diversity scenario is noted when the number of one species is 98 and the other species 1.

Taxonomic diversity (Phylogenetic): Taxonomic diversity (TD) is the genetic relationship that exists between various groups of species. This kind of diversity is represented using a hierarchical classification based on the phylogenetic evolution of different species of organisms.

The species richness depends largely on climatic conditions. The number of individuals of different species with in a region represents species evenness or species equitability. The product species richness and species evenness give species diversity of a region. When a species is confined entirely to a particular area, it is termed as endemic species. It describes the richness and variety of species in a particular region. It is defined as "the number of species per unit area." The richness of species explains the extent of biodiversity and provides several methods for comparing different sites. The richness of species depends on the conditions of climate.

3. Ecosystem Diversity

It defines the assemblage and Interaction of spices living composed and the physical environment a given area. It relates varieties of habitats, biotic communities' ecological processes in biosphere. It also tells about the diversity within the ecosystem. It is mentioned as Land escape diversity because it comprises placement and size of various ecosystems.

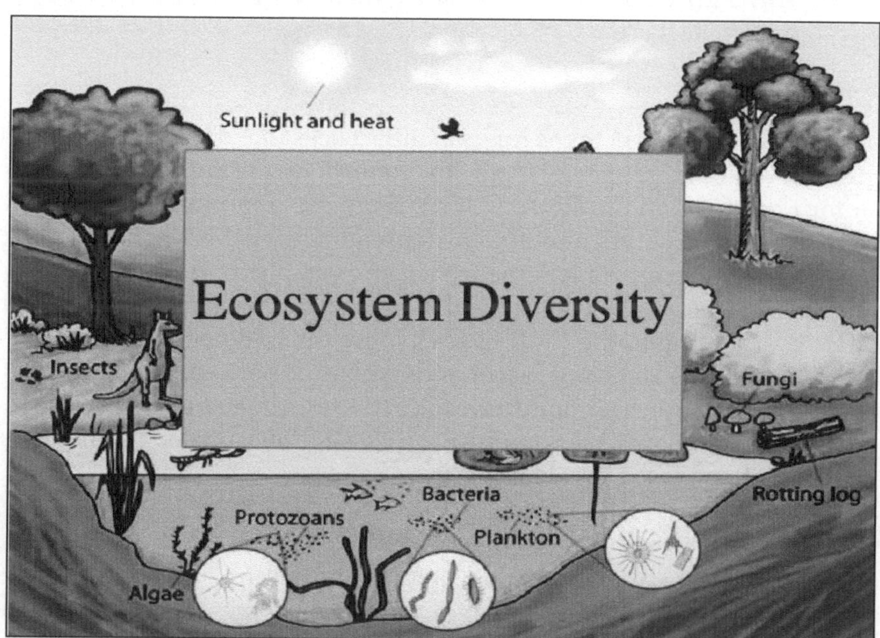

It is comparatively similar to species diversity excluding for the fact that it deals with the variations in ecosystems within a geographical location. From one region or country to another, there are diverse ecosystem or biomes. It describes the interaction and assemblage of species living together and the physical environment of that area. Ecosystem diversity relates to biotic communities, habitats and ecological processes in the entire biosphere. This diversity is also mentioned as landscape biodiversity because it consists of size and placement of different ecosystems. Ecological Diversity is much related to the species diversity. In our country, you can find more ecosystem diversity compared to other countries. We have several biomes like alpine meadows, rainforests, deserts, wetlands, and mangroves *etc.*

For example, the landscapes like grass lands, deserts, mountains *etc.* show ecosystem diversity. The ecosystem diversity is due to diversity of niches, trophic levels and ecological processes like nutrient cycling, food webs, energy flow, role of dominant species and various related biotic interactions. Such type of diversity can generate more productive and stable ecosystems or communities capable of tolerating various types of stresses *e.g.* drought, flood *etc.*

According to Whittaker (1965), the community diversities are of three types:

- ✰ α-*Diversity*: It tells the species diversity in a given community. It depends upon species richness and evenness.
- ✰ β-*Diversity*: It describes a range of communities due to replacement of species which arises due to the presence of different microhabitats, niches and environmental conditions.
- ✰ γ-*Diversity*: It describes diversity of habitat over a total land escape or geographical area.

4. Global Diversity

Global diversity refers to the range of differences that depict the composition of a group of two or more species in a global context. According to the International Union for Conservation of Nature (IUCN), there are approximately 1. 5 million different species of plants and animals spread across the world. 70 per cent represent animal species while 22 per cent represent plants. It is amazing to know that the total number of plant and animal species on this planet is about 1.5 million. This data is according to the IUCN (2004). More than 70 per cent of the species recorded are animals and plants account for about 22 per cent; 70 per cent of the animals are insects. However, one drawback of these estimates is that they do reveal any data for prokaryotes. This is because the conventional taxonomic methods are insufficient to find these microbial species.

4.2 Biogeographical Classification of India

India is one of the world's most biologically and culturally diverse countries. It is also one of the poorest in terms of per capita income. India is one of the world's 'mega diversity' countries. It is ranked ninth in the world in terms of higher plant species richness. At the ecosystem level, India is also well-endowed, with ten distinct biogeographic zones.

As per available data, the varieties of species living on the earth are 1753739. Out of the above species, 134781 are exist in in India although surface area of India is 2 per cent of the earth's surface. Wild life Institute of India has divided it into ten biogeographical regions and twenty five biotic provinces.

Biogeographical regions are:

- ☆ Trans Himalayas,
- ☆ Gangetic plain,
- ☆ Desert,
- ☆ Semiarid zone;
- ☆ Western Ghats;
- ☆ Deccan peninsula,
- ☆ North eastern zone,
- ☆ Coastal lands
- ☆ Himalayas,
- ☆ Islands.

It also contains two of the world's 25 biodiversity hotspots, because of their extremely high levels of species-richness and threatened status. India is considered to be the centre of origin for the following crop species pigeon pea, eggplant cucumber, possibly cotton and sesame. But for millennia, numerous other crop species have been presented to India and adapted to localised conditions.

India is one of the twelve mega diversity nations of the world due to the following reasons:

- ☆ Several crops arose in the country and spread throughout the world.
- ☆ There is wide variety of domestic animals like cows, buffaloes, goats, sheep, pigs, horses *etc.*
- ☆ It has 7.3 per cent of the global fauna and 10.88 per cent of global flora as per the data collected by Ministry of Environment and forest.
- ☆ It has 350 different mammals, 1200 species of birds- 453 different reptiles, 182 amphibians and 45,000 plants spices.
- ☆ There are a number of hot spots (namely Eastern Ghats, Western Ghats, North Eastern hills *etc.*).
- ☆ The marine biota includes sea weeds, fishes, crustaceans, molluscs, corals, reptiles *etc.*
- ☆ It has 10 different biogeographical regions and 25 biotic provinces having varieties of lands and species.

- ✰ In addition to geographical distribution, geological events in the land mass provide high level of biological diversity.
- ✰ It has 50,000 known species of insects which include 13,000 butterflies and moths.

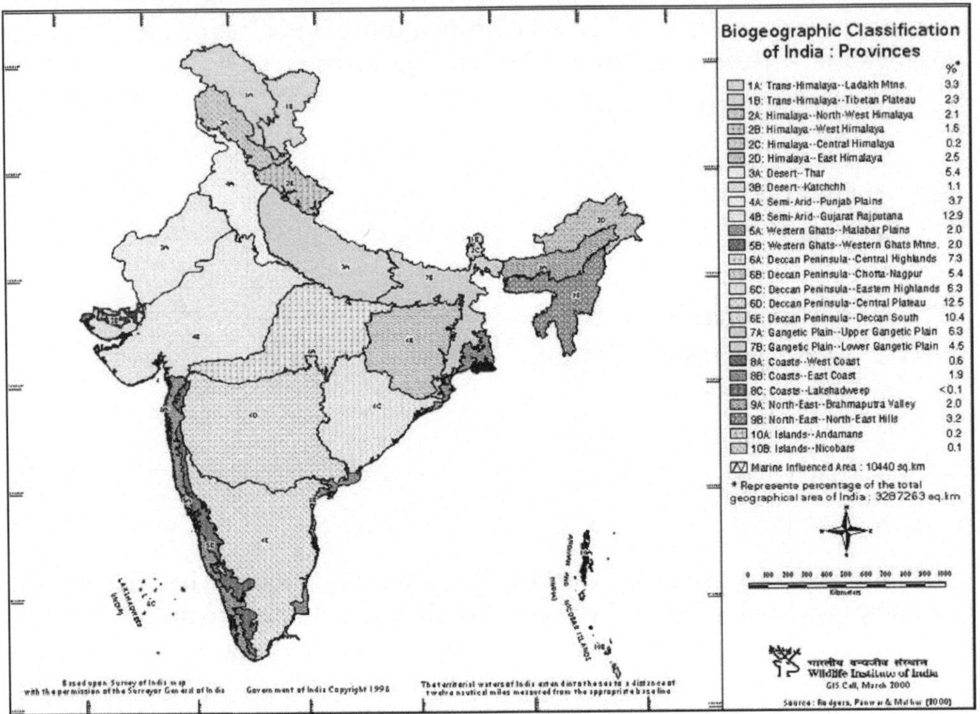

4.3 Value of Biodiversity

Biodiversity provides a variety of environmental services from its species and ecosystems that are essential at the global, regional and local levels. Biodiversity is essential for preserving ecological processes, such as fixing and recycling of nutrients, soil formation, circulation, and cleansing of air and water, global life support, maintaining the water balance within ecosystems, watershed protection, maintaining stream and river flows throughout the year, erosion control and local flood reduction. Food, clothing, housing, energy, medicines are all resources that are directly or indirectly linked to the biological variety present in the biosphere.

Some will argue that conservation can and should be based on economics. One view is that supply and demand determines resource use and prevents over-exploitation. As resources are depleted and supply lessens, price rises. Once a resource becomes very scarce, it is no longer profitable to harvest. The strong view is that markets prevent over-exploitation. A less strong view is that rising prices at least give an early warning of resource depletion.

With the growing realization that species are being lost and ecosystems degraded, other economic arguments have gained credence. One argument is that the economic value of yet-undiscovered resources (new medicines, new foods) is the best reason to save biodiversity. Another view is based on calculating the full value of all ecosystem goods and services as a way to demonstrate how much actual value nature provides.

Biodiversity or biological diversity simply means the variety and variability among living organisms and the ecological complexes in which they occur. Such variety refers to the variety at the species, genetic and ecosystem level. The value of biodiversity is classified into:

- ✰ Direct Values
- ✰ Indirect Values

Direct Values

These are those ways by which we can directly use biodiversity for our benefit. For example we can use plants as food of for deriving medicines in the laboratory. Economic value and recreational value comes under this category.

Direct values are further classified into:

Consumptive use Value: Consumptive use value is the value put on the products of nature which are consumed directly without passing through a market. For example, if we use firewood by cutting down a tree or consume an animal after hunting it. The consumptive value can be assigned to goods such as fuel woods, leaves, forest products *etc.* which may be consumed locally and do not figure in national and international market.

When direct consumption requires recreation, as in sport fishing and game viewing, the consumptive value is the whole recreational experience. Consumptive value seldom appears in national income accounts, but could be easily included in measures such as GDP. It is valued from the cost if resource was sold at market value, rather than being consumed.

High consumptive use values on resources may lead to the following problems:

- ✰ Over-exploitation of wildlife in developing countries
- ✰ Loss of traditional controls on hunting and
- ✰ Loss of wildlife populations at productive levels.
- ✰ Consumptive use value benefits the communities closest to the resource if harvested sustainably and managed efficiently.

A straight consumptive use is the direct utilization of timber, food, fuel wood and fodder by local communities. The diversity of organisms provide food, clothing, shelter, medicines, proteins, enzymes, papers, sports goods, musical instruments, beverages, narcotics, pets, zoo specimens, tourism and raw material for business prospects *etc.*

Productive use value: Productive use value is the value put on the products of nature which are consumed after passing through a market. For example, if we buy fish from the market then it will have productive use value. Productive use value refers to products that are commercially harvested (sold in a market). Its value is estimated at the production end rather than retail end by adding an inflated cost to the finished product. Productive use value is often the only value of biological resource reflected in national income accounts and may have a major impact on the national economy.

Timber, fish, honey, construction materials, mushrooms, fruits, medicinal plants and game meat sold in a market have productive use value.

This category comprises of marketable goods. The biotechnologist uses bio-rich areas to prospect and search for potential genetic properties in plants or animals that can be used to develop better varieties of crops for use in farming and plantation programs or to develop better live stock. To the pharmacist, biological diversity is the raw material from which new drugs can be identified from plant or animal products. To industrialists, biodiversity is rich storehouse from which to develop new products. For the agricultural scientist, the biodiversity is the basis for developing better crops. A variety of industries, like pharmaceuticals are highly dependent on identifying compounds of great economic value from the wide variety of wild species of plants located in undisturbed natural forests called "biological prospecting".

Indirect Values or Non-Consumptive Value

These are those ways by which we don't physically use a plant or animal, but by virtue of its existence it provides services that keep the ecosystem healthy. Indirect values would include ethical or moral value, existence value, ecological value, aesthetic value, cultural or spiritual value, option value and scientific or educational value.

Indirect values provide economic benefits without being harvested and do not appear in GDP. However, they are crucial to other natural products which influence the GDP. These values involve functions performed by biodiversity which are not of any use. Ex: Ecological Processes *etc.* Direct values are often derived from indirect values because plants and animals are supported by the services provided by their environments. Many classes of plant and animal species are consumed by tribal and non-tribal communities.

Examples include:

- Ecological functions
- Flood and storm protection
- Waste assimilation
- Microclimatic functions
- Nutrient cycles

- Photosynthesis
- Carbon stores
- Soil protection, *etc.*

Social Value

The loss of biodiversity straight effects the social life of the country perhaps through influencing ecosystem functions (energy flow and biogeochemical cycle). Social value of biodiversity prospecting motivated habitat conservation in some areas, as old-style societies valued it as a resource. Ecosystem people value biodiversity as a part of their livelihood as well as through cultural and religious sentiments. This can be easily unspoken by observing detrimental effects of global warming and acid rain which cause an unfavourable alteration in logical processes.

A great variety of crops have been cultivated in old-fashioned agricultural systems and permitted a wide range of produce to be grown and advertised throughout the year and acted as an insurance against the failure of one crop. In recent years, farmers have begun to receive economic incentives to grow cash crops for national or international markets, rather than to supply local needs. This has resulted in local food shortages, unemployment, landlessness, and increased vulnerability to drought and floods.

Aesthetic Value

Aesthetic values such as refreshing fragrance of the flowers, taste of berries, softness of mossed, melodious songs of birds, *etc.* compel the human beings to preserve them. Biodiversity is a direct source of pleasure and aesthetic satisfaction – its contribution to quality of life, outdoor recreation and scenic enjoyment. They provide opportunities for recreational activities such as hiking, canoeing, bird watching, river rafting, rock climbing, and trekking, parasailing, bird watching and nature photography.

The earth's natural beauty with its colour and hues, thick forest, and graceful beasts has inspired the human beings from their date of birth to take essential steps for its maintenance. The designing of thousands of new horticultural species, wild life conservation, landscape luxury, national parks, zoological and botanical gardens, snake, crocodile, butterfly parks, and biotechnologically operated novel curios species added to the existing aesthetics. Similarly botanical and zoological gardens are the means of biodiversity conservation and are of aesthetic values.

Ethical Value

Ethical values related to biodiversity upkeep are based on the importance of protecting all forms of life against illegal activities like cloning of animals, smuggling of valuable biodiversity instances, biopiracy, illicit trade *etc.* In India, several generations have preserved nature through local traditions. However, instant benefit rather than ethics appears to be modern man's objective.

Biodiversity must be seen in the light of holding ethical value. Since man is the most intelligent amongst the living organisms, it should be major responsibility and moral obligation of man to preserve and conserve other organisms which will directly or indirectly favour the existence of the man.

Ecological Value

Biodiversity holds great ecological value because it is crucial to uphold the ecological balance. Every species plays a unique role in the ecosystem. Through this role it upholds the ecological balance. Thus, the ecosystems don't get disrupted. So, even if we do not use a plant or animal for making products in our industries, by virtue of its very existence in the wild it delivers us with many important services. These services preserve ecological balance and the ecosystem. Some of these services are:

Waste Management: Nature has a unique way of managing wastes. The waste of one organism becomes food for another organism. So, wastes don't accrue. For example, forests absorb greenhouse gases like carbon dioxide during photosynthesis. This helps to decrease global warming. Forests also donate to precipitation due to transpiration. Many other plants and animals help to keep forests healthy. So, the entire biodiversity contributes towards maintaining climate stability.

Any disturbance in the delicately fabricated ecological balance maintained by different organisms, will lead to severe problems, which may threaten the survival of human beings.

Economic Value

Biodiversity has inordinate economic value because economic development be contingent upon efficient and economic management of biotic resources. We depend deeply on biological products for our subsistence. Biodiversity has economic value because it is a source of significant products. Some of these products are:

- *Food supplies*: Agriculture, the very basis of human survival, depends on plants and animals.
- *Source of medicines*: A large number of medicines are obtained from plants and animals. Cinchonas, Belladonna are important medicinal plants. Snake venom is used in making medicines.

Source of raw materials for industries.

It supports the economy of a country. Industries and agriculture produce revenue or income. They also engender employment. In fact the economy of many countries is deeply dependent on biodiversity.

So, it is highly vital for the human beings to take care of their surrounding species and make best use of their service, for better economic development. Thus, it is rightly told, survival of the man depends upon the survival of the biosphere.

Option Value

Keeping future possibilities open for their use is called 'option value'. It is impossible to predict which of our species or traditional varieties of crops and domestic animals will be of greatest use in the future. Important ecosystem services and uses for plants and animals are still unknown and await discovery. There are many plants and animals which have not yet been discovered or even if they have been exposed we do not know if they can be of any use to us. This untapped potential is referred to as option value. For instance, there might be a plant or animal which we can use in the future to find a treatment for corner. If we destroy biodiversity then we lose this chance of finding a cure for cancer. Thus biodiversity has great potential of being useful to us in the future. It develops valuable if boards are based on policy of obtaining wealth from wastes.

4.4 Biodiversity at Global, National and Local Levels

The present geological era is perhaps the richest in biological diversity. About 2.1 million species have been identified till date, while many more species are believed to exist. According to UNEP (1993-94) (UN convention on environment protection) estimate, the total number of species that might exist on Earth range between 9.0 – 52 million.

Invertebrate animals and plants make-up most of the species. About 70 per cent of all known species are invertebrates (animals without backbones such as insects, sponges, worms, *etc.*); while, about 15 per cent are plants. Mammals, the animal group to which man belong, comprise a comparatively small number of species. Of all the world's species, only 10 to 15 per cent live in North America and Europe. By contrast, the centres of utmost biodiversity tend to be in the tropics. Conservative estimates of the existing biodiversity is ten million species, but if estimates for insects are correct then it could be around 30 million species, we have till now enlisted about 1.4 million species. It includes among others about 98 per cent birds, 95 per cent reptiles and amphibians, 90 per cent fish and about 85 per cent higher plants known to exist on this Earth.

National and Local Level

India has over 108,276 species of bacteria, fungi, plants and animals already identified and described. Out of these, 84 per cent species constitute fungi (21.2 per cent), flowering plants (13.9 per cent), and insect (49.3 per cent). In terms of the number of species, the insects alone constitute nearly half of the biodiversity in India.

These species occur on land, fresh and marine waters, or occur as symbiosis in mutualistic or parasitic state with other organisms. In the world as a whole, 16, 04,000 species of Monera, Protista, Fungi, Plantae and Animalia have been described so far. However, it is predictable that at least 179, 80,000 species exist in the world, but as a working figure 122, 50,000 species are careful to be near reality. Percentage of Different Biota in India is 10th among the plant rich countries of

#	Name	Location
1	*California Floristic Province*	California, USA
2	*Madrean pine-oak woodlands*	Southern USA
3	*Mesoamerica*	Central Mexico, Belize, Guatemala, Nicaragua, and Northern Costa Rica
4	*Caribbean Islands*	East of Central America
5	*Atlantic Forest*	Parts of Brazil, Argentina, and Paraguay
6	*Cerrado*	Central Brazil
7	*Chilean Winter Rainfall-Valdivian Forests*	Central North of Chile, to the Western regions of Argentina
8	*Tumbes-Choco-Magdalena*	Pacific Coast of South America and the Galapagos Islands
9	*Tropical Andes*	Part of the Andes Mountains in South America
10	*Mediterranean Basin*	Surrounds the Mediterranean Sea
11	*Cape Floristic Region*	Southern tip of South Africa
12	*Coastal Forests of Eastern Africa*	Eastern Coast of Africa
13	*Horn of Africa*	Northeastern Africa
14	*Madagascar*	Southeast Coast of Africa
15	*Indian Ocean Islands*	Comoros, Mauritius, and Seychelles, surrounding Madagascar
16	*Maputaland-Pondoland-Albany*	Southeastern coast of South Africa
17	*Succulent Karoo*	Coastal region of South Africa
18	*Mountains of Central Asia*	Central Asia: extends through Afghanistan, China, Kazakhstan, Kyrgyzstan, Tajikistan, and Uzbekistan
19	*Eastern Himalaya*	Parts of China, Bhutan, India, Tibet, and Myanmar
20	*Indo-Burma*	Parts of Bangladesh, India, Myanmar, China, Cambodia, Vietnam, Thailand, and Malaysia, Hainan Island and Andaman Island
21	*Western Ghats*	Indian Peninsula
22	*Sri Lanka*	South of India
23	*East Melanesian Islands*	North East of Australia
24	*New Caledonia*	South Pacific Ocean
25	*New Zealand*	Southwest Pacific Ocean
26	*Philippines*	Southeast Asia
27	*Polynesia-Micronesia*	Southern Pacific Ocean
28	*Southwest Australia*	Southwest tip of Australia
29	*Sundaland*	Southeastern Asia comprising the Malay peninsula, Borneo Island, Java Island, and Sumatra Island as well as their smaller surrounding islands
30	*Wallacea*	Eastern Indonesia
31	*Japan*	Northen Pacific Ocean
32	*Mountains of Southwest China*	Includes Tibet, Sichuan, Qinghai, Gansu, and Myanmar
33	*Caucasus*	Border between Europe and Asia, separating the Black and Caspian seas
34	*Irano-Anatolian*	Parts of Armenia, Azerbaijan, Georgia, Iraq, Iran, Turkey, and Turkmenistan
35	*Forest of East Australia*	Eastern coast of Australia
36	*Guinean Forests of West Africa*	Coastal western africa

the world, fourth among the Asian countries, eleventh according to the number of endemic species of advanced vertebrates (amphibians, birds and mammals), and tenth in the world as far as richness in mammals is concerned. Out of the 10 'Hot spots' recognised in the world, India has four. These are Eastern Himalaya, North East India, Western Ghats and Andaman and Nicobar Islands.

The crops which first produced in India and spread during the world include rice, sugarcane, Asiatic vegans, jute, mango, citrus, and banana, several species of millets, spices, medicinal, aromatics and ornamentals. India ranks sixth among the centres of diversity and origin in terms of agro-biodiversity.

Various Patterns of Biodiversity

Species are not consistently dispersed over the Earth; diversity varies importantly from place to place. In terms of the number of basic kinds of organism and number of species of each kind, biodiversity has varied markedly through geological time; and, in terms of present day species richness, biodiversity differs importantly between one part of the earth and another. The present global patterns in biodiversity indicate that the species richness tends to vary geographically according to a series of fairly well defined rules. For example, in case of terrestrial environments:

- ☆ Warmer areas hold more species than the colder areas;
- ☆ Wetter areas hold more species than the drier ones;
- ☆ Larger areas hold more species than the smaller areas;
- ☆ Areas of varied climate and topography hold more species than the areas of uniform climate and topography;
- ☆ Areas at lower altitude (elevation) hold more species than the high altitude areas; and
- ☆ Less seasonal areas hold more species than the highly seasonal areas.
- ☆ Similarly, in case of pelagic marine species, there tends to be more species in warmer and less seasonal waters, *i.e.* at lower latitudes.

In a much more simplified way, it can be said that there are much more species, both per unit area and overall, in the tropics than in temperate regions and for more in the latter (temperate regions) than in Polar Regions. The moist tropical forests, in over-all, are the most species rich areas or environments on earth. Though they cover just about 7 per cent of the world's surface; but it is estimated that they may hold more than 90 per cent of the world's species, if the yet unknown tropical forest micro fauna (mainly insects) are accepted. If small tropical forest insects are discounted, then the areas that may be similarly rich in species are:

- ☆ Coral reefs; and
- ☆ Areas of Mediterranean climate in South Africa and Western Australia.
- ☆ These areas are rich in species, especially for flowering plants (angiosperms).

Latitudinal Gradients

Biodiversity differs with latitude and altitude of the place. If the environmental conditions are positive for the speciation, you will find more diversity in such areas. Biodiversity be contingent upon the latitudinal gradients of a place. They are reduced from the equator towards the poles. For example, Colombia (near the equator) has 1400 species of birds whereas New York (41° N) has 105 species. According to ecologists, communities with more species tend to be more steady than those with fewer species.

4.5 India as a Mega-Diversity Nation

India is one of the world's 'mega diversity' countries. It is graded ninth in the world in terms of advanced plant species richness. At the ecosystem level, India is also affluent, with ten distinct biogeographic zones. India is known for its rich heritage of Biological diversity, having already documented over 91,000 species of animals and 45,500 species of plants in its 10 biogeographic regions. Nearly 6,500 native plants are still used conspicuously in indigenous healthcare systems. Biodiversity is not dispersed evenly across the globe. Certain countries, lying typically in the tropics, are branded by high species richness and more number of endemic species-called Mega variety nations About 19 countries/Nations belongs to this group as on today.

India is one of them. India a mega diversity country with only 2.4 per cent of the land area of the world, but accounts for 7.8 per cent of the recorded species of the world. India also boasts of rich marine biodiversity, along the coastline of 7516.5 km with exclusive economic zone of 202 million sq.km, supporting the most productive ecosystems such as mangroves, estuaries, lagoons and coral reefs. The amount of zooplankton recorded is about 16,000 species. The benthic fauna largely consists of polychaeta (62 per cent), crustacean (20 per cent) and

Statistical records of biological diversity in India

TAXA	*Number of species*		
	India [endemics]	*World*	*Percentage of India to the world*
Bacteria / Virus	850	8,050	10.60
Algae	6,500	40,000	16.25
Fungi	14,500	70,000	20.70
Lichens	2,070	13,500	15.00
Bryophyta	2,850	14,500	19.70
Pteridophyta	1,200 [200]	10,000	12.00
Gymnosperms	48	650	7.40
Angiosperms	17,500 [4,950]	250,000	7.00
Protists	2,577 [640]	31,250	8.24
Mollusca	5,072 [967]	66,535	7.62
Arthropoda	69,903	9,70,670	7.20
Other invertebrates	8,329	87,121	9.56
Protochordata	119	2,106	5.65
Pisces	2,546 [78]	21,723	11.72
Amphibia	240 [123]	5,150	4.66
Reptilia	450 [182]	5,817	7.94
Aves	1,232 [60]	9,026	13.66
Mammalia	397 [44]	4,629	8.58
Total	1,36,830	16,29,158	7.46

molluscs (18 per cent) with the biomass of about 12 g per sq. Metre. Over 30 species of marine algae and 14 species of sea grass have been stated. There are over 45 species of mangrove plants. Over 342 species of corals belonging to 76 genera have been reported and about 50 per cent of the world's reef building corals are found in India.

- ☆ India is one of the twelve mega diversity nations of the world due to the following reasons:
- ☆ It has 7.3 per cent of the global fauna and 10.88 per cent of global flora as per the data collected by Ministry of Environment and forest.
- ☆ It has 350 different mammals, 1200 species of birds- 453 different reptiles, 182 amphibians and 45,000 plants spices.
- ☆ It has 50,000 known species of insects which include 13,000 butterflies and moths.
- ☆ It has 10 different biogeographical regions and 25 biotic provinces having varieties of lands and species.
- ☆ In addition to geographical distribution, geological events in the land mass provide high level of biological diversity.
- ☆ Several crops arose in the country and spread throughout the world.
- ☆ There is wide variety of domestic animals like cows, buffaloes, goats, sheep, pigs, horses *etc.*
- ☆ The marine biota includes sea weeds, fishes, crustaceans, molluscs, corals, reptiles *etc.*
- ☆ There are a number of hot spots (namely Eastern Ghats, Western Ghats, North Eastern hills *etc.*).
- ☆ Some of the salient features of India's biodiversity are as under:
- ☆ India has two major realms called the Palaearctic and the Indo Malayan; and three biomes, namely the tropical humid forests, the tropical dry deciduous forests and the warm desert/semi-deserts.
- ☆ India has ten biogeographic regions, namely the Trans-Himalayan, the Himalayan, the Indian desert, the semi-arid zone, the Western Ghats, the Deccan Peninsula, the Genetic plain, North-East India, the Islands, and the coasts.
- ☆ India is one of the 12 mega-diversity nations of the world.
- ☆ India is one of the 12 centres of origin of cultivated plants.

There are two hotspots that spread into India. There are the Western Ghats/Sri Lanka and the Indo-Burma region (covering the Eastern Himalayas). Further these hotspots are included amongst the top eight most important or hottest hotspots.

India has 26 documented endemic centres that are home to nearly a third of all the flowering plants (angiosperms) identified and described to date.

Recorded plant species

Taxonomic group	Number of species		% of the world flora
	World	India	
Angiosperms	250,000	17,500	7.0
Gymnosperms	650	48	7.4
Pteridophytes	10,000	1,200	12.0
Bryophytes	14,500	2,850	19.7
Lichens	13,500	2,075	15.0
Fungi	70,000	14,500	20.7
Algae	40,000	6,500	16.3
Virus/Bacteria	8,050	850	10.6
Total	**406,700**	**45,523**	**11.8**

Source: India's Third National Report to CBD, 2006.

Recorded animal species

Taxonomic group	Number of species		% in India
	World	India	
PROTISTA (Protozoa)	31,250	2,577	8.24
ANIMALIA			
Mesozoa	71	10	14.08
Porifera	4,562	500	10.70
Cnidaria	9,916	842	8.49
Ctenophora	100	12	12.00
Platyhelminthes	17,500	1,622	9.22
Nemertinea	600		
Rotifera	2,500	330	13.20
Gastrotricha	3,000	100	3.33
Kinorhyncha	100	10	10.00

Some of the habitats richest in biodiversity, such as tropical rainforests, are being destroyed because of human activities. The obliteration of a habitat could lead to extinction of species that lived there and can live nowhere else.

During last few decades India has cut down at least 50 per cent of its forests, polluted over 70 per cent of its water bodies, built or cultivated over much of its grasslands, and degraded many coastal areas.

At least 10 per cent of India's recorded wild flora (mainly flowering plants) and possibly a larger percentage of its wild fauna are threatened, although how many are on the verge of extinction, no one known.

Based on a survey of about two-third of the geographical area of the country, the Ministry of Forests and Environment (MOEF) reports that India has at present 89,317 species of fauna and 45,364 species of flora representing about 7.31 per cent of the world fauna and 10.88 per cent the world flora described so far. In plants, the species richness is high in angiosperms, bryophyte and petridophyta, and in the family orchidaceous. In animals, arthropod (insects) are predominant.

India is also rich in agro-biodiversity. There are 167 crop species and wild relatives. Further, India is considered to be the centre of origin of 30,000 to 50,000 varieties of rice, pigeon-pea, mango, turmeric, ginger, sugarcane, gooseberries, *etc.* and ranks seventh in terms of contribution to world agriculture.

Country		Number of Mammals	Number of Birds	Number of Reptiles	Number of Amphibians	Number of Angiosperms
1	Australia	255 (210)	650 (355)	686 (616)	197 (169)	23,000 (14,458)
2	Brazil	428 (131)	622 (>191)	467 (172)	516 (294)	55,000 (17,500)
3	China	394 (77)	1195 (99)	270 (133)	265 (175)	27,000 (10,000)
4	Columbia	359 (28)	1721 (>142)	383 (97)	407 (367)	45,000 (16,000)
5	Ecuador	280 (21)	1447 (37)	345 (114)	358 (138)	15,000 (4,500)
6	India	372 (44)	1224 (52)	408 (187)	246 (110)	17,000 (5,400)
7	Indonesia	515 (201)	1519 (397)	600 (150)	270 (100)	20,000 (16,000)
8	Madagascar	105 (77)*	250 (103)	269 (74)	144 (76)	10,000 (9,200)
9	Malaysia	293 (27)	1200 (11)	294 (68)	171 (57)	15,000 (7,250)
10	Mexico	449	1010	717	282	25,000
11	Peru	361 (46)	1703 (109)	297 (98)	251 (>89)	20,000 (5,356)
12	Zaire	409	1086	280	216	10,000

4.6 Hot-Spots of Biodiversity

"Hot Spots refer to the area with high density of diversity or mega diversity which are threatened at present." Biodiversity hotspots are defined as regions "where exceptional concentrations of endemic species are undergoing an exceptional loss of habitat". The concept of biodiversity hotspots was developed by the Norman Myers in 1988 when he identified that the tropical forest losing its plants species as well as habitat. IUCN prepares 'Red Data Book'.

There are 34 areas around the world which are fit as Biodiversity hotspots. These hotspots represent only 2.3 per cent of the total Earth's land surface. These hotspots are significant because Biodiversity underpins all life on Earth. Without species, there would be no air to breathe, no food to eat, no water to drink. There would be no human society at all. And as the places on Earth, where the most biodiversity is under the most threat, hotspots are critical to human survival.

In 1988, British ecologist Norman Myers published a seminal paper classifying 10 tropical forest "hotspots." These regions were characterized both by exceptional levels of plant endemism and serious levels of habitat loss. There are around 16 hot spots, out of which two are located in India by the name Western Ghats and North-East Himalayas. These hot spots are determined on the basis of four factors –

- ☆ Degree of expectation
- ☆ Degree of endemism
- ☆ Number of species diversity

Degree of threat to habitat as a result of fragmentation and degradation.

The hotspots remain important in our work for two important reasons:

Biodiversity reinforces all life on Earth. Without species, there would be no air to breathe, no food to eat, no water to drink. There would be no human society at all. And as the places on Earth where the most biodiversity is under the most threat, hotspots are dangerous to human survival.

The map of hotspots overlays extremely well with the map of the natural places that most advantage people. That's because hotspots are among the richest and most important ecosystems in the world – and they are home to many vulnerable populations who are unswervingly reliant on on nature to survive. By one estimate, despite comprising 2.4 per cent of Earth's land surface, forests, wetlands and other ecosystems in hotspots account for 35 per cent of the "ecosystem services" that vulnerable human populations depend on.

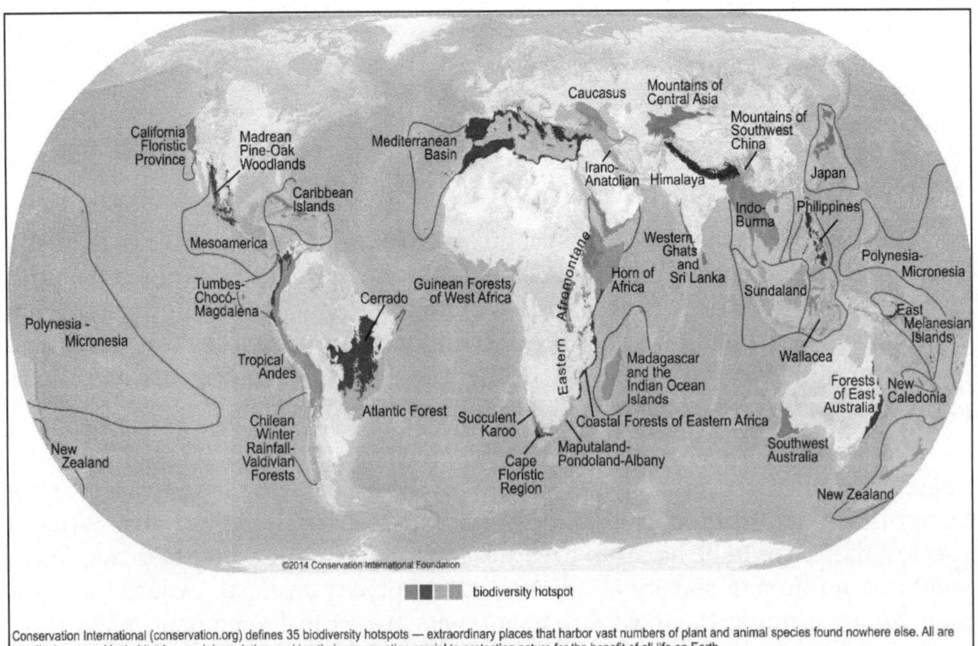

Conservation International (conservation.org) defines 35 biodiversity hotspots — extraordinary places that harbor vast numbers of plant and animal species found nowhere else. All are heavily threatened by habitat loss and degradation, making their conservation crucial to protecting nature for the benefit of all life on Earth.

4.7 Threats to Biodiversity

Any disturbance in a natural ecosystem tends to decrease its biodiversity. Waste produced due to surge in human population and industrialization plunders the environment and leads to reduced diversity in biological species. Any alteration in the system leads to a major inequity and looms the normal ecological cycle. The core risk to biodiversity on the planet, and therefore a threat to human welfare, is the mixture of human population growth and the resources used by that population. The human population needs resources to survive and grow, and many of those resources are being removed unsustainably from the environment. The three greatest nearby threats to biodiversity are habitat loss, poaching of wildlife, and man-wildlife conflicts. A major cause of extinction, anthropogenic (human-caused) climate change, has not yet had a large impact, but it is predicted to become significant during this century. Global climate change is also a consequence of human population needs for energy and the use of fossil fuels to meet those needs. Environmental issues, such as toxic pollution, have specific targeted effects on species, but are not generally seen as threats at the magnitude of the others.

More specifically, the threats are inter alia due to the following:

- ☆ Large-scale development projects such as mining and dam and road construction.
- ☆ Conversion of biodiversity-rich ecosystems, such as tropical forests to farmlands and industrial and residential sites.
- ☆ Poaching of wildlife and over-harvesting of forest products.

While there has been no complete assessment of biodiversity loss, three of four mammal species have been lost since 1950, and so also 15-20 plant species have develop non-existent. By and large over ten per cent of India's flowering plant species are endangered with extinction. Of particular seriousness is the loss of agro-biodiversity.

In one district of the state of Andhra Pradesh, Odisha and West Bengal, 95 per cent of rice varieties previously cultivated are no longer found. Although the reasons are various, this situation is primarily due to the spare of low-input poly-cultural agricultural systems with higher-input monocultures.

Habitat Loss

Ecosystem conversion and ecosystem squalor contribute to habitat disintegration. Habitat loss from exploitation of resources, agricultural conversion, and development is the largest factor contributing to the loss of biodiversity. When such habitats are lost due to deforestation and other anthropogenic activities such as mining, the individual environments are unable to provide shelter, food, water, or breeding grounds for the living organisms. Deforestation is a direct cause of extinction and loss of biodiversity. An estimated 18 million acres of forest are lost each year, due in part to logging and other human practices, destroying the ecosystems on which many species be contingent. Tropical rainforests in particular, such as the Amazon, hold a high percentage of the world's known species, yet the regions themselves are in decline due to humans.

Habitat loss may either be produced by natural events like natural disasters and geological events or anthropogenic activities like deforestation and man-induced climate change.

In the process of habitat squalor, the organisms that were once living in a particular area or region are displaced and are forced to relocate; thus resulting in biodiversity reduction.

Indeed, man-made efforts are the prime reasons for habitat loss. At present, the repetition of clearing out ecosystems for agriculture conversion and industrial expansion continues to displace organisms of their natural habitat. Other activities include logging and mining.

Habitat obliteration can affect ecosystems other than forests. Rivers and streams are significant ecosystems and are frequently the target of environment alteration. Damming of rivers affects flow and access to habitat. Altering a flow regime can reduce or remove populations that are adapted to seasonal changes in flow. For example, a projected 91 per cent of river ways in the United States have been modified with damming or stream bank modification. Many fish species in the United States, especially rare species or species with restricted distributions, have seen declines caused by river damming and habitat loss. Research has confirmed that species of amphibians that must carry out parts of their life cycles in both aquatic and terrestrial habitats are at greater risk of population declines and extermination because of the increased likelihood that one of their habitats or access between them will be lost. This is of particular concern because amphibians have been declining in numbers and going extinct more rapidly than many other groups for a variety of possible reasons.

Solution

- ✰ Curbing the problem of deforestation and habitat loss mostly depend on policy and implementation of laws.
- ✰ For one, corporations and companies can take up the wise practice of refusing to use paper and timber products that encourage deforestation.
- ✰ Likewise, awareness creation should be done so that consumers can desist from supporting companies or manufacturers that use timber and paper to make their products and most importantly, those that utilize unsustainable natural resource extraction or manufacturing processes.
- ✰ Governments and regulatory agencies should also take the lead in ensuring the enactment of stronger forest protection laws and policies.
- ✰ Individuals and organizations can also participate by support environmental conservation through charities and creating awareness.

- The solutions to deforestation mostly lie in policy – companies and corporations can adopt best practices and refuse to use timber and paper suppliers that contribute to deforestation.
- In the same vein, conscious consumers can refuse to patronize companies that do, and put pressure on retailers that employ unsustainable manufacturing methods. Individuals can also participate in land preservation through charities and private corporations.
- Ultimately, however, international governments need to enact stronger, scientific forest protection laws.

Poaching of Wildlife

Poaching refers to killing animals or commercial hunting. It contributes to loss of biodiversity. The main reasons for poaching are hunting for food, trade (market demand for skins and bodies of the animals) and lack of enforcement control.

Elk, musk deer, argali, wolves, and bears are traded regularly into China. Marmot pelts are being traditional goods for illegal trade to Russia for many years because of high demand on the Russian market. Saiga is being hunted for its horns that are valued in Chinese medicine. Large falcon trading (ger falcon, saker falcon and peregrine falcon) exclusively connected with demand for hunting birds in Arab countries remains at previous levels of approximately 100 birds a year.

Snow leopards are under threat due to its fantastic skin and any other part of the body good for Asian medicine. Besides there is a conflict between a snow leopard and local herders who kill the predator for attacking the cattle in the corrals and pastures.

Stronger enforcement and alternative income options for improvement of livelihood are the most effective measures against the threat of poaching.

Poaching can be of two types listed below:

- Subsistence poaching: This refers to killing animals for survival.
- Commercial poaching: This refers to hunting animals in order to sell their products.

Factors Influencing Poaching

- Human population: Increased human population in India has led to pressure on forest resources, leading to degradation of wildlife habitats
- Commercial activities: Although a ban has been imposed internationally on the trade of products of endangered species, there is a continued smuggling of wildlife products. Since trading of such products is highly profitable, poachers continue to hunt endangered animals and smuggle their fur, skin and tusks to other countries.
- A number of wildlife species are becoming extinct due to poaching and hunting.

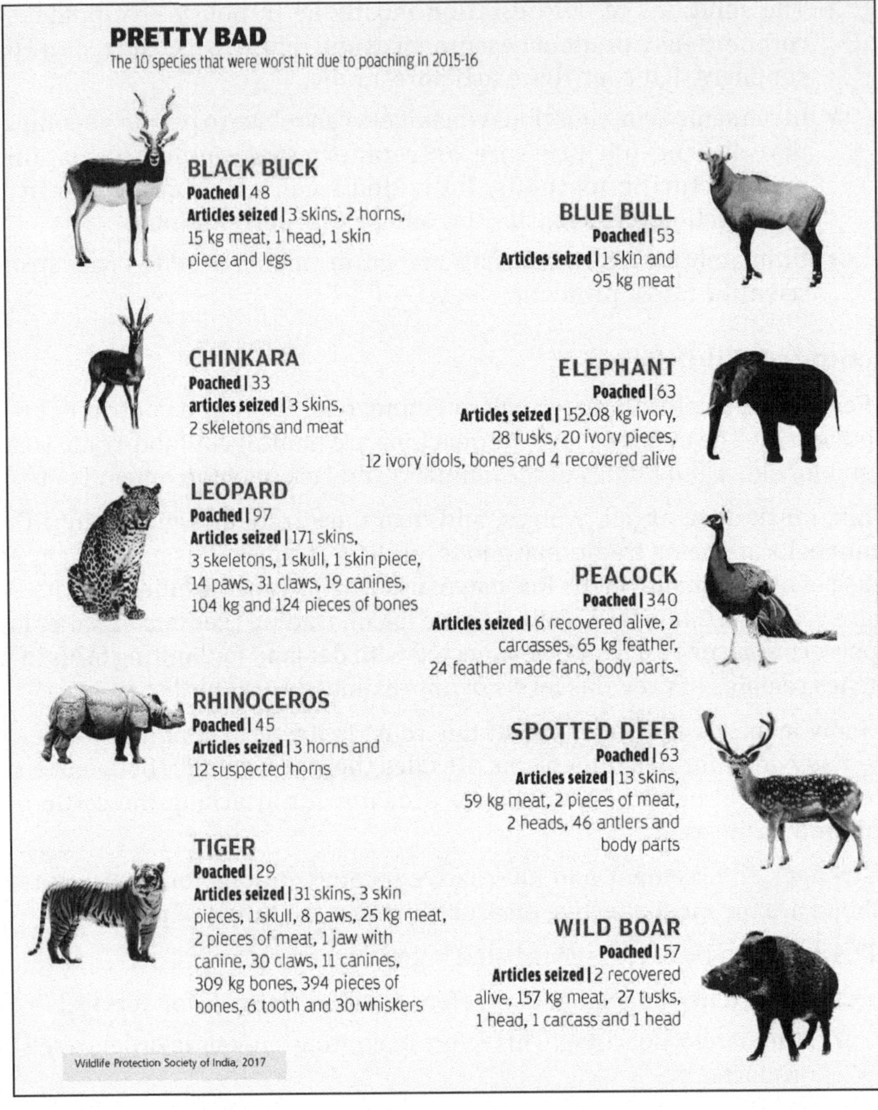

Man-Wildlife Conflicts

Man-wildlife conflicts arise, when wildlife starts causing immense damage and danger to man. Under such conditions it is very difficult for the forest department officials to convince the affected villagers to gain the villagers support for wildlife conservation.

Factors Influencing Man-animal Conflicts

- ☆ Shrinking forest cover compels wildlife to move outside the forest
- ☆ Human encroachment into forest area induces a man-wildlife conflict

- Injured animals have a tendency to attack man
- Wild animals venture out of the forest area in search of food
- Villagers' set-up electric wiring around their fields. This injures animals (Elephants) who suffer pain and get violent.
- Cash compensation paid by the government is not enough.
- Garbage near human settlements or food crops attracts wild animals.

Infestation by alien species, such as the Codling Moth, is also a major danger to BC ecosystems. The deliberate and unintentional introductions of a wide variety of species to ecosystems in which they do not belong have caused in ecosystems that differ radically in structure and function from those originally present. Exotic species are characteristically presented into ecosystems without their co-evolved predators and parasites, which allows an alien invader to out-compete native species with similar ecological requirements. This introduction can cause major threats to the native species as they often become subjected to great competition for resources, disease, and predation. When these species have successfully colonized the area, they are already called "invasive" ones.

Solution

- The problem of man-wildlife conflicts majorly due to invasive alien species is global and for this reason, it requires intervention at the local, state and international level.
- Local authorities and states need to establish systems to manage and prevent invasive alien species through risk assessments as a strategy of predicting the possibility of species becoming invasive.
- The assessments should also aim at determining the potential ecological damages and put in place effective preventive measures to counter the likely environmental impacts.

Therefore, creating systems to stop the outline of invasive alien species even before it happens, quickly removing newly detected invaders, and efficiently monitoring new invasions are the most efficient strategies. International bodies and scientist can assists in research and information quantification by using more creative means such as Google street view and other progressive technological techniques.

Solutions include creating systems to prevent introduction of invasive species in the first place, effectively monitoring for new infestations and swiftly eradication newly detected invaders.

4.8 Endangered and Endemic Species of India

4.8.1 Endangered Species

Endangered or threatened species is one whose number has been abridged to a dangerous number. Unless it is protected and conserved, it is in instant danger of extinction. Endangered species are animals or plants which occur in very less number and if not conserved properly, will extinct. Endemic species are plants or animals exist in some particular regions and nowhere in the world. Unless conserved, these also will vanish from earth. This article lists the Endangered and Endemic Species in India.

India is home to dissimilar types of animals, birds and fishes which include some significant farm animals like goats, poultry, cows, buffaloes, pigs *etc.* The country is also a habitat for wild animals like Bengal tigers, deer, wolves, pythons, Indian lions, bears, snakes, and monkeys, many types of bison, Asian elephants and antelope species. India is one of the mega diverse countries out of a total of seventeen mega diverse countries in the world. These seventeen mega diverse countries, including India, are the habitats of around 60 – 70 per cent of the world's biodiversity. The Western Ghats, The Eastern Himalayas and Indo–Burma are the three biodiversity hotspots out of total 34 in the whole world.

India has 6.5 per cent of the world's total wildlife species according to a report which was published by the United Nations Office on Drugs and Crime (UNODC) that includes 7.6 per cent of all mammals and 12.6 per cent of all bird species.

According a report issued by the International Union for Conservation of Nature (IUCN) Red List in 2014, 15 species of birds, 12 species of mammals, and 18 species of reptiles and amphibians have joined the critically endangered list.

4.8.1.1 Reasons for Endangerment

Loss of habitat is one of the primary reasons for the endangerment of species. Today, human intervention plays a main role in the destruction of the natural landscape. Human activities like removal of trees that provide both food and shelter for countless number of species, mining and agriculture.

Over hunting and poaching has a very unhelpful and disastrous effect on the number of animals and fishes all over the world.

Pollution like air pollution, water pollution and waste pollution, especially in the form of plastic plays a very dominant role in the endangerment of animal species. Pollution not only causes health hazards for humans, but it affects the animals also.

In a vigorous and enthusiastic environment there is always a precise balance between the number of predators and their prey animals. The marauders who are natural enemies of their prey animals choose the old and sick preys as they cannot keep up with their group. In this scenario the relationship between them is totally healthy as the predators only eat these prey animals that are already nearing the end of their life. But the problems become more apparent when the predators wander in such an area where they will get only a few numbers of their prey animals.

To save the animals from poaching and hunting they are often kept in a sanctuary and shelter. Though for some animals it has proved to be very much beneficial, there are other animals too who suffer and are in trouble for being endangered. The main two reasons are overcrowding and overgrazing. Usually there are too many animals that are enclosed in small areas. These animals often eat the same grass and trees in a limited area whereas in natural surroundings the grazing animals keep on changing the landscape while eating and keep on moving for most of the time. But in an enclosed and small area they eat from the same plants repeatedly which finally destroys the plants due to overstress.

Factors Affecting Endangered Species

- Human beings arrange wastes comprehensively in nature thus polluting the air, land and water. These pollutants enter the food chain and accumulate in living creatures resulting in death.
- Over-exploitation of natural resources and stealing of wild animals also leads to their extinction.
- Climate change brought about by build-up of greenhouse gases in the atmosphere. Climate change threatens organisms and ecosystems and they cannot adjust to the changing environmental conditions leading to their death and extinction.
- An international treaty to help protect endangered wildlife is, "Convention on International Trade in Endangered Species 1975" (CITES). This treaty is now signed by 160 countries.
- CITES lists 900 species that cannot be commercially traded as live specimens or wildlife products as they are in danger of extinction.
- CITES restricts trade of 2900 other species as they are endangered.

4.8.1.2 Endangered Animal Species

Snow Leopard

These beautiful cats have thick fur to protect them from the elements. Their paws are large like snowshoes, allowing them to gracefully walk through the snow. Unfortunately, these graceful creatures are hunted for their fur and by farmers when snow leopards prey on their livestock.

Bengal Tiger

They are found in tropical rainforests, grasslands and swamps of Southern India. These beautiful cats are fierce, eating up to 60 pounds of meat in one sitting from large game. Bengal tigers are under attack from humans for their pelts, habitat destruction and human fear of attack. Although most Bengal tigers will leave humans alone, some sick tigers become man-eaters, stalking local villagers instead of wildlife.

Ganges River Dolphin

The Ganges river dolphin is a rare freshwater dolphin living only in the Ganges river of Western India. Growing up to about 9 feet in length and about 370 pounds, this large dolphin eats turtles, fish and birds native to the Ganges. Unfortunately due to hunting, accidental death in fishing nets and the high levels of pollution in the Ganges, this dolphin is endangered.

Hangul

Kashmir Stag or Hangul is one of the greatest unfavourably endangered species found in the temperate grasslands of western Himalayas. Dachigam National Park in Kashmir signifies one such grassland habitat that supports Hangul, a highly endangered and the only subspecies of the Red deer (*Cervus elaphus*) to be found in India, which is now confined only to the Kashmir Valley.

Nilgiri Tahr

Nilgiri Tahr, a mountain goat, is the highly endangered flagship species happen on the crest lines and ridge forests of the southern Western Ghats. The ideal habitat of this species is the rocky outcrops adjacent to the shola-grasslands and other ridge forests. Only less than 2000 individuals of this species is remaining in the wild in the whole world with the major population confined to Eravikulam National Park in Kerala and Grizzled Giant Squirrel Wildlife Sanctuary in Tamil Nadu.

Asian Wild Buffalo

The Asian wild buffalo (*Bubalus bubalis*) has been chosen as endangered by the IUCN and included in Schedule I of the Wildlife (Protection) Act, 1972. The wild buffalo was once widely distributed over the tracts of tall grasslands and riverine forests in India and Nepal. The present population of wild buffalo in its entire range is estimated to be lower than 2,000 individuals.

Malabar Civet

The Malabar large spotted civet (*Viverra civettina* Blyth, 1862) was once a common species in the coastal districts of Malabar and Travancore in southwest India in the low elevation moist forests of the Western Ghats. By the late 1950s it was reported to be almost 'extinct'. None were seen for a long period of time until 1987, when it was rediscovered about 60 km east of Calicut in Kerala. Extensive deforestation has reduced the Malabar civets.

Other important endangered species are:

- Tortoise, Green sea Turtle, Gharial, Python (Reptiles)
- Peacock, Siberian White Crane, Pelican, Indian Bustard (Birds)
- Hoolock gibbin, Lion-tailed Macaque, Capped mokey, Golden monkey (Primates)
- *Rauvolfia serpentina* (medicinal plant), Sandal wood tree, *etc.*

4.8.1.3 Endangered Plant Species

India, the home of Ayurveda was once measured the epicentre of herbal bounty. According to legends, it grew the Sanjivani – the herb that could cure all ailments and counter all poisons. Traditionally, India's flora has been one of the richest across the globe. This can be attributed to the sheer range of soil, climatic conditions, topography and ecology present in the nation. According to studies, over 49,000 species of plants are to be found in the nation, that is, about 12 per cent of the total plant varieties known to mankind today. This includes some 16,000 species of flowering plants. This accounts for about 6 to 7 per cent of the total flowering plant species found in the world. Among all the plants found in India, scientific studies and research prove that about 3,000 plant species possess medicinal qualities and have been used in traditional remedies.

Here are some of the well-known endangered species of plants in India that are in urgent need of conservation:

- *Amentotaxus assamica* (Assam catkin yew) – Arunachal Pradesh
- Coniferous trees found only in the Delei Valley and Turoo Hills of Arunachal Pradesh. Low rates of reproduction have endangered the plant species.
- Ilex khasiana – Shillong (Meghalaya)
- Shrubs found only in Shillong Peak in Meghalaya. Only about three to four plants remaining.

Diospyros celibica (Ebony) – Karnataka

Well-known for its dark and high quality wood. Highly sought-after for furniture and wood work. Indiscriminate cutting has depleted the region covered by these trees.

Pterocarpus Santalinus (Red Sandalwood) – Eastern Ghats

This is a rare form of sandalwood found only in south-eastern India. Well-known for its medicinal properties. Endangered due to logging and loss of habitat.

Chlorophytum tuberosum (Musli) – Tamil Nadu

A flowering plant found only in the interiors of Africa and India. It is a well-known Ayurvedic medicine that gives vitality and strength. Indiscriminate harvesting has endangered the plant.

Kingiodendron Pinnatum (Malabar Mahagony) – Kerala, Karnataka, Tamil Nadu

Another tree best known for its wood. Logging, exploitation and destruction of habitat have led to low reproduction.

Actinodaphne lawsoni – Kerala

A canopy or sub-canopy tree that grows in high altitude forests. Harvested extensively for its medicinal properties. It is endemic and endangered due to habitat loss.

4.8.1.4 Conservation Ways of Endangered Species

If pollution can be controlled all across the globe then it can have a major positive impact on animals, fish and birds all over the world.

To save the endangered animals from extinction, a number of breeding programmes have been introduced. The government, NGOs and the other corporate bodies should come forward for this noble cause as this programme involves dedicated and special people and of course a lot of money.

Reintroducing the endangered animals to the wild once their numbers increase has become successful in some cases though all species have not done well.

If hunting and poaching can be controlled then there can be a significant change in the numbers of the endangered animals.

Endemic Species

Endemic species are animals or plants happen only in some specific areas and nowhere else in the world. In India, endemic species are frequently in Himalaya and Western Ghats. Endemic species are those that are found in just one area and nowhere else in the world. For instance, kangaroos are formerly endemic to Australia and are found nowhere else in the world. The cases where they have been dotted outside their natural environment is due to humans presenting them when the animal was in imprisonment.

Endemic species are those which are only found in a given region or location and nowhere else in the world. So the region which the species is endemic to ends up being called the "endemic site", a "national endemic", a "geographical range endemic", or a political region endemic depending on the location.

But as we see endemic plants and animals are unique to a particular geographical region; they are incredibly unique and more vulnerable to extinction. As a result, special efforts are required to conserve them. Endemism, or the incidence of endemic animals and plants, is more common in some regions than in others. In isolated environments such as the Hawaiian Islands, Australia, and the southern tip of Africa, as many of 90 per cent of naturally occurring species are endemic. In less remote regions, including Europe and much of North America, the percentage of endemic species can be very small.

Because endemic animals and plants by sense have partial geographic ranges, they can be chiefly vulnerable to human invasion and habitat destruction. Island species are chiefly vulnerable because islands usually lack large predators, and many island endemics changed without defences against predation. Cats, dogs, and other carnivores presented by sailors have destroyed many island endemics. The flora and fauna of Hawaii, exceptionally rich before Polynesians arrived with pigs, rats, and agriculture, were severely depleted because their range was limited and they had nowhere to retreat as human payment advanced. Tropical rain forests, with extraordinary species variety and high rates of endemism, are also vulnerable to human invasion. Many of the species detached daily in Amazonian rain forests are locally endemic, so that their whole range can be eliminated in a short time.

For instance, Endemic species, such as the tortoises of the Galápagos and the lemurs of Madagascar can be found small islands. Big islands also deliver the same separation but on a larger scale. Antarctica Hawaii and Australia are all huge land masses where we can find a lot of endemic species. Kangaroos, koalas, and polar bears are all endemic to these places.

The Redwood Forest on the West Coast of the United States has become endemic as it is now nearly totally partial to California. While there was a time when Redwoods used to cover much of the United States but have been destroyed by classification and are now limited to a small conservation area.

Diseases, on the other hand, can also be endemic. An endemic disease may be geographically isolated or it may be isolated to a certain group. Malaria is an example of an endemic disease because it is mostly limited to small pockets of infection in Africa.

Importance of Endemic Species

Endemic species are important for a number of reasons.

1. First, since endemic species have a generally restricted distribution, threats to endemics carry more risk of extinction than for broadly distributed species.
2. Second, when an endemic plant species becomes extinct, it takes with it between 10 and 30 endemic animal species.
3. Third, endemics, by definition, are highly adapted to their home range. As conditions within their range change, whether from anthropogenic or natural causes, their adaptations can function as a source of competitive strength or weakness.

In other words, some endemics of distinct regions may function as a sort of "collective insurance" for continued genetic diversity in the face of rapid changes, while others are at greatest risk of extinction as conditions change. Thus, endemic species are a focus for the conservation of biological diversity or biodiversity.

4.8.2.2 Endemic Species of India

A list of the endemic species of India is mentioned below:

- **Asiatic Lion, Gir Forest:** Asiatic Lion is also known as the Indian Lion and can be only found in and around Gir Forest National Park of Gujarat. These are listed as endangered species. These are one of the five big cats found in India, the others being Indian Leopards and Bengal Tigers.
- **Kashmir Stag, Kashmir Valley:** Also known as Hangul, Kashmir Stag is found in the dense forests of Dachigum National Park, Kasmir Valley and Chamba district, Himachal Pradesh.

- **Lion-Tailed Macaque, Western Ghats:** It is the rarest and the most threatened and endangered primate species found only in the Western Ghats of Southern India.
- **Purple Frog, Western Ghats:** The purple frog, also known as Pignose frog is only found in the rainforests of western ghats in India. It spends most of its life underground.
- **Sangai Deer, Loktak Lake:** It is also known as Brow Antlered Deer exclusively found in Keibul Lamjao National Park of Manipur. This park is a marshy wetland located in the southern parts of Loktak lake.
- **Nilgiri Tahr, Nilgiri Hills:** It is a wild sheep species, endangered and endemic to the Nilgiri Hills of Western Gats.

Other endemic species of India include:

- Pygmy Hog, Assam
- Bronzeback Vine Snake, Western Ghats
- Nilgiri Blue Robin, Nilgiri Hills
- Malabar Civet, Western Ghats
- Anaimalai Gliding Frog, Anaimalai Hills
- Namdapha Flying Squirrel, Arunachal Pradesh
- Indian Giant Squirrel
- Bonnet Macaque

Examples of endemic Flora species are

- Sapria Himalayana
- Ovaria Lurida
- Nepenthis khasiana *etc.*
- Nilgiri langur: The Nilgiri langur was first found in 1829.
- Asiatic lion: The Asiatic lion was first found in 1826 by the Austrian zoologist.
- One-horned rhinoceros: The one-horned rhinoceros were found in 1798.
- Mugger crocodile: The mugger crocodile was found in 1831 by Rene Lesson.
- White-rumped vulture: The white-rumped vulture was found in 1876 by Peter coal.
- Blackbuck: The Blackbuck was found in 1756 by Duke stark.
- Lion-tailed macaque: The long-tailed macaque was found in 1821 by sir Thomas.
- Niligiri blue robin: The Nilgiri blue robin was found in 1844 by Thomas C. Jerdon.

- ☆ Sangai deer: The Sangai deer was found in 1898 in the Manipur state only.
- ☆ Kashmir stag: The Kashmir stag was founded in 1906 by Tomy Charles.

4.9 Conservation of Biodiversity

Conservation and sustainable use of biodiversity have been an essential part of Indian philosophy. The diverse eco-climatic conditions joined with sole geological and cultural features have subsidised to an amazing variety of environments, which port and withstand huge biological diversity at all levels. With only 2.4 per cent of world's land area, India books for 7-8 per cent of recorded species of the world. In terms of species richness, India ranks seventh in mammals, ninth in birds and fifth in reptiles.

In the modern world, the conservation of biodiversity and its supportable use is becoming progressively significant and documented worldwide. The year 2010 marked the implementation of Aichi Targets as the global goal at the tenth Conference of the Parties to the Convention on Biological Diversity (COP10) held in Japan. In 2015, the 2030 Agenda for Sustainable Development was adopted by the United Nations General Assembly and the Sustainable Development Goals (SDGs) were announced, which included goals for biodiversity. In 2019, the Intergovernmental Science-Policy Platform on Biodiversity and Ecosystem Services* (IPBES) released their first global assessment which tinted the opposing effects of human activities on the ecosystems. The valuation reported that around one million animal and plant species are now endangered with extermination and that the health of ecosystems on which we and all other species depend is deteriorating more rapidly than ever. Initiatives for the conservation of biodiversity are becoming an inseparable issue when companies try to promote sustainable management.

Conserving biodiversity has economic, social, and cultural values. Conservation of biodiversity is vital to the biological and cultural legacy of many people and the dangerous components of healthy ecosystems that are used to provision economic and social developments. Moreover, it is used to uphold the earth's genetic library from which society has derived the foundation of its agriculture and medicine. The twenty-first century is foretold to be an era of bio-economy driven by advances of bioscience and biotechnology. Bio-economy may become the fourth economy form after agricultural, industrial, and information technology economies, having far-reaching impacts on sustainable development in agriculture, forestry, environmental protection, industry, food supply, health care and other micro-economy aspects. Thus, a strategic vision for conservation and sustainable use of biodiversity in importance the 21st century is of far-reaching implication for sustainable development economy and society.

4.9.1 Types of Biodiversity

Biodiversity is the diversity of plants, animals and other living creatures on Earth. From the minimum micro-organism to the major mammal, biodiversity forms a web of life, creating the ecosystems we live in and be contingent upon for our survival, and providing us with food, water, medicine and many other important ecological services.

Biodiversity is being lost at alarming rates - between 1,000 and 10,000 times faster than the natural extinction rate, owing to problems such as habitat obliteration and climate change. A comprehensive range of ecosystem facilities are tainted as a result. The loss of biodiversity has overwhelming special effects on local and native communities around the world who depend on it for their livelihoods. Biodiversity is significant for the upkeep of a healthy environment and used for straight human benefits like food, medicine, and energy. It is also used for recycling of diverse vital elements, for vindication of pollution, for defence of watersheds, to mitigate soil erosion and to control excessive variations in climate and catastrophic events. For instance, biodiversity provides different services free of charge worth of billions dollar every year for crucial well-being of the society.

Some of these services are providing clean water and air, soil formation and protection, pollination, pest control, food, fuel, fibres, medicine, and construction and industry raw materials. Agricultural biodiversity is another significant constituent of biodiversity, which has a straighter link to the well-being and livelihood of mankind than other forms of biodiversity. Food plant and animal species have been collected, used, domesticated, and improved through traditional systems of selection over many generations.

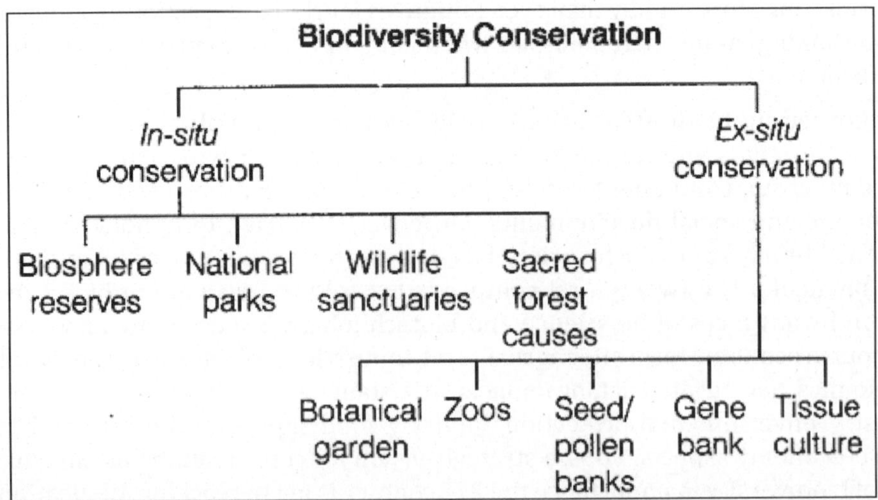

In situ Conservation

In situ conservation, which is also known as "on-site conservation", mentions to the conservation of wild species in their usual habitats and environment. It targets to conserve the natural habitats of the living creatures and uphold and recuperate wild species, particularly the endangered species. It is the procedure of conserving all the living species, especially the wild and endangered species in their natural habitats and environment. *In situ* conservation of Biodiversity includes biosphere reserves, national parks, wildlife sanctuaries, *etc.*

In situ conservation is defined as conservation of ecosystems and natural habitats, the conservation of viable populations of the species in their natural surroundings and, in the case of the refined species, in the surrounds where they have advanced their characteristic properties. *In situ* conservation can be done in farmers' fields, in pasture lands, and in protected areas. For cultivated species, *in situ* conservation concerns the maintenance of the local intra- and inter-population diversity available in various ecological and geographical sites. Thus, it allows ongoing host-parasite coevolution, which is likely to provide material resistance to pests and diseases, and CBD documented it as a main approach to conserve biodiversity. However, *in situ* conservation has certain boundaries like more difficult access to breeders requiring the application of its complimentary technique. For instance, some of the natural habitats or wild habitats are very risky when compared to relatively safe captive environment.

This includes the establishment of

- ✰ National parks and sanctuaries
- ✰ Biosphere reserves
- ✰ Nature reserves
- ✰ Reserved and protected forests
- ✰ Preservation plots
- ✰ Reserved forests

Ex situ Conservation

Ex situ conservation, which is also recognized as off-site conservation, refers to the conservation of endangered species in the false or man-made habitats that copy their natural habitats, *e.g.* zoo, aquarium, botanical garden *etc.* It is the process of conserving all the living species in the artificial habitats that imitate their natural habitats. *Ex situ* Conservation of Biodiversity includes aquariums, botanical gardens, zoos, Cryopreservation DNA banks, *etc.*

It offers less flexibility to the animals as it is lesser in area than the area of *in situ* conservation. This method of conservation is appropriate for the animals which are not found in abundance. *Ex situ* conservation is a method of conservation of biological diversity outside its natural habitats, directing all levels of biodiversity such as genetic, species, and ecosystems. Its idea was established earlier before its authorised acceptance under the Convention on Biological Diversity signed in 1992 in Rio de Janeiro.

In broad, *ex situ* conservation is practical as an additional measure to addition *in situ* conservation, which refers to upkeep of biological diversity in its natural habitats. In some cases, *ex situ* management will be dominant to a conservation strategy and in others it will be of secondary importance. Approximately, *ex situ* conservation comprises a diversity of activities, from handling captive populations, education and raising awareness, supporting research initiatives and collaborating with *in situ* efforts. It is used as valued tools in studying and conserving biological resources (plants, animals, and microorganisms) for different purposes through dissimilar techniques such as zoos, captive breeding, aquarium, botanical gardens, and gene banks.

Types of Ex Situ Conservation

Zoos

Zoos or zoological gardens or zoological parks in which animals are limited within presences or semi-natural and open areas, showed to the public, and in which they may also breed.

Captive Breeding

Captive breeding is an essential part of the general conservation action plan for a species that helps to stop extinction of species, subspecies, or population. It is a concentrated organisation practice for endangered individuals, populations, and species by anthropogenic and natural factors. In small and fragmented populations, even if the human caused intimidations could be magically reversed, the species would still have a high probability of extinction by random demographic and genetic events, environmental variations, and catastrophes.

Aquarium

An aquarium is a non-natural habitat for water-dwelling animals. It can also be used to house amphibians or large marine mammals and plant species for tourist attractions. It is usually found in zoos or marine parks with different size. The 15,750 labelled type of freshwater fish include around 25 per cent of living vertebrate species diversity and a key for global economic and nutritional resources of which more than 11 per cent is endangered (60-extinct, 8-extinct in the wild and 1679-threatened).

Botanical Gardens

Botanical gardens contain of living plants, grown out of doors or under glass in greenhouses and conservatories. They are used to grow and display plants primarily for scientific and educational purposes. They also include herbarium, lecture rooms, laboratories, libraries, museum and experimental or research plantings.

Gene Banks

Genome resource banking is another management technique used for biodiversity conservation. Different types of gene banks have been established for the storage of biodiversity, depending on the type of materials conserved. These include seed banks (for seeds), field gene banks (for live plants), in vitro gene banks (for plant tissues and cells), pollen, chromosome, and deoxyribonucleic acid (DNA) banks for animals (living sperm, eggs, embryos, tissues, chromosomes, and DNA) that are held in short term or long term laboratory storage; usually cryopreserved or freeze-dried.

4.10 Agrobiodiversity

Agrobiodiversity is the consequence of natural selection procedures and the cautious selection and creative developments of farmers, herders and fishers over millennia. Agrobiodiversity is a vital sub-set of biodiversity. Many people's food and livelihood security depend on the continuous management of numerous biological resources that are significant for food and agriculture. Agricultural biodiversity is a comprehensive term that includes all mechanisms of biological diversity of significance to food and agriculture, and all components of biological diversity that establish the agricultural ecosystems, also named agro-ecosystems: the diversity and inconsistency of animals, plants and micro-organisms, at the genetic, species and ecosystem levels, which are vital to endure key purposes of the agro-ecosystem, its construction and processes.

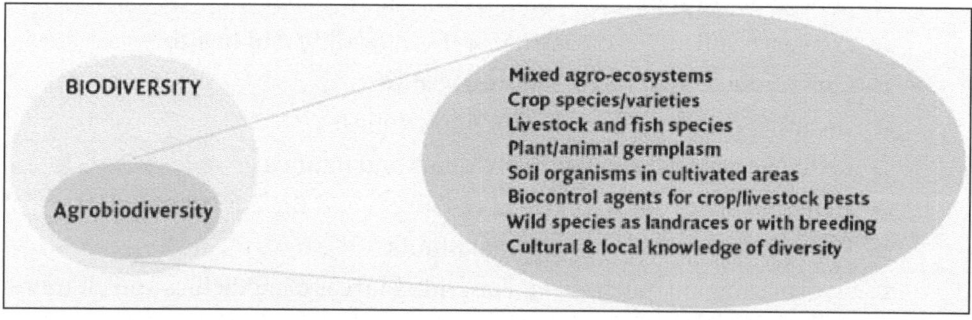

Agricultural biodiversity is the consequence of the connections among hereditary resources, the environment and the management systems and practices used by farmers. This is the result of both natural selection and human creative industrialised over times.

Agrobiodiversity is significant not only for environmental constancy but it also additions the local living. The level of dependence of the locals on agrobiodiversity is even more concentrated in an area where there is no logged government forests or endangered areas. A total of 145 households were plotted to explore the agrobiodiversity status in the farmlands of Ullapara upazilla in Sirajganj District, Bangladesh, where endangered areas do not exist. A total of 37 tree species belonging to 22 families including timber, fruits, fodder, fuel wood, and multi-purpose tree species (MPTS) were logged. The most shared species was Eucalyptus camaldulensis (83 per cent) followed by Swietenia macrophylla (60 per cent). The regular tree density for the whole study area was 87.35 ± 9.28 trees per ha. The nasty worth of Shannon–Wiener index and consistency of species were 3.33 and 2.12, respectively, with larger farms having the uppermost tree class variety and bordering farms needing the lowest. Actual farm size and the species diversity were found definitely correlated (Spearman's $\rho = 0.33$, $n = 145$, $p \leq 0.01$). Five types of cereal crops and two types of shared vegetables were grown. Mean production of all agriculture crops was the uppermost on the large farms. Chicken and duck were the chief livestock in the study area. Farmland, like other agroforestry systems, supports a level of tree class diversity along with crop and livestock diversity which may donate to biodiversity conservation in least forest areas and supply diverse products and make income for farmers as incomes of their being.

The Role of Agrobiodiversity

Experience and research have shown that agrobiodiversity can:

- ✰ Increase productivity, food security, and economic returns
- ✰ Reduce the pressure of agriculture on fragile areas, forests and endangered species
- ✰ Make farming systems more stable, robust, and sustainable
- ✰ Contribute to sound pest and disease management
- ✰ Conserve soil and increase natural soil fertility and health
- ✰ Contribute to sustainable intensification
- ✰ Diversify products and income opportunities
- ✰ Reduce or spread risks to individuals and nations
- ✰ Help maximize effective use of resources and the environment
- ✰ Reduce dependency on external inputs
- ✰ Improve human nutrition and provide sources of medicines and vitamins, and
- ✰ Conserve ecosystem structure and stability of species diversity.

There are several distinctive features of agrobiodiversity, compared to other components of biodiversity:

Agrobiodiversity is actively managed by male and female farmers; many components of agrobiodiversity would not endure without this human interference; local knowledge and culture are integral parts of agrobiodiversity management;

Many frugally important agricultural systems are based on 'alien' crop or livestock species introduced from away (for example, horticultural production systems or Friesian cows in Africa). This makes a high degree of interdependence between countries for the hereditary resources on which our food systems are based;

As regards crop diversity, diversity within species is at least as important as diversity between species; Because of the degree of human management, conservation of agrobiodiversity in production systems is inherently linked to sustainable use - preservation through establishing protected areas is less relevant; and

In industrial-type agricultural systems, much crop diversity is now held *ex situ* in gene banks or breeders' materials rather than on-farm.

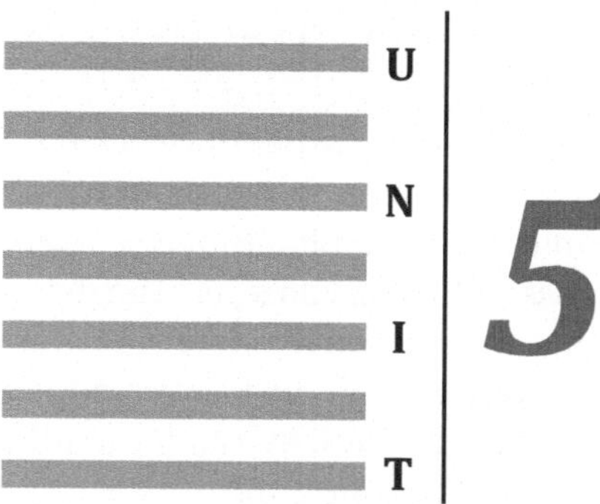

Environmental Pollution

Introduction

Pollution is the outline of pollutants into the environment that reason damage or uneasiness to humans or other living organisms, or that damage the environment, which can come in the form of chemical substances, or energy such as noise, heat or light. Pollutants can be unsurprisingly happening substances or energies, but are reflected pollutants when in excess of natural levels.

Pollution is the introduction of damaging materials into the environment. These destructive resources are called pollutants. Pollutants can be natural, such as volcanic ash. They can also be shaped by human activity, such as garbage or runoff produced by factories. Pollutants damage the quality of air, water, and land. Many things that are useful to people produce pollution. Cars spew pollutants from their exhaust pipes. Burning coal to create electricity pollutes the air. Industries and homes generate garbage and sewage that can pollute the land and water. Pesticides–chemical poisons used to kill weeds and insects–seep into waterways and harm wildlife. All living things–from one-celled microbes to blue whales–depend on Earth's supply of air and water. When these resources are polluted, all forms of life are threatened.

Pollution may be a worldwide problem. Although urban areas are typically more polluted than the countryside, pollution can spread to overseas where no people live. For instance, pesticides and other chemicals are found within the Antarctic ice sheet. Within the middle of the northern Pacific, a large collection of microscopic plastic particles forms what's called the nice Pacific Garbage Patch. Air and water currents carry pollution. Ocean currents and migrating fish carry marine pollutants far and wide. Winds can develop material accidentally released from a setup and scatter it round the world. Smoke from a factory in one country drifts into another country.

It matters first and for many because it's negative impacts on crucial environmental services like the supply of fresh air and cleans water without which life on Earth as we all know it might not exist. People are the explanation we've got pollution. Pollution affects the environment because pollution can affect the living conditions of individuals and plants. Pollution can cause the environment to start out and fall. If we don't have clean water for plants and trees how are we visiting make paper and grow veggies for people to eat? This can be why the large idea affects the environment and folks are the most reason for pollution.

What is Environmental Pollution?

Environmental pollution takes place when the environment cannot process and neutralize harmful by-products of human activities (poisonous gas emissions) in due course without any structural or functional damage to its system.

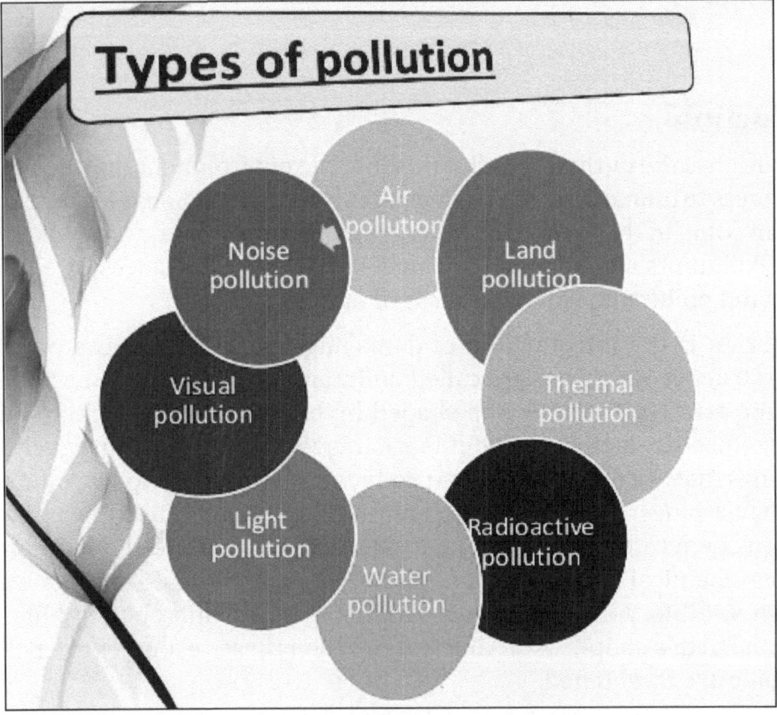

Environmental pollution is reaching worrying proportions worldwide. Urbanization and industrialization along with economic development have led to increase in energy consumption and waste discharges. The global environmental pollution, including greenhouse gas emissions and acid deposition, as well as water pollution and waste management is considered as international public health problems, which should be investigated from multiple perspectives including social, economic, legislation, and environmental engineering systems, as well as lifestyle habits helping health promotion and strengthening environmental systems to resist contamination.

Environmental pollutants have various adverse health effects from early life some of the most important harmful effects are perinatal disorders, infant mortality, respiratory disorders, allergy, malignancies, cardiovascular disorders, and increase in stress oxidative, endothelial dysfunction, mental disorders, and various other harmful effects. Though, short-term effects of environmental pollutants are usually highlighted, wide range of hazards of air pollution from early life and their possible implication on chronic non-communicable diseases of adulthood should be underscored. Numerous studies have exposed that environmental particulate exposure has been linked to increased risk of morbidity and mortality from many diseases, organ disturbances, cancers, and other chronic diseases. Therefore it is time to take action and control the pollution. Otherwise, the waste products from consumption, heating, agriculture, mining, manufacturing, transportation, and other human activities will degrade the environment.

5.1 Types of Pollution

As stated before, there are different types of pollution, which are either caused by natural events (like forest fires) or by man-made activities (like cars, factories, nuclear wastes, *etc.*) These are further classified into the following types of pollution:

- Air Pollution
- Water Pollution
- Soil Pollution
- Noise Pollution
- Marine Pollution
- Thermal Pollution
- Nuclear Hazards

5.2 Air Pollution

5.2.1 Definition

Air pollution are often outlined because the presence of toxic chemicals or compounds (including those of biological origin) within the air, at levels that cause a health risk. In a fair broader sense, pollution suggests that the presence of chemicals or compounds within the air that area unit typically not gift and that lower the standard of the air or cause prejudicial changes to the standard of life (such because the damaging of the layer or inflicting world warming).

Air pollution may be a mixture of solid particles and gases within the air. Automobile emissions, chemicals from factories, dust, and spore and mould spores could also be suspended as particles. Ozone, a gas, may be a major a part of pollution in cities. Once gas forms pollution, it is also referred to as air pollution. Pollution is currently thought of to be the world's largest environmental health threat, accounting for seven million deaths round the world per annum. Pollution causes and exacerbates variety of diseases, starting from respiratory illness to cancer, pulmonic sicknesses and cardiovascular disease. Outside pollution and stuff, one among its major elements, are classified as malignant neoplastic disease to humans by the International Agency for analysis on Cancer.

In accordance with recent estimates by the globe Health Organization, exposure to pollution is therefore a lot of necessary risk issue for major non-communicable diseases than antecedent thought. Pollution is that the largest contributor to the burden of sickness from the setting. the most substances moving health are: atomic number 7 oxides (NOx), sulphur oxides (SOx), gas and stuff with the latter - particularly stuff below two.5 microns (PM two.5) - being of greatest concern, as these little particles penetrate deep into the lungs, moving each the metabolic process and tube systems. Each extent and length of the exposure influence health outcomes.

History of Air Pollution

Along with superb technological advances, the economic Revolution of the mid-19th century introduced new sources of air and pollution. By the centre of the twentieth century, the results of those changes were starting to be felt in countries round the world. Within the Nineteen Sixties, AN environmental movement began to emerge that sought-after to stem the tide of pollutants flowing into the planet's ecosystems. Out of this movement came events like Earth Day, and legislative victories just like the Clean Air Act (1970) and therefore the Clean Water Act (1972).

In the latter a part of the thirteenth century, in an attempt to cut back pollution, England's King of Great Britain vulnerable Londoners with harsh penalties if they didn't stop burning sea-coal. However, the king's regulations–and those of subsequent leaders–had very little impact.

By the late eighteenth century and initial a part of the nineteenth century, coal came into large-scale use throughout the economic Revolution. The ensuing air pollution and soot had serious health impacts on the residents of growing urban centres. Within the nice air pollution of 1952, pollutants from factories and residential fireplaces mixed with air condensation killed a minimum of 4,000 folks in London over the course of many days. some years earlier, in 1948, severe industrial pollution created a deadly air pollution that physiological condition twenty folks in Donora, Pennsylvania, and made 7,000 a lot of sick. Acid rain, initial discovered within the decennium, was another downside ensuing from coal-powered plants. the discharge of human-produced sulphur and gas compounds into the atmosphere negatively wedged plants, fish, soil, forests and a few building materials.

Today, the leading reason behind pollution within the U.S. is cars that were initial factory-made within the U.S. by industrialist within the early twentieth century. Motorcar emissions conjointly increase the number of greenhouse gases within the atmosphere that successively contribute to heating.

Types of Pollutants

Primarily air pollutants can be caused by primary sources or secondary sources. The pollutants that are a direct result of the process can be called primary pollutants. A classic example of a primary pollutant would be the sulfur-dioxide emitted from factories

Secondary pollutants are the ones that are caused by the intermingling and reactions of primary pollutants. Smog created by the interactions of several primary pollutants is known to be as a secondary pollutant.

5.2.2 Causes of Air Pollution

1. The Burning of Fossil Fuels

Sulphur dioxide produced from the burning of fossil fuels like coal, petroleum and other sweatshop combustibles are one the main reason of air pollution. Pollution producing from vehicles including trucks, jeeps, cars, trains, airplanes cause a huge amount of pollution. We trust on them to fulfil our daily basic needs of transport.

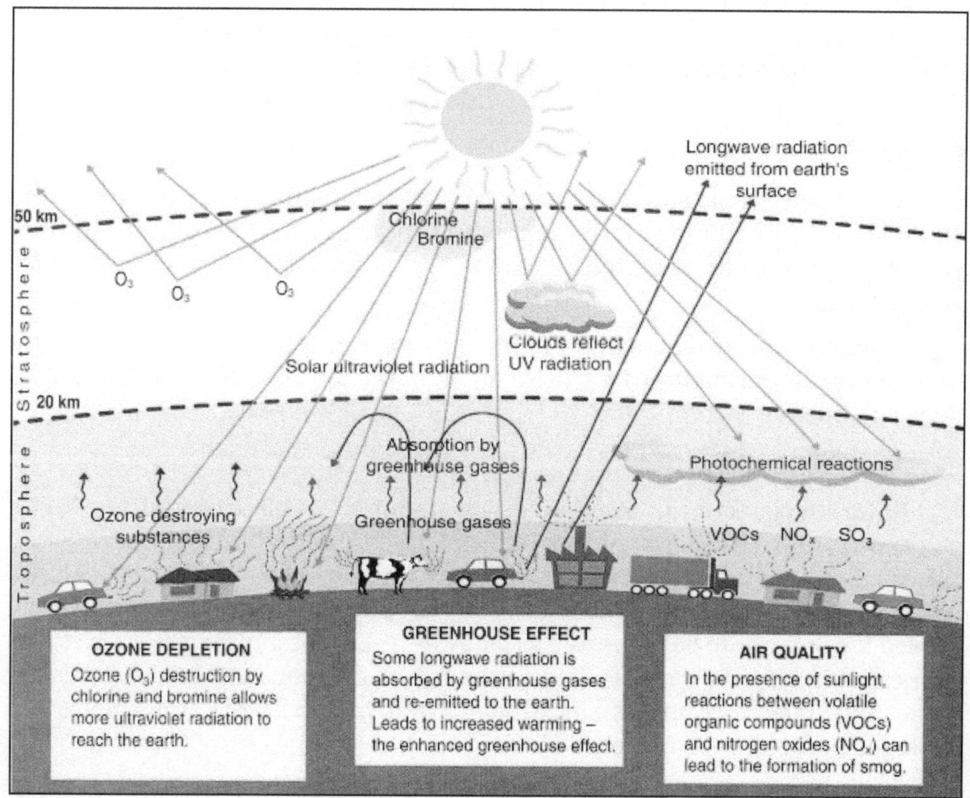

But, their abuse is killing our environment as unsafe gases are polluting the environment. Carbon Monoxide caused by improper or incomplete combustion and generally emitted from vehicles is another major pollutant along with Nitrogen Oxides that is produced from both natural and man-made processes.

2. Agricultural Activities

Ammonia is a very common by-product from agriculture-related activities and is one of the most hazardous gases in the atmosphere. Use of insecticides, pesticides, and fertilizers in agricultural activities has grown quite a lot. They emit harmful chemicals into the air and can also cause water pollution.

3. Dissipate from Factories and Industries

Manufacturing industries release a large amount of carbon monoxide, hydrocarbons, organic compounds, and chemicals into the air thereby depleting the quality of air. Manufacturing industries can be found at every corner of the earth and there is no area that has not been affected by it. Petroleum refineries also release hydrocarbons and various other chemicals that pollute the air and also cause land pollution.

4. Mining Processes

Mining is a process wherein minerals below the earth are extracted using large equipment. During the process dust and chemicals are released in the air causing massive air pollution. This is one of the reasons which is responsible for the deteriorating health conditions of workers and nearby residents.

5. Inside Air Pollution

Household cleaning products, painting supplies produce toxic chemicals in the air and cause air pollution. Have you ever noticed that once you paint the walls of your house, it creates some sort of smell which makes it literally impossible for you to breathe?

Suspended particulate matter popular by its acronym SPM, is another cause of pollution. Referring to the particles afloat in the air, SPM is usually caused by dust, combustion, *etc.*

6. Natural Disasters

Natural disasters can also reason air pollution to increase quickly. When volcanoes erupt, they eject volcanic ash and gases into the atmosphere. Volcanic ash can discolour the sky for months. After the eruption of the Indonesian volcano of Krakatoa in 1883, ash darkened the sky around the world. The dimmer sky caused fewer crops to be harvested as far away as Europe and North America. For years, meteorologists tracked what was known as the "equatorial smoke stream." In fact, this smoke stream was a jet stream, a wind high in Earth's atmosphere that Krakatoa's air pollution made visible.

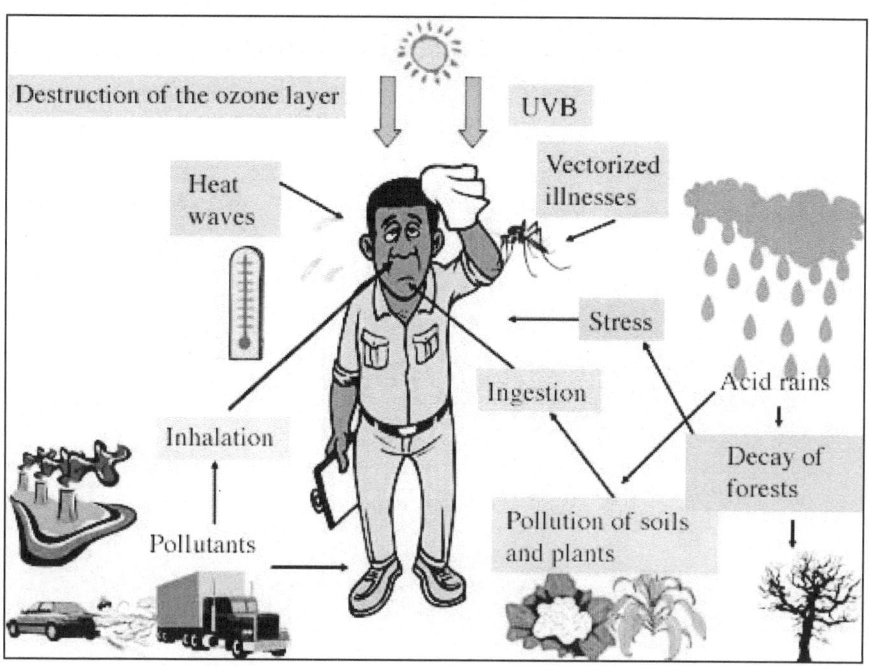

5.2.3 Effects of Air Pollution

1. **Respiratory and heart problems:** Smog makes breathing difficult, especially for children and older adults. Some cities that suffer from extreme smog issue air pollution warnings. The government of Hong Kong, for example, will warn people not to go outside or engage in strenuous physical activity (such as running or swimming) when smog is very thick.

   The effects of air pollution are alarming. They are known to create several respiratory and heart conditions along with Cancer, among other threats to the body. Several million are known to have died due to direct or indirect effects of Air pollution. Children in areas exposed to air pollutants are said to commonly suffer from pneumonia and asthma.

2. **Global warming:** Another direct effect is the immediate alterations that the world is witnessing due to global warming. With increased temperatures worldwide, increase in sea levels and melting of ice from colder regions and icebergs, displacement and loss of habitat have already signalled an impending disaster if actions for preservation and normalization aren't undertaken soon.

   Global warming is causing ice sheets and glaciers to melt. The melting ice is causing sea levels to rise at a rate of 2 millimetres (0.09 inches) per year. The rising seas will eventually flood low-lying coastal regions. Entire nations, such as the islands of Maldives, are threatened by this climate change.

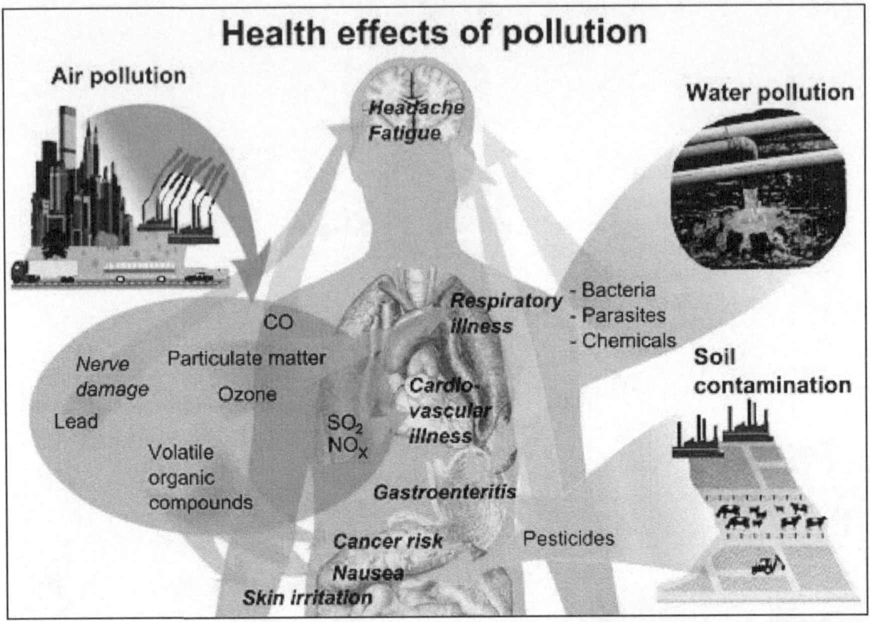

Global warming also contributes to the phenomenon of ocean acidification. Ocean acidification is the process of ocean waters absorbing more carbon dioxide from the atmosphere. Fewer organisms can survive in warmer, less salty waters. The ocean food web is threatened as plants and animals such as coral fail to adapt to more acidic oceans.

3. **Acid rain:** Harmful gases like nitrogen oxides and sulfur oxides are released into the atmosphere during the burning of fossil fuels. When it rains, the water droplets combine with these air pollutants, becomes acidic and then falls on the ground in the form of acid rain. Acid rain can cause great damage to human, animals, and crops.

   Acid rain can kill all the trees in a forest. It can also devastate lakes, streams, and other waterways. When lakes become acidic, fish can't survive. In Sweden, acid rain created thousands of "dead lakes," where fish no longer live.

   Acid rain also wears away marble and other kinds of stone. It has erased the words on gravestones and damaged many historic buildings and monuments. The Taj Mahal, in Agra, India, was once gleaming white. Years of exposure to acid rain has left it pale.

4. **Eutrophication:** Eutrophication is a condition where a high amount of nitrogen present in some pollutants gets developed on sea's surface and turns itself into algae and adversely affect fish, plants and animal species. The green coloured algae that are present on lakes and ponds is due to the presence of this chemical only.

5. **Effect on wildlife:** Just like humans, animals also face some devastating effects of air pollution. Toxic chemicals present in the air can force wildlife species to move to a new place and change their habitat. The toxic pollutants deposit over the surface of the water and can also affect sea animals.
6. **Depletion of the ozone layer:** Ozone exists in the Earth's stratosphere and is responsible for protecting humans from harmful ultraviolet (UV) rays. Earth's ozone layer is depleting due to the presence of chlorofluorocarbons, hydro chlorofluorocarbons in the atmosphere. As the ozone layer will go thin, it will emit harmful rays back on earth and can cause skin and eye related problems. UV rays also have the capability to affect crops.

Other harmful and disastrous effects of Air Pollution include:

1. It grounds depletion of ozone layer due to which ultraviolet radiations can reach the earth and cause skin cancer, damage to eyes and resistant system.
2. It reasons acid rain which damages crop plants, trees, buildings, monuments, statues and metal structures and also makes the soil acidic.
3. It causes greenhouse effect and global warming which leads to excessive heating of earth's atmosphere, further leading to weather variability and rise in sea level. The increased temperature may cause melting of ice caps and glaciers, resulting in floods.
4. Air pollution from certain metals, pesticides and fungicides causes serious ailments.
5. Lead pollution causes anaemia, brain damage, convulsions and death.
6. Certain metals cause problem in kidney, liver, circulatory system and nervous system.
7. Fungicides cause nerve damage and death.
8. Pesticides like DDT(Dichloride diphenyl trichloroethane) which are toxic enter into our food chain and gets accumulated in the body causing kidney disorders and problems of brain and circulatory system.
9. The table below shows the types of health effects experienced by the most common pollutants at elevated levels:

Pollutant	*Health effects at very high levels*
Nitrogen Dioxide, Sulphur Dioxide, Ozone	These gases irritate the airways of the lungs, increasing the symptoms of those suffering from lung diseases
Particles	Fine particles can be carried deep into the lungs where they can cause inflammation and a worsening of heart and lung diseases
Carbon Monoxide	This gas prevents the uptake of oxygen by the blood. This can lead to a significant reduction in the supply of oxygen to the heart, particularly in people suffering from heart disease

5.2.4 Control Measures of Air Pollution

Different techniques are used for controlling air pollution caused by 'gaseous pollutants' and that caused by 'particulate pollutants'.

Methods of Controlling Gaseous Pollutants

The air pollution caused by gaseous pollutants like hydrocarbons, sulphur dioxide, ammonia, carbon monoxide, *etc.* can be controlled by using three different methods-Combustion, Absorption and Adsorption.

- ☆ Combustion-This technique is applied when the pollutants are organic gases or vapours. The organic air pollutants are subjected to 'flame combustion or catalytic combustion' when they are converted to less harmful product carbon dioxide and a harmless product water.
- ☆ Absorption-In this method, the polluted air containing gaseous pollutants is passed through a scrubber containing a suitable liquid absorbent. The liquid absorbs the harmful gaseous pollutants present in air.
- ☆ Adsorption-In this method, the polluted air is passed through porous solid adsorbents kept in suitable containers. The gaseous pollutants are adsorbed at the surface of the porous solid and clean air passes through.

Methods of Controlling Particulate Emissions

The air pollution caused by particulate matter like dust, soot, ash, *etc.* can be controlled by using fabric filters, wet scrubbers, electrostatic precipitators and certain mechanical devices.

- ☆ Mechanical Devices-It works on the basis of following:
- ☆ Gravity-In this process, the particulate settle down by the action of gravitational force and get removed.
- ☆ Sudden change in the direction of air flow-It brings about separation of particles due to greater momentum.

Fabric Filters

The particulate matter is passed through a porous medium made of woven or filled fabrics.

- ☆ The particulate present in the polluted air are filtered and gets collected in the fabric filters, while the gases are discharged.
- ☆ The process of controlling air pollution by using fabric filters is called 'bag filtration'.

Wet Scrubbers

They are used to trap SO_2, NH_3 and metal fumes by passing the fumes through water.

Electrostatic Precipitators

When the polluted air containing particulate pollutants is passed through an electrostatic precipitator, it induces electric charge on the particles and then the aerosol particles get precipitated on the electrodes.

Some other methods of controlling Air Pollution:

- ☆ Tall chimneys should be installed in factories.
- ☆ Better designed equipment and smokeless fuels should be used in homes and industries.
- ☆ Renewable and non-polluting sources of energy like solar energy, wind energy, *etc.* should be used.
- ☆ Automobiles should be properly maintained and adhere to emission control standards.
- ☆ More trees should be planted along roadsides and houses.
- ☆ Use public mode of transportation: Encourage people to use more and more public modes of transportation to reduce pollution. Also, try to make use of carpooling. If you and your colleagues come from the same locality and have same timings you can explore this option to save energy and money.
- ☆ Conserve energy: Switch off fans and lights when you are going out. A large number of fossil fuels are burnt to produce electricity. You can save the environment from degradation by reducing the number of fossil fuels to be burned.

5.3 Water Pollution

5.3.1 Definition

Water pollution is defined as the attendance in groundwater of poisonous chemicals and biological agents that surpass what is logically found in the water and may stance a threat to human health and/or the environment. Moreover, water pollution may contain of chemicals introduced into the water bodies as a result of various human activities. Any quantity of those chemicals pollutes the water, regardless of the harm they may pose to human health and the environment.

What is Water Pollution?

It is the contamination of water bodies such as oceans, seas, lakes, rivers, groundwater, and aquifers by pollutants. The contaminants can end up in the water by the direct or indirect application. It is the second most prevalent type of environmental pollution, after air pollution. Below is an insight into the different types of water pollution and its effects.

Water pollution is any pollution of water with chemicals or other foreign substances that are harmful to human, plant, or animal health. These pollutants include fertilizers and pesticides from agricultural runoff; sewage and food processing waste; lead, mercury, and other heavy metals; chemical wastes from industrial discharges; and chemical contamination from hazardous waste sites. Worldwide, nearly 2 billion people drink contaminated water that could be harmful to their health.

When toxic substances enter lakes, streams, rivers, oceans, and other water bodies, they get dissolved or lie suspended in water or get deposited on the bed. This results in the pollution of water whereby the quality of the water deteriorates, affecting aquatic ecosystems. Pollutants can also seep down and affect the groundwater deposits.

History of Water Pollution?

Just like air, water is under assault from numerous types of pollution. For centuries, humans unknowingly contaminated sources of drinking water with raw sewage, which led to diseases such as cholera and typhoid. According to a CNN report, one gram of human excrement contains approximately "10 million viruses, 1 million bacteria, 1,000 parasite cysts and 100 parasite eggs." Today, over 1 billion people worldwide lack access to safe water and every 15 seconds somewhere on the planet, a child dies from a water-related disease, according to Water Partners International (www.water.org).

Water pollution strengthened with the advent of the Industrial Revolution, when factories began releasing pollutants directly into rivers and streams. In 1969, chemical waste released into Ohio's Cuyahoga River caused it to burst into flames and the waterway became a symbol of how industrial pollution was destroying America's natural resources. In 2007, CNN reported that "up to 500 million tons of heavy metals, solvents and toxic sludge slip into the global water supply every year. In the developing world [according to UNESCO] as much as 70 per cent of industrial waste is just dumped untreated into the rivers and lakes. China is a perfect case in point. According to Greenpeace, around 70 per cent of China's lakes and rivers are now polluted from industrial waste, leaving 300 million people 'forced to rely on polluted water supplies.'" Water sources are also contaminated by rain runoff from such things as oil-slick roads; construction, mining and dump sites; and livestock wastes from farm operations. Leaky septic tanks, pesticides and fertilizers are among the other sources that can contaminate groundwater. Over half the American population (including the majority of those living in rural areas) relies on groundwater for drinking water, according to The Groundwater Foundation (www.groundwater.org), which also notes that the largest use for groundwater is crop irrigation.

In 1972, Congress passed the Clean Water Act to reduce water pollution. Various pieces of anti-pollution legislation have followed since that time and today the U.S. has relatively clean, safe drinking water compared with much of the world. However, water pollution is still a problem. In 2006, the Environmental News

Service (ENS) reported that "more than 62 per cent of industrial and municipal facilities across the country discharged more pollution into U.S. waterways than their Clean Water Act permits allowed between July 2003 and December 2004." The ENS also noted that over 40 per cent of American waterways were unsafe for swimming and fishing. Additionally, water resources face an ongoing threat from man-made environmental disasters such as the 1989 Exxon Valdez oil spill, during which approximately 11 million gallons of crude oil were accidentally dumped into the sea off Alaska's Prince William Sound. The disaster, which created a 3,000-square-mile oil slick, instantly killed hundreds of thousands of birds, fish and other wildlife and devastated the area for years afterward.

Water contamination is categorised based on the source or the nature of the water body it affects. Some of the most common pollution types are:

5.3.2 Causes of Water Pollution

Water pollution can come from a variety of sources. Pollution can enter water directly, through both legal and illegal discharges from factories, for example, or imperfect water treatment plants. Spills and leaks from oil pipelines or hydraulic fracturing (fracking) operations can degrade water supplies. Wind, storms, and littering–especially of plastic waste–can also send debris into waterways.

Some of the causes of water pollution are:

Sewage

With billions of people on the planet, positioning of sewage waste is a main problem. According to 2015 figures from the World Health Organization (the most recent available at the time this article was updated in 2019), some 2.1 billion people (28 per cent of the world's population) don't have admission to safe drinking water, while 2.3 billion (30 per cent of the world's population) don't have proper hygiene (hygienic toilet facilities); although there have been great developments in securing access to clean water, relatively little growth has been made on improving global sanitation in the last decade. Sewage removal touches people's instant environments and leads to water-related illnesses such as diarrhoea that kills 525,000 children under five each year. In industrialised countries, most people have flush toilets that take sewage waste rapidly and hygienically away from their homes.

More than 80 per cent of the world's wastewater flows back into the setting without being preserved or reused, according to the United Nations; in some least-developed countries, the figure tops 95 percent.

Agricultural Run off

The use of land for agriculture and the practices followed in cultivation greatly affect the quality of groundwater. Intensive cultivation of crops causes chemicals from fertilizers (*e.g.* nitrate) and pesticides to seep into the groundwater, a process commonly known as leaching.

Industrial Effluents

Wastewater from manufacturing or chemical processes in industries contributes to water pollution. Industrial wastewater usually contains specific and readily identifiable chemical compounds. Most major industries have treatment facilities for industrial effluents. But this is not the case with small-scale industries, which cannot afford enormous investments in pollution control equipment as their profit margin is very slender.

Suspended Matter

Some pollutants do not dissolve in water and are too large to mix with water molecules. They comprise can, straws, and other large objects. When deferred, they form a layer on the superficial of the water, preventing oxygen diffusion leading to oxygen depletion pollution.

Some of the particulate substance may relax at the lowest of the lake, ocean, or river, affecting the life on the floor of the river, lake, or ocean. In more or less cases, the material can comprise of harmful toxins.

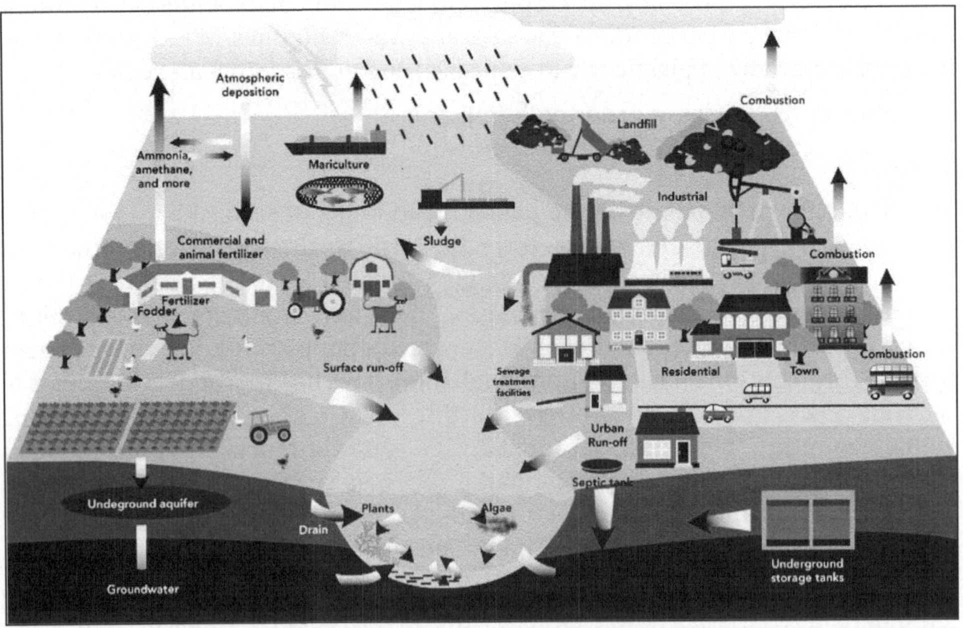

Plastics

While plastics are not toxic in quite the same way as poisonous chemicals, they nevertheless present a major hazard to seabirds, fish, and other marine creatures. For example, plastic fishing lines and other debris can strangle or choke fish. (This is sometimes called ghost fishing.) About half of all the world's seabird species are known to have eaten plastic residues. In one study of 450 shear-waters in the

Eutrophication

At the point when new water is misleadingly enhanced with supplements, it brings about an irregular increment in the development of water plants. This is known as eutrophication. The release of waste from enterprises, farming, and urban networks into water bodies for the most part extends the organic limits of amphibian frameworks. Compound run-off from fields likewise adds supplements to water. Abundance supplements cause the water body to get gagged with natural substances and creatures. At the point when natural issue surpasses the limit of the miniaturized scale living beings in water that separate and reuse the natural issue, it supports quick development, or blossoms, of green growth. At the point when they kick the bucket, the remaining parts of the green growth add to the natural squanders as of now in the water; in the long run, the water gets insufficient in oxygen. Anaerobic life forms (those that don't expect oxygen to live) at that point assault the natural squanders, discharging gases, for example, methane and hydrogen sulfide, which are hurtful to the oxygen-requiring (high-impact) types of life. The outcome is a putrid, squander filled waterway. This has just happened in such places as Lake Erie and the Baltic Sea, and is a developing issue in freshwater lakes all over India. Eutrophication can deliver issues, for example, awful tastes and scents just as green filth green growth. Additionally the development of established plants expands, which diminishes the measure of oxygen in the most profound waters of the lake. It additionally prompts the passing of all types of life in the water bodies.

North Pacific, over 80 per cent of the birds were found to contain plastic residues in their stomachs. In the early 1990s, marine scientist Tim Benton collected debris from a 2km (1.5 mile) length of beach in the remote Pitcairn islands in the South Pacific. His study recorded approximately a thousand pieces of garbage including 268 pieces of plastic, 71 plastic bottles, and two dolls heads.

Types of Water Pollution

Surface Water Pollution

Covering about 70 per cent of the earth, surface water is what fills our oceans, lakes, rivers, and all those other blue bits on the world map. Surface water from freshwater sources (that is, from sources other than the ocean) accounts for more than 60 per cent of the water delivered to American homes. But a significant pool of that water is in peril. According to the most recent surveys on national water quality from the U.S. Environmental Protection

Agency, nearly half of our rivers and streams and more than one-third of our lakes are polluted and unfit for swimming, fishing, and drinking. Nutrient pollution, which includes nitrates and phosphates, is the leading type of contamination in these freshwater sources. While plants and animals need these nutrients to grow, they have become a major pollutant due to farm waste and fertilizer runoff. Municipal and industrial waste discharges contribute their fair share of toxins as well. There's also all the random junk that industry and individuals dump directly into waterways.

Groundwater Pollution

It occurs when the hazardous chemicals and particles applied on the surface by humans seep into the ground by rainwater. The contaminants pollute under-groundwater features such as underground rivers and waterbeds. As a result, wells and boreholes might become contaminated. The most common cause of this type of pollution is the use of pesticides and fertiliser on farms.

Chemical Pollution

Chemicals are the most common type of water contaminants. They affect both surface and under-groundwater bodies. Industries and farming activities are the leading cause.

Solvents and metals used in industries can pollute rivers and lakes. Weed, insects, and fungi control in farms using pesticides is the other cause of soil contamination. Chemical contamination also results from petroleum spills.

Nutrients Pollution

Even though nutrients are essential for plant and aquatic life, an excess of it is dangerous. Wastewater and fertilisers have a high content of nutrients required for plant growth. Consequently, they cause rapid and uncontrolled growth of vegetation and algae on the water surface when they end up in the water.

It leads to clogging of water filters and contamination of drinking water. It also uses up all the oxygen leading to the destruction of marine life.

Oxygen Depletion Pollution

Aquatic microorganisms thrive on biodegradable substances. When many of these materials get into the water, the number of microorganisms increases. They use up all the oxygen in the water. The depletion of oxygen leads to the death of aerobic microorganisms but promotes the thriving of anaerobic organisms.

Certain anaerobic microorganisms contaminate the water by producing toxins such as sulphides and ammonia. All these are harmful to humans and aquatic life.

Microbiological Pollution

It is a natural type of contamination of water, as it results from natural existing microorganisms such as protozoa, viruses, and bacteria. Water containing some of these microorganisms can cause diseases such as cholera and bilharzia. The effects of microbiological pollution are common in areas where people drink untreated water.

5.3.3 Effects of Water Pollution

Contamination of water influences both the humans and amphibian life. Most water sources near urban communities and urban focuses are contaminated by trash and dumping of synthetic compounds, lawfully or unlawfully. The following are a portion of the basic just as antagonistic impacts of dirtying water bodies. Water contamination can bring about human medical issues, harmed natural life, and long haul biological system harm. At the point when agrarian and mechanical spill over floods conduits with over-abundance supplements, for example, nitrogen and phosphorus, these supplements frequently fuel green growth sprouts that at that point make no man's lands, or low-oxygen regions where fish and other oceanic life cannot flourish anymore.

A few people accept contamination is an unpreventable consequence of human action: they contend that on the off chance that we need to have manufacturing plants, urban areas, ships, vehicles, oil, and beach front hotels, some level of contamination is practically sure to result. As it were, contamination is a fundamental wickedness that individuals must endure in the event that they need to gain ground. Luckily, not every person concurs with this view. One explanation individuals have woken up to the issue of contamination is that it brings expenses of its own that subvert any monetary advantages that come to fruition by dirtying.

Take oil slicks, for instance. They can occur if big haulers are excessively inadequately worked to endure mishaps adrift. In any case, the monetary advantage of settling on big hauler quality brings a financial cost when an oil slick happens. The oil can appear on close by sea shores, crush the biological system, and seriously influence the travel industry. The primary issue is that the individuals who bear the expense of the spill (normally a little waterfront network) are not the individuals who caused the issue in any case (the individuals who work the big hauler). However, apparently, each and every individual who puts gas (petroleum) into their vehicle–or uses practically any sort of oil fuelled transport–adds to the issue here and there. So oil slicks are an issue for everybody, not simply individuals who live by the coast and big hauler works.

Sewage is another genuine case of how contamination can influence all of us. Sewage released into seaside waters can appear on sea shores and cause a well-being peril. Individuals who wash or surf in the water can become sick on the off chance that they swallow dirtied water–yet sewage can have other destructive impacts as well: it can harm shellfish, (for example, cockles and mussels) that develop close to the shore. Individuals who eat harmed shellfish chance experiencing an intense–and here and there deadly–disease called incapacitated shellfish harming. Shellfish is no longer gotten along numerous shores since it is essentially excessively contaminated with sewage or poisonous synthetic squanders that have released from the land close by.

The impacts of water contamination are obliterating to individuals as well as to creatures, fish, and flying creatures. Dirtied water is unsatisfactory for drinking, entertainment, agribusiness, and industry. It decreases the tasteful nature of lakes and waterways. All the more genuinely, tainted water pulverizes oceanic life and lessens its conceptive capacity. In the long run, it is a peril to human well-being. It's not possible for anyone to get away from the impacts of water pollution.

Effects on Human Beings

Life is a cycle, and human misconduct often comes back to haunt you. Adding pollution to water bodies has affected the human family in many ways. According to a 2017 WHO report, 2.1 billion people do not have access to safe water. In 2019, it explained that 785 million people lack vital drinking water. Germs found in water, in the form of germs that cause diseases and germs from human and animal waste, are a major cause of illness from contaminated drinking water. Diseases spread by unsafe water include cholera, giardia and typhoid. Even in wealthy countries, accidental or illegal evacuation from sanitation facilities, as well as from farms and urban areas, contributes to harmful pathogens on waterways. Thousands of people across the United States are getting sick every year with Legionnaires disease (a powerful form of lung disease that is established in water sources such as cool towers and tap water), with cases ranging from California's Disneyland to Manhattan's Upper East Side.

Globally, dirty water also affects the health value of the disease. At least 2 billion people drink water from contaminated sources, according to the World Health Organization, and that water can transmit dangerous diseases such as cholera and typhoid.

Effects on the Environment

Contamination matters since it hurts nature on which individuals depend. The earth isn't something removed and separate from our lives. It is anything but a pretty shoreline many miles from our homes or a wild scene that we see just on TV. The earth is everything that encompasses us that gives us life and well-being. Wrecking the earth at last diminishes the nature of our own lives–and that, most egotistically, is the reason contamination should matter to us all.

So as to flourish, sound biological systems depend on an intricate trap of creatures, plants, microbes, and parasites–all of which connect, straightforwardly or by implication, with one another. Damage to any of these living beings can make a chain impact, risking whole oceanic conditions.

At the point when water contamination causes an algal sprout in a lake or marine condition, the multiplication of recently presented supplements animates plant and green growth development, which thusly diminishes oxygen levels in the water. This shortage of oxygen, known as eutrophication, chokes out plants and creatures and can make "no man's lands," where waters are basically without life. In specific cases, these unsafe algal sprouts can likewise create neurotoxins that influence untamed life, from whales to the ocean turtles.

Death of Aquatic Life

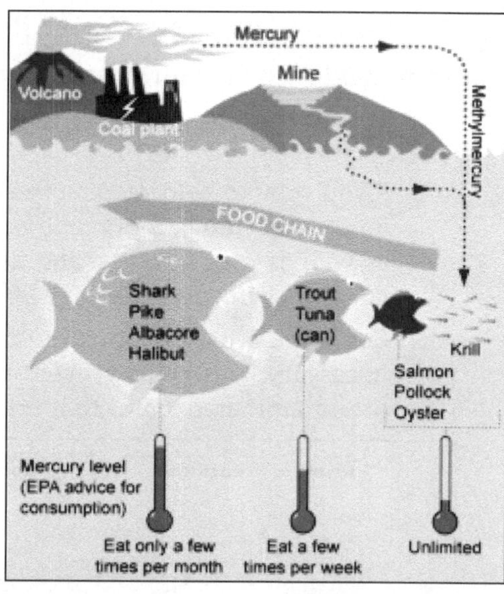

Animals and plants that be contingent on water for life are the most pretentious by contaminated water. Statistics from the Centre for Biological Diversity on the effects of the Deep Horizon spill provides a useful sight of the impact of pollution on aquatic life. In the report, the 2010 spill on the Gulf of Mexico harmed over 82,000 birds, 25,900 marine animals, 6165 sea turtles, and an unidentified number of fish and invertebrates.

Chemicals and heavy metals from manufacturing and municipal wastewater pollute waterways as well. These pollutants are toxic to aquatic life–most often dipping an organism's life span and ability to copy–and make their way up the food chain as predator eats prey. That's how tuna and other big fish accrue high quantities of toxins, such as mercury.

Food Chain Commotion

Pollution disturbs the food chain by touching the toxins from one level in the chain to higher levels. In some cases, pollution can wipe out an entire part of the food chain. Such affect the other organisms by either causing excessive growth, in case the predator dies or death (if it wipes out the prey).

Destruction of Ecosystems

Nutrient pollution from upstream of land movements down and pollutes the larger water body. It in turns, promotes the algae growth and consequently growth of many more water creatures takes place. The extra growth of algae touches the fish and other aquatic organisms by fascinating and plummeting their oxygen supply. Fishes also die off due to clogging of their gills as a result of algal growth. Thus, the aquatic ecosystems are affected very adversely disturbing all the food chains. The introduction or elimination of certain microorganisms distorts the ecosystem. Nutrient effluence, for instance, leads to an increase in algae, which reduces the water of oxygen, thus leading to the demise of fish and other aquatic life.

Effects on Fauna

Animals, both aquatic and terrestrial, are prone to risks generated from the wastewater. In a classic case of marine pollution in recent time, 16000 miles of a US coastline was affected by an oil spill. In this case, death of many animals has been reported and there was great loss to aquatic fauna. Animals are also affected by the solid waste thrown into the water bodies, as it harm them in many ways.

5.3.4 Control Measures of Water Pollution

Water pollution in natural water bodies can be documented and counted on the basis of various parameters, such as, melted oxygen (DO), biochemical oxygen demand (BOD), coliform organisms, pH *etc.*

- ☆ If you want to help keep our waters clean, there are many things you can do to help. You can prevent water pollution of nearby rivers and lakes as well as groundwater and drinking water by following some simple guidelines in your everyday life.
- ☆ Conserve water by turning off the tap when running water is not necessary. This helps prevent water shortages and reduces the amount of contaminated water that needs treatment.

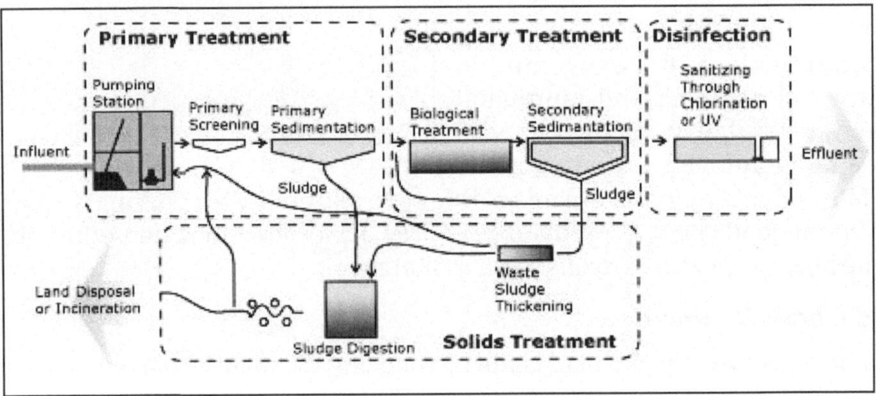

- ☆ Be careful about what you throw down your sink or toilet. Don't throw paints, oils or other forms of litter down the drain. Use environmentally household products, such as washing powder, household cleaning agents and toiletries.
- ☆ Take great care not to overuse pesticides and fertilisers. This will prevent runoffs of the material into nearby water sources.
- ☆ By having more plants in your garden you are preventing fertiliser, pesticides and contaminated water from running off into nearby water sources.
- ☆ Don't throw litter into rivers, lakes or oceans. Help clean up any litter you see on beaches or in rivers and lakes, make sure it is safe to collect the litter and put it in a nearby dustbin.
- ☆ There is no easy way to solve water pollution; if there were, it wouldn't be so much of a problem. Broadly speaking, there are three different things that can help to tackle the problem–education, laws, and economics–and they work together as a team.

- ☆ Education: Making people aware of the problem is the first step to solving it.
- ☆ Raise the Voice: One of the most effective ways to stand up for our waters is to speak out in support of the Clean Water Rule, which clarifies the Clean Water Act's scope and protects the drinking water of one in three Americans.
- ☆ Laws: There are many laws that protect the world's oceans, rivers and lakes from unnecessary water pollution. Each continent and country may differ in which laws they enforce but they aim to have the same overall positive influence.
- ☆ Water Treatment Technologies:
  - ❒ Drinking water treatment technology
  - ❒ Wastewater treatment technology
  - ❒ Drinking water treatment technology

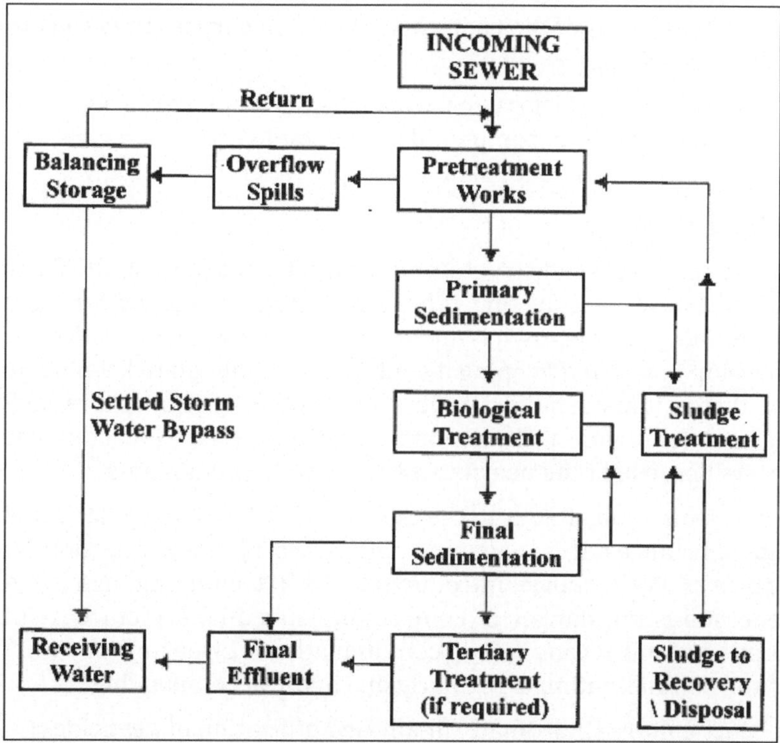

The various methods or the techniques which may be adopted for purifying the public water supplies are:

1. Screening
2. Plain sedimentation

3. Sedimentation aided with coagulation
4. Filtration
5. Disinfection
6. Softening
7. Miscellaneous treatments such as fluoridation, re-carbonation, liming, desalination *etc.*

5.3 Soil Pollution

5.3.1 Definition

Soil contamination is characterized as the nearness of harmful synthetic concoctions (poisons or contaminants) in soil, in sufficiently high fixations to represent a hazard to human well-being or potentially the biological system. On account of contaminants which happen normally in soil, in any event, when their levels are not sufficiently high to represent a hazard, soil contamination is still said to happen if the degrees of the contaminants in soil surpass the levels that ought to normally be available.

Soil contamination is characterized as the development in soils of relentless poisonous mixes, synthetic compounds, salts, radioactive materials, or sickness causing operators, which affect plant development and creature well-being.

What is Soil Pollution?

With the ascent of solid structures and streets, one piece of the Earth that we once in a while observe is the dirt. It has a wide range of names, for example, soil, mud, and ground. Nonetheless, it is unquestionably imperative to us. The plants that feed us develop in soil and keeping it sound is fundamental to keeping up a lovely planet. Be that as it may, similar to every single other type of nature, soil likewise experiences contamination. The contamination of soil is a typical thing nowadays, and it occurs because of the nearness of man-made components.

Soil contamination alludes to the tainting of soil with irregular convergences of harmful substances. It is a genuine natural worry since it harbors numerous well-being risks. For instance, introduction to soil containing high convergences of benzene builds the danger of contracting leukemia. A picture itemizing the discolouration of soil because of soil contamination is given underneath. The main driver of soil contamination is regularly one of the accompanying:

- ✰ Horticulture (inordinate/ill-advised utilization of pesticides)
- ✰ Exorbitant mechanical action
- ✰ Poor administration or wasteful removal of waste

The difficulties looked in soil remediation (cleaning of soil) are firmly identified with the degree of soil contamination. The more prominent the tainting, the more noteworthy the prerequisite of assets for remediation.

5.3.2 Causes of Soil Pollution

All soils, whether polluted or unpolluted, contain a variety of compounds (contaminants) which are naturally present. Such contaminants include metals, inorganic ions and salts (*e.g.* phosphates, carbonates, sulphates, nitrates), and many organic compounds (such as lipids, proteins, DNA, fatty acids, hydrocarbons, PAHs, alcohols, *etc.*). These compounds are mainly formed through soil microbial activity and decomposition of organisms (*e.g.*, plants and animals). Additionally, various compounds get into the soil from the atmosphere, for instance with precipitation water, as well as by wind activity or other types of soil disturbances, and from surface water bodies and shallow groundwater flowing through the soil. When the amounts of soil contaminants exceed natural levels (what is naturally present in various soils), pollution is generated. There are two main causes through which soil pollution is generated: Anthropogenic (man-made) Causes, and Natural Causes.

Anthropogenic Causes (Man-made)

Anthropogenic (man-made) soil pollution creates in numerous kinds of procedures, some deliberate (industrial) and some accidental. Human-caused soil pollution can work in conjunction with natural procedures to increase the toxic pollution levels in the soil. We also damage the soil as part of our ordinary lives. It can be through indecorous disposal of toxic waste that ends up in landfills or bodies of water, or human waste that goes through our sewer system and ends up in our soil. Either way, humans play a major role in causing soil pollution.

Almost all cases of soil pollution are anthropogenic in nature. A variety of human activities can lead to the contamination of soil. Some such processes are listed below:

- ☆ The demolition of old buildings can involve the contamination of nearby soil with asbestos.
- ☆ Usage of lead-based paint during construction activities can also pollute the soil with hazardous concentrations of lead.
- ☆ Spillage of petrol and diesel during transportation can contaminate soils with the hydrocarbons found in petroleum.
- ☆ Activities associated with metal casting factories (foundries) often cause the dispersion of metallic contaminants into the nearby soils.
- ☆ Underground mining activities can cause the contamination of land with heavy metals.
- ☆ Improper disposal of highly toxic industrial/chemical waste can severely pollute the soil. For example, the storage of toxic wastes in landfills can result in the seepage of the waste into the soil. This waste can go on to pollute groundwater as well.
- ☆ Chemical pesticides contain several hazardous substances. Excessive and inefficient use of chemical pesticides can result in severe soil pollution.

- ✰ Sewage produced in urbanized areas can also contaminate soil (if not disposed of correctly). These wastes may also contain several carcinogenic substances.
- ✰ Other forms of waste that can pollute soil include nuclear waste, e-waste, and coal ash.

Chemical Spills and Leaks

Accidental spills and escapes are common when substances are stowed, elated, and used. The chemicals free into the ground during such coincidences are damaging and contaminate the soil. For instance, when oil is elated through tubes from oil fields to processing plants, oil might leak through holes in the pipe and pollute the soil in the surrounding area. Other chemicals harm the composition of the soil and make it easier to erode by water and air. Plants absorb many of these pesticides and when they decompose, they reason soil pollution since they become a part of the land.

- ✰ Polycyclic aromatic hydrocarbons (often abbreviated to PAHs) are organic compounds that:
- ✰ Contain only carbon and hydrogen atoms.
- ✰ Contain more than one aromatic ring in their chemical structures.

Pesticides

Pesticides are substances (or mixtures of substances) that are used to kill or inhibit the growth of pests. Common types of pesticides used in agriculture include:

- ✰ Herbicides – used to kill/control weeds and other unwanted plants.
- ✰ Insecticides – used to kill insects.
- ✰ Fungicides – used to kill parasitic fungi or inhibit their growth.

However, the unintentional diffusion of pesticides into the environment (commonly known as 'pesticide drift') poses a variety of environmental concerns such as water pollution and soil pollution. Some important soil contaminants found in pesticides are listed below.

Agriculture

The use of chemicals as part of agriculture is at record levels. Whether it's to increase the production of crops or restrict the growth of harmful fungi, chemicals such as herbicides and pesticides are an integral part of our agricultural process. Modern agricultural practices pollute the soil to a large extent. With the advancing agro-technology, huge quantities of fertilizers, pesticides, herbicides and weedicides are added to increase the crop yield. Apart from these farm wastes, manure, slurry, debris, soil erosion containing mostly inorganic chemicals are reported to cause soil pollution

Urban Wastes

Urban wastes comprise of both commercial and domestic wastes consisting of dried sludge and sewage. All the urban solid wastes are commonly referred to as refuse. Constituents of urban refuse: This refuse consists of garbage and rubbish materials like plastics, glasses, metallic cans, fibres, paper, rubbers, street sweepings, fuel residues, leaves, containers, abandoned vehicles and other discarded manufactured products. Urban domestic wastes though disposed of separately from industrial wastes, can still be dangerous. This happens because they are not easily degraded.

Unintentional Oil Spills

Oil leaks can happen during storage and transport of chemicals. This can be seen at most of the fuel stations. The chemicals present in the fuel deteriorates the quality of soil and make them unsuitable for cultivation. These chemicals can enter into the groundwater through the soil and make the water undrinkable.

Acid Rain

Acid rain is caused when pollutants present in the air mix up with the rain and fall back on the ground. The polluted water could dissolve away some of the important nutrients found in soil and change the structure of the soil.

Industrial Activity

Industrial activity has been the biggest contributor to the problem in the last century, especially since the amount of mining and manufacturing has increased. Most industries are dependent on extracting minerals from the Earth. Whether it is iron ore or coal, the by-products are contaminated and they are not disposed of in a manner that can be considered safe. As a result, the industrial waste lingers in the soil surface for a long time and makes it unsuitable for use.

Natural Pollution of Soil

In some tremendously infrequent processes, some contaminants are obviously accrued in soils. This can happen due to the discrepancy statement of soil by the atmosphere. Additional manner in which this kind of soil pollution can happen is via the conveyance of soil pollutants with rain water. Apart from the infrequent cases when a natural accretion of chemicals leads to soil pollution, natural processes may also have an effect on the human unconfined toxic chemicals into the soil, overall lessening or cumulative the impurity toxicity and/or the level of pollution of the soil. This is likely due to the multifaceted soil environment, involving the presence of other chemicals and natural conditions which may interact with the released pollutants.

An example of natural soil pollution is the accumulation of compounds containing the perchlorate anion (ClO_4^-) in some dry, arid ecosystems. It is important to note that some contaminants can be naturally produced in the soil under the effect of certain environmental conditions. For example, perchlorates can be formed in soils containing chlorine and certain metals during a thunderstorm.

5.3.3 Effects of Soil Pollution

Soil pollution ports a comprehensive range of undesirable significances that distress plants, animals, humans, and the ecosystem as an entire. Since children are more vulnerable to diseases, contaminated soil stances a better danger to them. Soil pollution touches plants, animals and humans alike. While anybody is vulnerable to soil pollution, soil pollution belongings may differ based on age, general health status and other factors, such as the type of pollutant or contaminant inhaled or ingested. Though, children are typically more vulnerable to contact to pollutants, because they come in close interaction with the soil by playing in the ground; combined with lower edges for disease, this triggers higher risks than for adults.

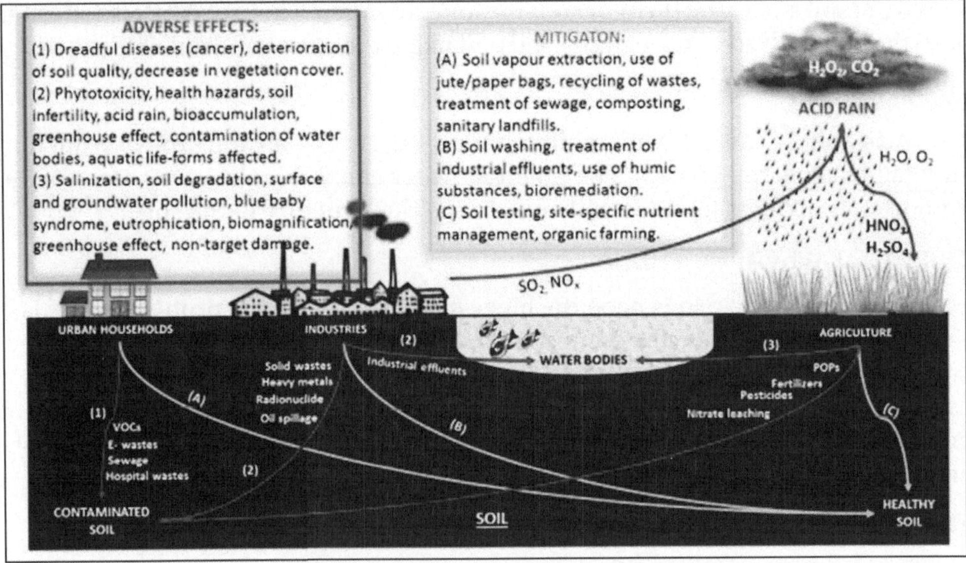

Some of the major effects of soil pollution are stated below:

1. Effect on Health of Humans

Soil contaminants can exist in all three phases (solid, liquid, and gaseous). Therefore, these contaminants can find their way into the human body via several channels such as direct contact with the skin or through the inhalation of contaminated soil dust.

The short term effects of human exposure to polluted soil include:

- ☆ Headaches, nausea, and vomiting.
- ☆ Coughing, pain in the chest, and wheezing.
- ☆ Irritation of the skin and the eyes.
- ☆ Fatigue and weakness.

A variety of long-term ailments have been linked to soil pollution. Some such diseases are listed below.

- ☆ Exposure to high levels of lead can result in permanent damage to the nervous system. Children are particularly vulnerable to lead.
- ☆ Depression of the CNS (Central Nervous System).
- ☆ Damage to vital organs such as the kidney and the liver.
- ☆ Higher risk of developing cancer.

2. Effect on Growth of Plants

The ecological balance of any system gets affected due to the extensive corruption of the soil. Most plants are powerless to adapt when the interaction of the soil changes so fundamentally in a small period of time. Fungi and germs found in the soil that quandary it composed instigate to weakening, which creates an additional problematic of soil erosion.

Since soil pollution is frequently escorted by a reduction in the obtainability of nutrients, plant life stops to flourish in such soils. Soils dirty with inanimate aluminium can show toxic to plants. Also, this type of pollution often upsurges the salinity of the soil, making it unwelcoming for the development of plant life.

Plants that are grown in polluted soil may accrue high attentions of soil contaminants through a process known as bioaccumulation. When these plants are spent by herbivores, all the accrued pollutants are approved up the food chain. This can consequence in the loss/extinction of many wanted animal species. Also, these contaminants can eventually make their way to the top of the food chain and manifest as diseases in human existences. The fertility slowly diminishes, making land unsuitable for agriculture and any local vegetation to endure. The soil pollution reasons outsized areas of land to become dangerous to health. Different deserts, which are suitable for its innate vegetation, such land cannot provision most forms of life.

3. Decreased Soil Fertility

The toxic chemicals current in the soil can reduction soil fertility and consequently decrease in the soil yield. The polluted soil is then used to produce fruits and vegetables which absences quality nutrients and may cover some venomous matter to reason thoughtful health problems in people consuming them. With the swift development of human population, we need all the food we can get. Chemicals used on soils decrease soil fertility so food production drops.

4. Toxic Dust

The emanation of toxic and foul gases from landfills contaminates the environment and reasons solemn belongings on the fitness of some people. The unfriendly smell causes tiresomeness to other people.

5. Changes in Soil Structure

The death of many soil organisms (*e.g.* earthworms) in the soil can lead to modification in soil structure. Separately from that, it could also force other predators to move to other places in search of food.

A number of ways have been suggested to control the present degree of pollution. Such efforts at cleaning up the environment need sufficiently of time and resources to be inclined in. Industries have been given regulations for the removal of dangerous waste, which goals at minimalizing the area that becomes contaminated.

Organic methods of farming are being supported, which do not use chemical-laden pesticides and fertilizers. Use of plants that can remove the pollutants from the soil is being fortified.

6. Effects on the Ecosystem

Soil pollution can reason an unevenness of the ecosystem of the soil. The soil is an important habitat and is the house of dissimilar type of microorganisms, animals, reptiles, mammals, birds, and insects. Thus, soil pollution can damagingly impact the lives of the living organisms and can result in the steady death of many organisms. It can reason health threats to animals cropping in the dirty soil or microorganisms exist in in the soil.

- ✰ Since the instable pollutants in the soil can be approved away into the atmosphere by winds or can seep into under-groundwater reserves, soil pollution can be a straight contributor to air and water pollution.
- ✰ It can also subsidise towards acid rain (by releasing huge quantities of ammonia into the atmosphere).
- ✰ Acidic soils are unwelcoming to several microorganisms that recover soil texture and help in the decay of biological matter.
- ✰ Crop yield is importantly pretentious by this form of pollution. In China, over 12 million tons of grain (worth approximately 2.6 billion USD) is found to be unfit for human ingesting due to pollution with heavy metals.

7. Inferior Crop Quality

It can reduction the quality of the crop. Regular use of chemical fertilizers, inorganic fertilizers, pesticides will decrease the fertility of the soil at a rapid rate and alter the structure of the soil. This will lead to decrease in soil quality and poor quality of crops. Over the time the soil will become less productive due to the accumulation of toxic chemicals in large quantity.

5.3.4 Control Measures of Soil Pollution

Develop Essential Legislation on Soil Pollution Control

Though determinations are being put in place to safeguard soil contamination control, the procedure ought to be faster by recruiting the essential legislation. The current recordings such as urban and rural planning, agricultural practices, and land management should be studied and efficient to include modern soil pollution control and prevention measures.

Legislatures on pesticide management, polluted sites management, and woodland soil management should as well be efficient. Also, events should be put in place to promote continuous improvement of soil pollution control standards.

Appropriate Solid Waste Management

It is significant to position of solid waste correctly by treated it before it's unconfined into the environment. Acidic and alkaline waste, for instance, can be counteracted before they are willing of to avoid soil contamination. Biodegradable waste should also be wrecked down in a skilful environment before it is unconfined into the environment. A great illustration is the proper behaviour of sewage sludge.

Firmly Control the Pollution of New Soil

It is inexpensive to avert soil pollution than it is to accomplish contaminated soil and thus, the suitable impartial should be defensive new soil. Before any new land is used, thorough research should be lead to ensure whether the land is suitable for the intended use. There should also be put in place severe measures against those who dump destructive or poisonous materials to the mudflat, sand, swampland, and salt marshes as well as those who release pollutants unlawfully.

Promote Objective Assessment and Accountability among all Stakeholders

To guarantee long term achievement in the deterrence and control of soil contamination, all stakeholders' role in the application of the soil contamination control should be clear. All tasks should be defined and inter-departmental collaboration and organisation should be strengthened to ensure seamless application and instant resolutions focused towards correct removal of garbage, proper treatment of sewage sludge, minimalized the use of agro-chemicals, reforestation, construction of drainage systems, and strengthening chemical use related laws.

5.4 Marine Pollution

Introduction

Covering more than 70 per cent of our planet, oceans are among the earth's most valued natural resources. They govern the weather, clean the air, help feed the world, and deliver a living for millions. They also are home-based to most of the life on earth, from tiny algae to the blue whale, the largest animal on the planet.

It is projected that a staggering 80 per cent of marine pollution originates on land. Land-based pollutants – such as agricultural run-off and nutrients from sewage outflows - are contributing to ocean 'dead zones' – areas which can no longer sustain life because they have low or zero oxygen. There are now some 500 of these dead zones around the world. Marine pollution has been an ever-present problematic since the arrival of large-scale agrarian activity and development. Though, significant laws and regulations at an international level to tackle the problem came only in the mid-twentieth century. During United Nations Conventions on the Law of the Sea in the early 1950s, the various stakeholders come together to deliberate and formulate laws relating to marine pollution. Till mid-twentieth century the popular of the scientists upheld that oceans were vast enough to be able to dilute the amount of pollution being exhausted into them, thus, seeing pollution inoffensive to the marine life.

5.4.1 Definition

Marine Pollution (UN definition) – "The introduction by man, directly, or indirectly, of substances or energy to the marine environment resulting in deleterious effects such as: hazards to human health, hindrance to marine activities, impairment of the quality of seawater for various uses and reduction of amenities."

Marine pollution refers to direct or indirect introduction by humans of substances or energy into the marine environment (including estuaries), resulting in harm to living resources, hazards to human health, hindrances to marine activities including fishing, impairment of the quality of sea water and reduction of amenities.

History of Marine Pollution

In 1926, the International Maritime Conference in Washington produced the first international convention relating to oil pollution, writes Captain M.M. ITTISO

General anxiety over oil pollution appears to have originated shortly after World War I. First the United States, and then the League of Nations undertook to obtain an explicit international agreement on measures to combat oil pollution. Prior to this time few states had sought to extend pollution legislation beyond their territorial sea, and no states had attempted to apply prohibitory regulations beyond this area. It can, therefore, be said that all prescriptive activity had taken the form of international agreement.

In 1926, the International Maritime Conference in Washington produced the first international convention relating to oil pollution. The problem was discussed in both technical and legal terms, but the convention failed to be ratified by any nation. Oil pollution, particularly on the shores of the Atlantic Ocean, was considerable during and after the Second World War, as a direct result of the many torpedoed, sunken and otherwise damaged vessels.

By the 1950s, a rising world economy, resulting in an ever-increasing demand for hydrocarbon fuel contributed to a renewed concern over the pollution problem. Nothing concrete was achieved until 1954 when the International Convention for the Prevention of Oil Pollution (OILPOL'54) was concluded in London and entered into force in 1958. This convention prohibited the international discharge of oil and oily mixtures from certain vessels in specific ocean areas.

Although OILPOL contributed to cleaner seas, it did not establish any enforcement system other than through the flag states. OILPOL was superseded by the International Convention for the Prevention of Pollution from Ships (MARPOL'73/78) after a 1973 conference in London and amended by the 1978 Protocol, in which the question of polluting substances other than oil was considered.

These included sewage, garbage, noxious liquid substances in bulk, harmful substances carried by sea in packaged form or in containers, cargo wastes and nuclear waste.

5.4.2 Causes of Marine Pollution

As massive as the Earth's oceans are, the total of pollution that arrives the water from both man-made and natural activities is affecting marine ecosystems through the world. Most of the pollutants that enter the ocean come from the land, according to the National Oceanographic and Atmospheric Administration. Animals often eat these contaminants which, in turn, means humans will as well. There are many causes of ocean pollution, including:

Sewage

Sewage is demarcated as the wastewater and its constituent filths that are elated in the sewer system. Some of the materials that are in the sewage waste are damaging and donate greatly to ocean pollution. These materials may cause thoughtful health problems to the aquatic beings once they devour them

Industrial Chemicals

Another main pollutant is the chemicals from businesses and from the manures and other farm products that are carried by run-off water into the ocean waters. Many businesses dump their waste materials and substances into the ocean waters. These chemicals pollute the ocean by altering the pH level of the waters. Most aquatic plants and animals cannot survive in adverse pH levels.

Nuclear Waste

Another major ocean pollutant is the nuclear waste, which is mostly produced from industrial, medical, and also scientific procedures that use radioactive material. The common industries that produce nuclear waste comprise power stations, the military, and the reprocessing plants. This energy enters the food chain through kelp and plankton, and once the marine animals consume these plants they become contaminated.

Plastics

Plastic pollution mainly includes the build-up plastic in the ocean waters and thus producing adverse belongings on marine organisms. Marine organisms are pretentious by the plastics through direct digestion of the plastic wastes and also ended contact to chemicals that are within the plastics. The ever-growing need of human population on plastic has full the oceans and the land, it covers of 80 per cent of the wreckage found in the oceans. Plastic dumped and found in the oceans are hazardous for the marine life forms and wildlife, as sometimes it stifles and obstructs them to death. The rising levels of plastic dumps found in the oceans are overpowering, ingesting, and entangling the life underwater as well as above it.

Acid Rain

Acid rain is not a main cause of ocean pollution, but it also donates to water pollution. Exploding volcanoes, fossil fuels, rotting vegetation, and nitrogen oxides when free into the atmosphere react with water and other substances in the air to form sulphuric and nitric acid. The wind setbacks these chemicals across the atmosphere, and when it rains, these chemicals find their way into the marine waters. Acid rain makes water acidic and thus abolishes the marine life as most aquatic organisms cannot live in acidic conditions.

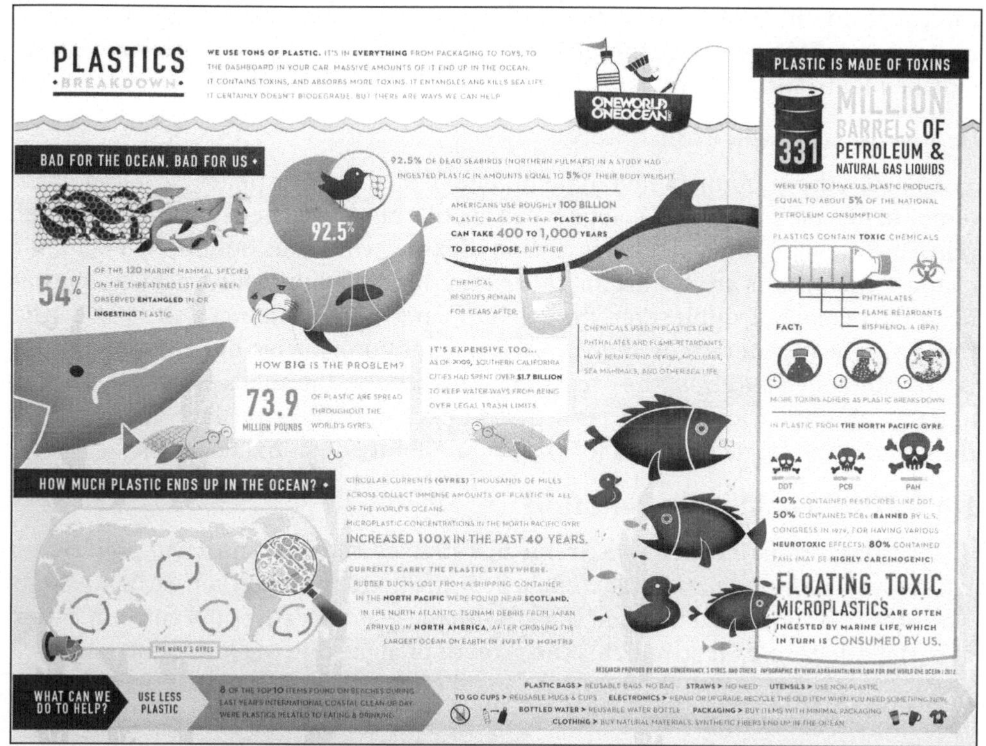

Oil Spills

Oil spillage is another main cause of ocean pollution in that the oil forms a layer on the water preventing oxygen circulation. Lack of oxygen in the ocean waters results in the destruction of marine life over a long period. Ships and platforms release large amounts of oil every year. However, oil isn't the only pollutant that comes from ships, which may also discharge fuel, plastic, and human waste. Crude oil is one of the most difficult to clean up. It is also toxic, suffocating, and devastating to marine life. Ships also cause noise pollution, disrupting the balance of life for marine animals such as dolphins and whales that use echolocation. Crates lost during storms, accidents, and other emergencies pollute the ocean as well.

Runoff from Land

The second major source of marine pollution is runoff from the land, from both agricultural and urban areas. Runoff from the land comes from both urban and agricultural areas. Often referred to as non-point source pollution, it can originate from various sources like cars and trucks, septic tanks, farms, and timber harvesting operations. Chemicals that end up on roads and highways flow over and under the ground with rain water, as do pesticides, fertilizers, and carbon-, nitrogen-, and phosphorous-rich particulates, eventually reaching the ocean. Inland mining can cause an influx of mineral and soil deposits. These travel through rivers and estuaries, making soil a real threat to marine ecosystems. Runoff can even smother marine plants and coral reefs.

Pollutants from the Atmosphere

Finally, the announcement of pollutants from the atmosphere is an important donor to marine pollution. Various types of pollutants can get in the water through rain. One menace, in particular, is carbon dioxide, which has built up with climate change. The oceans are absorbing the excess and becoming more acidic. This has been particularly troublesome for calcium carbonate structures such as corals, then, cannot regenerate or regrow. About 1 million species depend on thriving coral habitats. In addition, the excess CO_2 can dissolve the shells of various marine animals. Acidification, which puts these species at risk and has ramifications throughout the food chain and for the sustainability of global fisheries.

5.4.3 Effects of Marine Pollution

The pollution of water by extreme nutrients is known as nutrient pollution, a type of water contamination that touches the life under water. When extra nutrients like nitrates or phosphates get dissolved with the water it causes the eutrophication of surface waters, as it stimulates the growth of algae due to excess nutrients.

Effect of Toxic Wastes on Marine Animals

The oil spill is hazardous to marine life in several ways. The oil spilled in the ocean could get on to the gills and feathers of marine animals, which makes it difficult for them to move or fly properly or feed their children. The long term result on marine life can comprise cancer, failure in the reproductive system, behavioural changes, and even death. Animals higher up the food chain, such as seals, can have contamination levels millions of times higher than the water in which they live. And polar bears, which feed on seals, can have contamination levels up to 3 billion times higher than their environment.

Disruption to the Cycle of Coral Reefs

Oil spill floats on the surface of the water and stops sunlight from reaching to marine plants and affects the process of photosynthesis. Skin irritation, eye irritation, lung and liver problems can impact marine life over a long period of time.

Depletes Oxygen Content in Water

Most of the debris in the ocean does not decompose and remain in the ocean for years. It uses oxygen as it degrades. As a result of this, oxygen levels go down. When oxygen levels go down, the chances of survival of marine animals like whales, turtles, sharks, dolphins, penguins for a long time also goes down.

Failure in the Reproductive System of Sea Animals

Industrial and agricultural wastes include various poisonous chemicals that are measured hazardous for marine life. Chemicals from pesticides can accumulate in the fatty tissue of animals, leading to failure in their reproductive system.

Effect on Food Chain

Chemicals used in industries and agriculture get eroded into the rivers and from there are carried into the oceans. These chemicals do not get liquefied and sink at the lowest of the ocean. Small animals ingest these chemicals and are later eaten by large animals, which then touches the whole food chain.

Affects Human Health

Animals from impacted food chain are then eaten by humans which affects their health as toxins from these contaminated animals get deposited in the tissues of people and can lead to cancer, birth defects or long term health problems. The trash poses dangers to both humans and animals.

Other specific effects of ocean pollution on sea life include:

- ✰ Oil covering the feathers of birds and the gills of fish.
- ✰ Skin and eye irritation, and lung and liver problems, from oil deposits and by-products.
- ✰ Reproductive system failure from exposure to poisonous industrial and agricultural chemicals.
- ✰ The fatty tissues of fish can hold onto toxic substances which people then consume.

Type	*Primary Source/Cause*	*Effect*
Nutrients	Runoff approximately 50 per cent sewage, 50 per cent from forestry, farming, and other land use. Also airborne nitrogen oxides from power plants, cars *etc.*	Feed algal blooms in coastal waters. Decomposing algae depletes water of oxygen, killing other marine life. Can spur algal blooms (red tides), releasing toxins that can kill fish and poison people.
Sediments	Erosion from mining, forestry, farming, and other land-use; coastal dredging and mining	Cloud water; impede photosynthesis below surface waters. Clog gills of fish. Smother and bury coastal ecosystems. Carry toxins and excess nutrients.
Pathogens	Sewage, livestock.	Contaminate coastal swimming areas and seafood, spreading cholera, typhoid and other diseases.
Alien Species	Several thousand per day transported in ballast water; also spread through canals linking bodies of water and fishery enhancement projects.	Outcompete native species and reduce biological diversity. Introduce new marine diseases. Associated with increased incidence of red tides and other algal blooms. Problem in major ports.

Type	*Primary Source/Cause*	*Effect*
Persistent Toxins (PCBs, Heavy metals, DDT *etc.*)	Industrial discharge; wastewater discharge from cities; pesticides from farms, forests, home use *etc.*; seepage from landfills.	Poison or cause disease in coastal marine life, especially near major cities or industry. Contaminate seafood. Fat-soluble toxins that bio-accumulate in predators can cause disease and reproductive failure.
Oil	46 per cent from cars, heavy machinery, industry, other land-based sources; 32 per cent from oil tanker operations and other shipping; 13 per cent from accidents at sea; also offshore oil drilling and natural seepage.	Low level contamination can kill larvae and cause disease in marine life. Oil slicks kill marine life, especially in coastal habitats. Tar balls from coagulated oil litter beaches and coastal habitat. Oil pollution is down 60 per cent from 1981.
Plastics	Fishing nets; cargo and cruise ships; beach litter; wastes from plastics industry and landfills.	Discard fishing gear continues to catch fish. Other plastic debris entangles marine life or is mistaken for food. Plastics litter beaches and coasts and may persist for 200 to 400 years.
Radioactive substances	Discarded nuclear submarine and military waste; atmospheric fallout; also industrial wastes.	Hot spots of radio activity. Can enter food chain and cause disease in marine life. Concentrate in top predators and shellfish, which are eaten by people.
Thermal	Cooling water from power plants and industrial sites	Kill off corals and other temperature sensitive sedentary species. Displace other marine life.
Noise	Supertankers, other large vessels and machinery	Can be heard thousands of kilometres away under water. May stress and disrupt marine life.

5.4.4 Control Measures of Marine Pollution

Answers for marine pollution comprise preclusion and clean-up. Disposable and single-use plastic is plentifully used in today's society, from shopping bags to shipping wrapping to plastic bottles. Changing society's method to plastic use will be a long and frugally challenging process. Clean-up, in difference, may be unbearable for some items. Many types of debris (including some plastics) do not float, so they are lost deep in the ocean. Plastics that do float tend to collect in large "patches" in ocean gyres. Marine life is a significant part of our Eco-system and it is everyone's accountability to save marine animals and natural resources. We can make a difference working together to stop marine pollution. Some of the important preventive measures of marine pollution have been listed below:

Installing Overflow Pipe in Fuel Tank

When bunkering, if fuel is supplied to the fuel tank of our ship beyond capacity, there is a possibility of fuel flowing out of the ship through the air-vent of the fuel tank. In order to prevent this situation, we have installed an overflow pipe in the overflow Tank in order that fuel oil spilt out of the fuel tank flows into the overflow tank, and furthermore, a flow detection sensor in the pipe or a high-level alarm sensor in the tank in order to detect overflow immediately.

Indirect Cooling System (Central Cooling System)

It is a device for cooling the train and lubricating oil by circuitously exchanging heat with seawater via devoted freshwater. Use of this system stops leakages or overboard spills of lubricating oil due to trouble in the cooling system, because thermal exchange is not straight made between lubricating oil and seawater in this system.

Other Preventive Measures

- ☆ Stop using plastic and littering garbage as they not only choke up the drains but also releases into the oceans.
- ☆ Ensure that chemicals mentioned above are not used anywhere near the streams of water and try cutting down on the usage of such chemicals.
- ☆ For farmers, they need to switch from chemical fertilizers and pesticides and move towards the usage of organic farming methods.
- ☆ Use public transport and reduce the carbon footprint by taking small and substantial measures that will not help in reducing the pollution from the environment but will ensure a safe and healthy future for the upcoming generations.
- ☆ Prevent from any oil or chemical spill in the oceans and if in case there is an oil or chemical spill near you volunteer and help in cleaning out the ocean water.
- ☆ Volunteer or initiate beach clean-up activities and spread awareness about the same in the nearby vicinity.
- ☆ Get involved: there are probably cleanup efforts happening near you!
- ☆ Don't discard anything near the coast, when you go to the beach make sure you pick up after yourself.
- ☆ Ban toxic products from your boat-maintenance and don't throw anything overboard. Use and overflow system to avoid oil spills, and maintain your boat regularly to avoid leaks.
- ☆ Try to green your household and gardening chemical products, use them sparingly and wisely. Don't use fertilizers before it rains or pour oil or chemicals down the drain: they would just end up in the ocean!
- ☆ Try to avoid using single-serving plastic items and replace them by reusable items (cloth bags, reusable cups and silverware, non-plastic bottles).

- ☆ Refuse excess packaging, try to re-use and recycle as much as possible. Remember that very little of the plastic produced each year is actually recycled and much of it finds its way to the ocean.
- ☆ Make sure only rain goes down the storm water drain

5.5 Noise Pollution

Introduction

Noise pollution, unwanted or extreme sound that can have harmful belongings on human health and environmental quality. Noise pollution is usually shaped inside many industrial facilities and some other workplaces, but it also comes from highway, railway, and airplane traffic and from outdoor construction activities. Such sounds aren't just maddening. There is increasing evidence that long-term environmental noise above a certain level can have a negative effect on your health. These effects can be physical, mental and possibly even disrupt children's learning.

5.5.1 Definition

Noise pollution is usually defined as even exposure to raised sound levels that may principal to opposing effects in humans or other living organisms. According to the World Health Organization, sound levels less than 70 dB are not harmful to living organisms, irrespective of how long or consistent the exposure is.

By definition, noise pollution takes place when there is either an extreme amount of noise or an unfriendly sound that causes a provisional disturbance in the natural balance. This definition is typically appropriate to sounds or noises that are unusual in either their volume or their manufacture. Our environment is such that it has become problematic to escape the noise.

History of Noise Pollution

Throughout recorded history, mankind has been overwhelmed by a diversity of both natural and man-made ills. In the 21st Century we are experiencing the man-made plague of environmental noise from which there is virtually no escape, no matter where we are – in our homes and yards, on our streets, in our cars, at theatres, restaurants, parks, arenas, and in other public places. Notwithstanding efforts to regulate it, noise contamination has become an unfortunate fact of life worldwide. In a way that is similar to second-hand smoke, second-hand noise is an unwelcome airborne pollutant produced by others; it is imposed on us without our agreement, often against our wills, and at times, places, and capacities over which we have no switch.

There is rising evidence that noise effluence is not merely an irritation; like other forms of pollution, it has wide-ranging opposing health, social, and economic effects. A new hunt (September, 2006) of the National Library of Medicine data base for adverse health effects of noise revealed over 5000 citations, many of recent vintage. As the population grows and as sources of noise become more numerous and more powerful, there is increasing exposure to noise pollution, which has profound public health implications. Noise, even at levels that are not harmful to hearing, is perceived subconsciously as a danger signal, even during sleep. Thus, the purpose of this review is to précis what is known of these opposing health belongings and to encourage physicians, nurses, and other health professionals to join with groups around the country that are trying to restore the Constitutionally guaranteed right of domestic tranquillity. Noise Free America and the Noise Pollution Clearing-house are two such organizations. There are numerous Internet sites that contain relevant information about noise and the ongoing efforts to restore quiet in communities across the United States. The interested reader should consult Noise Off (www.NoiseOFF.org), The Noise Pollution Clearing-house (www.nonoise.org), Noise Free America (www.noisefree.org), or the League for the Hard of Hearing (www.lhh.org/noise) for additional information about this subject.

How Sound is Measured?

Sound and noise is unhurried in decibels (dB), and according to figures from the Centre for Hearing and Communication, continued exposure to noise levels gaging more than 85 dBA (adjusted decibels) will cause hearing loss over time.

Both the distance of contact to the sound, and the capacity (dBA) will tell you how harmful the noise is. dBA are expressions of the relative loudness of sounds in the air as picked up by the human ear. What this also means is that with these adjusted decibels, the values of low frequency sounds are reduced as the human ear is less sensitive at this frequency. This is different to regular unweighted decibels, where no correction is made for audio frequency.

Environmental Noise Levels Measured in dBA

This list will help illustrate the range of environmental noise levels that affect everyone in society today.

- 0 – the quietest sound audible via normal hearing
- 10 – normal breathing
- 20 – whispering from 5 feet away
- 30 – soft whispering
- 40 – quiet office
- 50 – rainfall/refrigerator
- 60 – regular conversation
- 70 – freeway traffic
- 85 – noisy restaurant/heavy traffic
- 100 – shouting directly into the ear
- 120 – thunder

According to statistics from the Dangerous Decibels, lengthy contact to sound over the threshold of 85 decibels can develop an issue, particularly when the contact time to the sound increases. With these figures in mind, it is important to note that the effects of noise pollution on humans are best understood by investigative the dissimilar types of noise an separate is bare to daily.

While sound is somewhat that is indispensable within our daily lives, noise is not. Sounds that cause uneasiness to the ears that produce unpleasant effects, or are unwelcome, are considered to be noise. Noise pollution generally interferes with normal activities, for example, conversing or sleeping. This type of pollution can also diminish or disrupt quality of life. Not all noise is considered to be pollution. If it happens on an intermittent basis, noise is usually termed as "nuisance."

Noise pollution is not only an issue that affects humans, it is also harming the world's whale population. These mammals, who use their ears more than their eyes, make calls that portable distances of 20 miles and more. They need to do this to connect, in much the same way as humans need to. It also helps avert them from becoming lost during migration. However, with today's heavy, and constant ship traffic, drowning out their voices, these majestic creatures are suffering in silence.

Measuring and Perceiving Loudness

Sound is measured in decibels (dB), and according to statistics from the Center for Hearing and Communication, continued exposure to noise levels measuring more than 85 dBA (adjusted decibels) will cause hearing loss over time.

Both the distance of exposure to the sound, and the volume (dBA) will tell you how harmful the noise is. dBA are expressions of the relative loudness of sounds in the air as picked up by the human ear. What this also means is that with these adjusted decibels, the values of low frequency sounds are reduced as the human ear is less sensitive at this frequency. This is different to regular unweighted decibels, where no correction is made for audio frequency.

Environmental Noise Levels Measured in dBA

This list will help illustrate the range of environmental noise levels that affect everyone in society today.

- ☆ 0 – the quietest sound audible via normal hearing
- ☆ 10 – normal breathing
- ☆ 20 – whispering from 5 feet away
- ☆ 30 – soft whispering
- ☆ 40 – quiet office
- ☆ 50 – rainfall/refrigerator
- ☆ 60 – regular conversation
- ☆ 70 – freeway traffic
- ☆ 85 – noisy restaurant/heavy traffic

- ☆ 100 – shouting directly into the ear
- ☆ 120 – thunder

According to statistics from the Dangerous Decibels, prolonged exposure to noise over the threshold of 85 decibels can become an issue, especially when the exposure time to the sound increases. With these figures in mind, it is important to note that the effects of noise pollution on humans are best understood by examining the different types of noise an individual is exposed to daily.

Types of Noise Pollution

There are various different types of noise pollution, including:

Man Made Noise – including traffic, subway noise, airports, constructions sites, transport, and so on. In fact, figures from Dangerous Decibels indicate that the sound of an idling bulldozer can reach as high as 85 decibels, and people who work closely with the machines can sustain irreversible damage to their hearing within just one day.

Other types of man-made noise pollution are frivolous in nature. Dangerous Decibels states that at maximum volume, personal music players can exceed 100 decibels, which can cause irreversible damage to the hearing of people who listen for only 15 minutes a day.

5.5.2 Causes of Noise Pollution

1. Industrialization

Factories that use weighty equipment emit large quantities of sound that are harmful to the people that work for the company and those living in the neighbourhood. Most of the businesses use big machines which are accomplished of creating a large quantity of noise. Apart from that, various apparatus like compressors, generators, exhaust fans, grinding mills also contributes in producing big noise. Consequently, you must have seen workers in these factories and industries wearing earplugs to minimalize the effect of noise. People who work with saws, mills, lathes, drill presses, punch presses, and progressive dies are at risk for hearing loss if they do not wear protective gear equipment.

Inhabitants in areas neighbouring businesses or industrial facilities report penetrating sound levels that can damage hearing ability. Industrial towns which are close to housing areas have habitually caused uneasiness and irritation to the residents. Within the industries, it is mandatory for workers to wear earplugs to shield them against the admission of loud noise.

2. Poor Urban Planning

In most of the emerging countries, poor urban development also plays a vital role. Vehicle-related sounds are a common reason of noise pollution, particularly in busy metropolises. Honking horns and loud consumes attribute to cerebral distress that could lead to road rage, headaches, and physical discomfort. Congested houses, large families sharing small space, fight over parking, recurrent fights over basic amenities leads to noise pollution which may disrupt the environment of society.

3. Social Events

Social events like worship places, parties, gigs and discos, and so on create a great deal of noise for the people residing in that area. Noise is at its peak in most of the social events. Whether it is marriage, parties, pub, disc or place of worship, people normally flout rules set by the local administration and create nuisance in the area. People play songs on full volume and dance till midnight which makes the condition of people living nearby pretty worse. In markets, you can see people selling clothes via making a loud noise to attract the attention of people. Though, the loud sounds from these social events are mainly termed as 'nuisance' as opposed to noise pollution. Still, they basket usual activates like conversation, sleeping, or upsets one's quality of life.

4. Transportation

Transportation causes of noise pollution predominantly encompass noise from traffic, rails, and aircraft. The numbers of automobiles on the roads are progressively becoming irresistible owing to automobile revolution in urban settings. A large number of vehicles on roads, airplanes flying over houses, underground trains produce heavy noise and people get it difficult to get accustomed to that. The high noise leads to a situation wherein a normal person loses the ability to hear properly. Public transport drivers and people that use buses, subways, and trains to commute to work are all exposed to high levels of noise pollution.

5. Construction Activities

Construction works, blasting, stone crushing, bulldozing, welding, automobile repair activities, quarrying and so on are various grounds of noise pollution. People involved in these activities at times have to wear earplugs to avert them from deafening exposure. Under construction activities like mining, building of bridges, dams, buildings, stations, roads, flyovers takes place in almost every part of the world. These construction activities take place every day as we need more buildings, bridges to accommodate more people and to reduce traffic congestion. The down point is that these construction tackle are too noisy. The residents of areas boarding where these activities take place usually complain of unpleasant and intense noise.

6. Household Chores

Household utilities are bases of noises and can cause trouble to the peace of the mind. Common sources household noise contaminants include domestic gadgets such as weight cookers, washing machines, mixer-grinders, desert coolers, air conditioners, sewing machines and vacuum cleaners. We people are enclosed by gadgets and use them extensively in our daily life. Gadgets like TV, mobile, mixer grinder, pressure cooker, vacuum cleaners, washing machine and dryer, cooler, air conditioners are minor donors to the amount of noise that is shaped but it disturbs the quality of life of your neighbourhood in a bad way.

These gadgets and household actions can be very noisy thus causing trouble to everyone in the neighbourhood. Statistically, household utilities are the main sources of neighbourhood noises. While this form of pollution may seem inoffensive, it, in fact, has far-reaching consequences. The opposing belongings on the health of the environment are fairly severe. Not only is the local wildlife pretentious by pollution but persons also face a number of glitches due to it.

7. Agricultural Machinery

Agricultural equipment like thrashers, tube wells, power saw, tractors, power tillers, and harvesters have proved to produce high levels of noise when under process. As part of agricultural mechanization to make work easier, they have likewise accomplished and noise generating sources.

8. Public Addresses and Alarm Systems

In some countries with lax noise pollution policies, people have the smallest reason for using loudspeakers. They are used for details including birth, marriage and death rites, commercial advertising, and spiritual functions. During elections and demonstrations, public address systems are also commonly used. Accordingly, public address systems in their own way contribute to noise pollution.

9. Defence Machinery

A good deal of noise pollution is added to the environment by military tanks, explosions, shooting practices, military airplane training, and rocket launching. High-pitched noises from sonic booms and jet engines have a deafening impact on the ears. Furthermore, these high-pitched noises have in some extreme cases shattered window panes and old dilapidated structures.

10. Storms

Storms like hurricanes, cyclones, tornadoes can be very overwhelming and can make huge noise contamination. The loud noises made by these storms can cause disturbance in your daily activities and may put intense heaviness on your ears. Similarly, fast and thunderstorms create high-pitched noises that may upset your hearing ability.

5.5.3 Effects of Noise Pollution

The effects of noise pollution on the health of humans can be vast – including having mental, behavioural, and physical effects. For usually apprehensive individuals, noise can make them feel even more nervous, angry, and irritable. Looking beyond these indications, the other long-term effects of noise pollution are often even worse.

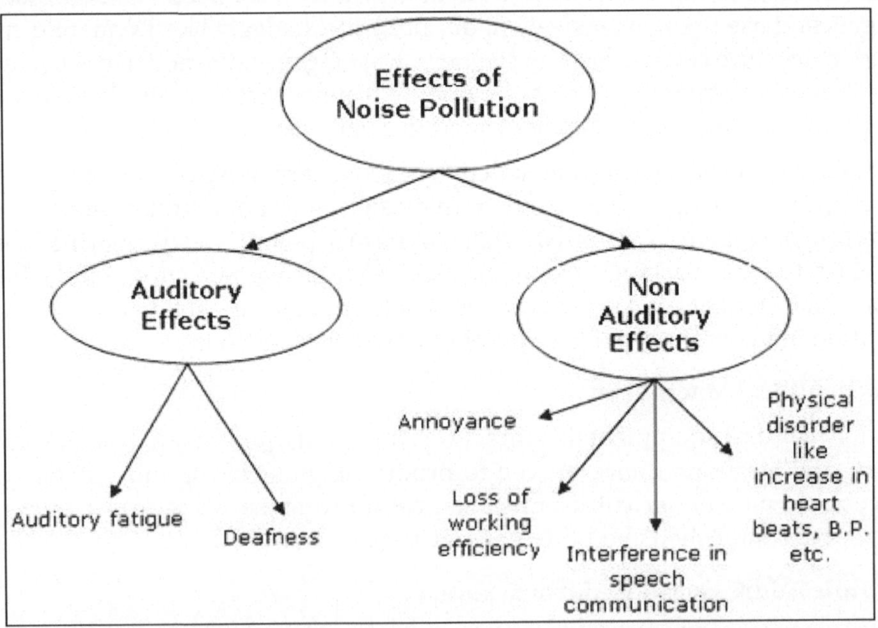

Some of the effects of Noise Pollution are briefly mentioned below:

Effects on Human Health

Whether we realize we are exposed to it or not, noise pollution can be dangerous to our health in various ways.

Hypertension is, in this case, a direct result of noise effluence caused raised blood levels for a longer period of time.

- ☆ **Health Issues**: Excessive noise pollution in working areas such as offices, construction sites, bars and even in our homes can influence psychological health.

Hearing loss can be straight caused by noise pollution, whether listening to loud music in your headphones or being exposed to loud drilling noises at work, heavy air or land traffic, or separate incidents in which noise levels reach dangerous intervals, such as around140 dB for adult or 120 dB for children.

Sleep disturbances are usually caused by constant air or land traffic at night, and they are a serious condition in that they can affect everyday performance and lead to serious diseases.

- ☆ **Child development.** Children appear to be more sensitive to noise pollution, and a number of noise-pollution-related diseases and dysfunctions are known to affect children, from hearing impairment to psychological and physical effects. Also, children who regularly use music players at high volumes are at risk of developing hearing dysfunctions. In 2001, it was estimated that 12.5 per cent of American children between the ages of 6 to 19 years had impaired hearing in one or both ears
- ☆ **Cardiovascular Issues:** Blood pressure levels, cardiovascular disease, and stress-related heart problems are on the rise. Studies suggest that high-intensity noise causes high blood pressure and increases heartbeat rate as it disrupts the normal blood flow. Bringing them to a manageable level depends on our understanding of noise pollution and how we tackle it. Dementia isn't necessarily caused by noise pollution, but its onset can be favoured or compounded by noise pollution.

Effects on the Economy

The World Health Organization estimates that one out of three people in Europe is harmed by traffic noise. More than the purely medical effects of noise pollution on the individual, there is a significant social and economic impact. Since noise pollution leads to sleep disturbance, it affects the individual's work performance during the day, it leads to hypertension and cardiovascular disease and costs the health system additional time and money, and it negatively affects school performance in children.

5.5.4 Control Measures of Noise Pollution

Noise control events can be comprised in the design of the installation area and the design and building of a new workplace. Where a plant is designed for a particular workplace, designers should design the plant to eliminate or control noise emissions where there is a risk to a person's hearing.

Noise can be controlled by using the following hierarchy of controls.

Elimination

Elimination is a process that eradicates the noise basis it is the most operative way to avert risks to workers and should always be measured when presenting a new work process, selecting new work gear and scheming the layout of the workplaces. Instances of noise elimination will include evading the use of loud processes or equipment, abolition of influences between solid objects or surfaces, outsourcing the noisy work procedures and moving the noisy processes away from other work doings.

Advance planning and presenting an appropriate buying or hire policy are vital to dipping the level of noise at work. As at an early stage how the novel work process or new equipment would work without revealing the labours to extreme noise is the greatest cost-effective and long-term amount trades can take to decrease overall noise levels. Before acquiring new machinery, its noise levels should be careful this can be attained by liaising with and obtaining info from the builder or supplier of the plant or machinery. This may include installation orders, upkeep preparations and likely noise levels under the exact circumstances in which the apparatus will be operated.

Substitution

Substitution is a process of substituting noisy machinery or apparatus with lower replacements. When elimination is not possible, replacement of the noisy machinery or equipment for softer ones may be the next-best substitute to defend labours from contact to noise.

Employers should always reflect substitute apparatus and work procedures which would make the job fewer loud. It's also noteworthy to retain up-to-date with the appropriate values and industry good repetition. Performing a task contrarily can also defend the workers from noise contact for instance, the use of welded in its place of engrossed building in fabrications and the use of hydraulic procedures to curve physical crops less noise than beating.

Engineering Controls

Engineering panels are all about making vicissitudes to processes, apparatus or gear so that the workers are bare to less noise. For instance, using screens, barriers, enclosures and porous materials help to reduce workers' noise exposure.

Administrative Controls

Administrative controls are the way work is prearranged to decrease either the amount of workers who are bare or the length of time they are exposed to noise. Administrative controls should be used when it is not possible to decrease noise exposure through removal, substitution or engineering noise control measures.

Personal Protective Equipment

Personal protective equipment protects the users from any opposing effects on hearing caused by exposure to high levels of noise. It is the last option in the hierarchy of control and should be used as a last option after all pains to remove or decrease the noise levels have been pushed through mechanical and structural means.

All hearing defence must be proficient of plummeting the noise experience to the safe exposure levels and should be made available for workers to use. It is important to make sure that the hearing protection chosen to protect the workers is suitable for individuals working environment and compatible with other individual protective apparatus being used, for example hard hats, dust masks, eye protection *etc.* It is good practice to offer different types of guards so that workers can choose ones which suit them better.

Selecting the Hearing Protection Device

There is no solitary ideal type of hearing protection for all persons and circumstances. No matter how good the hearing defence is, it's not the best control as its efficiency is dependent on its correct use, disorder and whether it fits appropriately.

Hearing protection must be reviewed periodically for any impairment and replaced if it's not of a disposable nature. Employers should make sure that hearing protection is available and well upheld, while employees are responsible for reporting any damage, defects or loss.

Reviewing Control Measures

Implemented noise control events should be studied frequently and if essential studied, to make sure they work as deliberate. Employers should make sure that:

- all noise-related hazards have been identified, assessed and controlled
- control measures introduced are working effectively
- control measures introduced have not created any new problems
- the engineering controls put in place have made the job safer
- safety procedures are being followed adequately by the workers
- training and instruction provided to workers on how to work safely has been successful

5.6 Thermal Pollution

The term thermal pollution has been used to designate the harmful belongings of intense waste release by various power plants. It means the damage of quality and worsening of marine and earthly environment by various manufacturing plants like thermal, atomic, nuclear, coal-fired plants, oil field generators, factories, and mills. Simply put, thermal pollution comes from hot water or cold water being dumped into a body of water. Bodies of water naturally tend to dissipate the heat gained from warm currents, underwater hot springs, and from the sun. Thermal pollution is called that because it overcomes the natural temperature control mechanisms that work in the water. The unexpected temperature change poses a health risk to a wide array of aquatic and amphibious creatures.

5.6.1 Definition

Thermal pollution is distinct as rapid increase or decrease in temperature of a usual body of water which may be ocean, lake, river or pond by human effect. This generally befalls when a plant or ability takes in water from a usual reserve and puts it back with a changed temperature. Typically, these amenities use it as a cooling method for their machinery or to help better produce their products.

The broadest definition of thermal pollution is the degradation of water quality by any process that changes ambient water temperature. Thermal pollution is usually associated with increases of water temperatures in a stream, lake, or ocean due to the discharge of heated water from industrial processes, such as the generation of electricity. The real definition of thermal pollution is "the harmful release of heated liquid into a body of water or heat released into the air as a waste product of business."

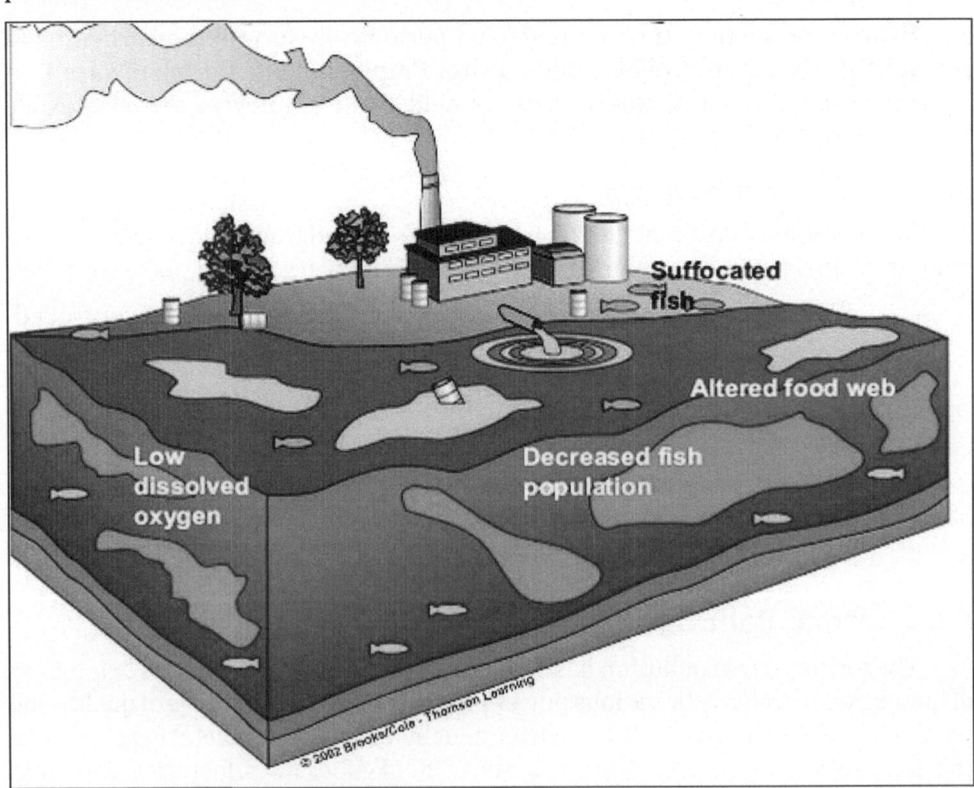

5.6.2 Causes of Thermal Pollution

Human alterations to natural temperature rules of freshwater homes cause thermal pollution. Thermal pollution is any nonconformity from the natural temperature in a habitat and can range from raised temperatures related with industrial chilling activities to discharges of cold water into watercourses below large impoundments. While the main issue is the activity of different industries, this type of pollution can occur from various other sources as well. Here is a look at some different ways:

1. Power, Industrial, and Manufacturing Plants

When plants pull water from nearby sources to keep their machines cool and then release it back to its original source, the water is expressively warmer than it was when initially pulled. The process alters oxygen levels and causes impairment to the wildlife that lives in the water.

Not only does it affect wildlife, but it also kills vegetation that grows alongside water. By destroying natural habitats, marine animals suffer. Industrial sewages including drainage from research organisations, hospitals, sugar industrial companies, explosives, and atomic trials produce a lot of unutilized heat and traces of heat fascinating materials into neighbouring water resources. This results in raised fevers of the water bodies along with plain drops in the levels of melted oxygen. The result is the demise of numerous fish and other aquatic life forms.

Power, production and manufacturing plants top at clearing lots of heat into close water resources. The processes of power cohort and production emit lots of heat. So, to cover the amount of heat produced, factories and power plants draw water from the nearby water resources to maintain smooth operations after which the water is liquidated back into the ocean at elevated temperatures.

2. Erosion from Soil

Thermal pollution can as well be understood as a consequence of soil erosion. This is because continued soil erosion causes siltation and sedimentation thus levitation the water levels which enlarges the surface area of the water bare to sunlight. The more the exposure to sun's heat, the higher the water temperatures become. Even though working plants are the most important cause of thermal pollution, soil erosion is additional primary cause. Soil erosion is the breakdown of the top layers of soil, and it can occur obviously or from human activity, like clearing trees and flora to build houses or businesses.

The erosion of soil causes water levels to rise and be even more bare to sunlight. Over time, the higher temperatures in the water cause problems for the biomes because of the breakdown of oxygen. Stream-side erosion may also remove vegetation cover along the streams further exposing the water to the sun's thermal radiation. The high temperatures can have destructive effects on the aquatic community.

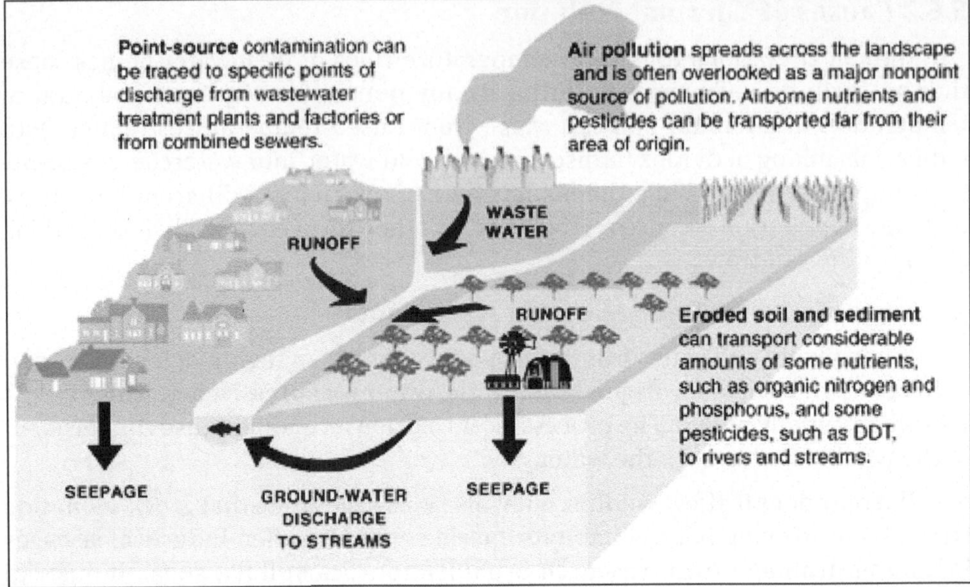

3. Water Overflow

Another cause of thermal pollution comes from the runoff water that comes from the pavement. During hot days and months, urban surfaces including parking lots, macadam roads, and paths usually heat up by absorbing the thermal radiation from the sun. When it rains, the urban storm water runoff absorbs the heat and discharges it into the nearby sewer systems and water bodies.

4. Deforestation

One of the advantages of trees and large plants is that they provide shade. For many bodies of water, the nearby trees and plants help keep sunlight from outstanding directly on the water. Deforestation is the procedure of eliminating a large number of trees and plants from an area. When it happens near bodies of water, it causes the temperature of the water to increase rapidly. Forests and vegetation cover is responsible for reflecting back and absorbing the sun's heat. Besides, it also cushions against the accumulation of intense heat on rivers, ponds, canals, and lakes. Thus, deforestation leaves the water sources directly unprotected to sunlight and licences intense sun's heat that would have otherwise been absorbed by trees or vegetation cover to accrue in water sources.

Deforestation has been one of the leading culprits in global warming because of the augmented greenhouse gasses from the lack of oxygen, and also the thermal pollution it causes that destroys marine ecosystems. Also, atmospheric temperatures are heightened due to the thermal effect of CO_2 greenhouse gases which could have been absorbed by trees. Heightened atmospheric temperatures mean compounded temperature levels of the water resources.

5. Natural Disasters and Causes

Geothermal activity and active volcanoes below the oceans and seas can persuade natural underground heating of the water bodies. Hot rocks and active lava have the potential of heating and raising water body temperatures. The increase in temperatures merely translates to important harmful influences on the marine life and the environment in general.

Volcanoes or other geothermal activity on the ocean floor can cause a fast increase in water temperature, and so can lightening striking water. While naturally happening thermal pollution is unavoidable, the maximum important causes of it are human-made.

5.6.3 Effects of Thermal Contamination

The effects of thermal pollution are varied, but in short, thermal pollution indemnities water ecosystems and reduces animal populations. Plant species, algae, bacteria, and multi-celled animals all reply inversely to important temperature changes. Creatures that cannot adapt can die of various causes or can be forced out of the area. Reproductive glitches can further reduce the diversity of life in the polluted area.

1. Reduction in DO (Dissolved Oxygen) Levels

Warm water holds less oxygen than cool water. If the oxygen level drops animals that cannot move to another area may begin to die. In deeper bodies of water, the injection of warm water can keep oxygen from scattering into deep water, which is potentially good for bacteria but dangerous for aquatic animals. The decreased oxygen can cause algae blooms that pose a threat to aquatic plants and animals. This algae bloom problem is probably the most common and best-known side effect of thermal pollution.

2. Slowing of Metabolism

Water temperatures usually change gradually as the seasons change. During cold seasons, when fewer food sources are available, fish breakdowns slow. When manufacturing releases heat up the water, fish breakdowns also speed up, causing undernourishment due to insufficient food sources. Mass fish kills can occur as a result.

3. Oxygen Reduction

Organic matter rots faster in warmer temperatures, reducing dissolved oxygen from the water and cumulative bacteria levels. This causes an excess of organic nutrients. Aquatic plants grow at a faster rate in warmer waters, subsequent in overpopulation and shorter life spans. Algae blooms occur, choking fish gills and clouding the water. As the algae growth increases, other aquatic plants are unable to perform photosynthesis because of a lack of light. Corals and other marine life may die because of oxygen depletion.

4. Involuntary Relocation

Fish and amphibians may move away from the warm water to a more-suitable location, disrupting the ecosystem for animals that remain. Birds may also be forced to leave in search of areas with more food. Plants and certain animals will be stuck in the area, which can lead to huge losses. Migration away from the polluted area contributes to a dramatic loss of biodiversity at sites where thermal pollution happens.

5. Cool Water Releases

Most thermal pollution is caused by the discharge of warmer water, but cooler water discharges also have a detrimental effect on aquatic life. Cold waters from the bottoms of deep reservoirs are emptied into warmer rivers. Plankton cannot survive the cooler temperatures. Many fish, their eggs and their fry may die. As some organisms die, others reproduce and take over, potentially changing the area's ecosystem.

6 Upsurges Metabolic Rate

Thermal pollution increases the metabolic rate of organisms as increasing enzyme activity occurs that causes organisms to consume more food than what is normally required, if their environment were not changed. It disrupts the stability of food chain and alter the balance of species composition.

7 Disturbs Generative Systems

A significant halt in the reproduction of marine wildlife (although this may be true, reproduction can still occur between fish – but the likelihood of defects in new-borns is significantly higher) can happen due to increasing temperatures as reproduction can happen within certain range of temperature. Excessive temperature can cause the release of immature eggs or can prevent normal development of certain eggs.

8. Biodiversity Damage

Whether marine animals and organisms opt to move away or are killed because of thermal pollution, biodiversity is affected. Animals, plants, and other organisms which can adapt more easily to temperature shifts will survive while other ones will not.

5.6.4 Control Measure of Thermal Pollution

Once governments and communities documented the damaging belongings of thermal pollution, conservation efforts began. Here is how some corporations can last operating without destroying ecosystems:

- ☆ Treating heated water before releasing it back into nearby water sources.
- ☆ Installing cooling tower or cooling ponds where the built-up heat can be released through evaporation or heat transfer.

- Recycling heated water, a process called cogeneration. The recycled water is used for industrial heating or other domestic operations.
- Directing storm water runoff to basins that are located underground. Cities that are looking to operate more environmentally friendly do this process to help with their green-living infrastructure.
- Storm water basins above ground to help reduce thermal pollution but are not necessarily as efficient as underwater systems because above-ground systems are exposed to more heat.
- **Use less electricity** –The use of less electricity will lead to less workload on power plants and these plants will not have to use their machines too much, meaning controlled use of water as coolant.
- **Use of Healthier Technologies** –Use of better technologies is strongly recommended for solving the problem of thermal pollution. There are technologies available which help in the cooling down of machines. If machines will be cooled down with the help of technologies, the use of water as coolant will come to a much reduced level.
- **Holding back the water for good** –Storing the heated water for a particular time will help in bringing back the high temperature of water to a normal level.
- **Plantation of more trees upon the banks of rivers, seas and other water bodies** –Planting more trees also helps in controlling the problem of soil erosion because the strong roots of trees hold the soil firmly and stop it from erosion. generations to plant more and more trees.
- **Artificial Lakes** –These artificial lakes or ponds generally contain two ends. From one end, the hot water is transferred into the lake; it is processed through evaporation or other technique and finally, when it cools down, it is taken out from the other end. The evaporated heat dissolves in the air.
- **Recycling used water** – Smart people always find intelligent solutions for even the most difficult of problems. If people start working upon the ideas of recycling the used water in plants and factories, the problem of thermal pollution will definitely be lessened to a significant extent. Every plant or industry should make it a rule that water used as coolant will not be spilled back into water bodies.
- **Co-generation** – Co-generation is also a delightful idea to battle thermal pollution. In the process of co-generation, the useless heat from hot water can be recycled and used smartly in many tasks by industries.
- **Cooling towers** – Cooling towers is also a good idea when talking about the solutions for thermal pollution. Use of water from water systems for cooling systems for cooling purposes, with subsequent return to the water way after passage through a condenser, is called cooling process.

5.7 Nuclear Hazards

Radioactive pollution is a special type of physical pollution which is related to all major life supporting systems like air, water and soil. I have tried to explain this article that how to treat these radioactive substances, its sources, its effects, control and disposal of these radioactive substances. Radioactive (nuclear) pollution is a special form of physical pollution related to all major life-supporting systems – air, water and soil. It is always convenient to discuss radioactive pollution separately because its nature of contamination is different from other types of pollution. Its effects are also of special kinds.

5.7.1 Definition

Any unsolicited effect caused to the environment due to dangerous materials or radiations is called nuclear pollution. Major basis is the Nuclear power plants. If suggestions of the dangerous materials are present in the water that is released from the plant, it will cause nuclear pollution. Release of energies can also cause this kind of pollution.

It affects almost all life forms in the nearby environment. From planktons to Human beings nothing is spared. To be more exact, the energies can cause changes that lead to cancer, and the dose of radiation or the level of contamination controls lethality or how deadly it is.

5.7.2 Causes of Nuclear Hazards

Most activities that involve radioactive materials have possible to pollute the environment these include:

Nuclear Weapons Testing

Beginning with the Second World War when Japan was passive after the use of the nuclear bombs on the cities of Hiroshima and Nagasaki, countries have been in the race to grow their own nuclear arms, in the name of protection, but more to threaten rival nations. These were led by US, Russia, Britain, France and China. Nowadays N. Korea, Iran and many of the emerging countries are armed to build these weapons as well.

Nuclear Power Plants

Intense Nuclear energy from radioactive fuel is used to heat water to steam. The steam is then used to turn the turbines that in turn work the producers to produce power. Small amounts of radiation are released during this process into the water which may then position of lengthily causing nuclear pollution.

Improper Disposal of Spent Nuclear Fuel

The favourite dumping ground of many countries was the Pacific Ocean. Greenpeace an organisation dedicated to preservative the environment and saving the earth from pollution has brought attention to this activity and opposes it with passion.

Some of the other sources or causes of nuclear hazards/pollution are mentioned below:

The sources of radioactivity include both natural and man-made.

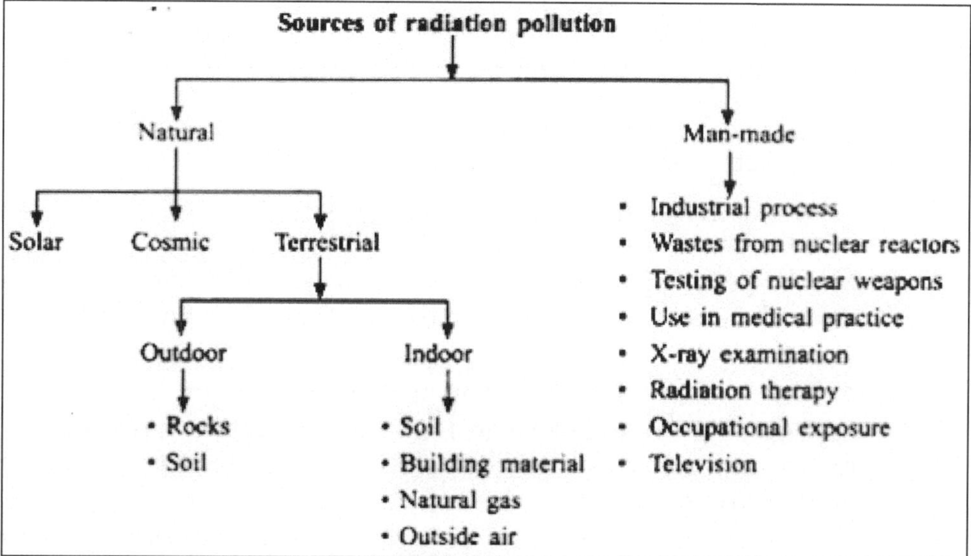

5.7.3 Effects of Nuclear Hazards

The effects of radioactive pollutants depend upon half-life, energy releasing volume, rate of diffusion and rate of statement of the pollutant. Various atmospheric conditions and climatic conditions such as wind, temperature and rainfall also control their effects. Negative influence on health and environment

Effects on Human Health and Environment

Nuclear accidents may produce outcome which can pollute water supplies for years after the incident. The organism lives in water show sensitivity to the radiations. The 1986 explosion of a nuclear generator in Chernobyl (Ukraine) created a large radioactive cloud which polluted existing water supplies and produced contaminated rain in nearby countries. The radiations which are harmful influence nature and occur in the coastal areas. The fishes and water polluted by the radiation. Nuclear radiation can contaminate soil, leading to plants which contain radiation and pose a health threat to individuals. Researchers found that current soil samples and local foods, including coconut meat, contained radiation levels significant enough to pose a health risk to individuals. Radiation also damages chromosomes. It increases the frequency of chromosomal aberrations and causes genetic mutations. Such genetic changes may adversely affect plant metabolism or change their characteristics in subsequent generations.

All organisms are affected from radiation pollution, and the effects are extremely dangerous. The effects may be somatic (individual exposed is affected) or genetic (future generations) damage.

5.7.4 Control Measures of Nuclear Hazards

On one hand, the peaceful uses of radioactive materials are so wide and effective that modern civilization cannot go without them; on the other hand, there is no cure for radiation damage. Thus the only option against nuclear hazards is to check and prevent radioactive pollution. For this:

- ✰ Laboratory generated nuclear wastes should be disposed off safely and scientifically.
- ✰ Nuclear power plants should be located in areas after careful study of the geology of the area, tectonic activity and meeting other established conditions.
- ✰ Appropriate protection against occupational exposure
- ✰ Leakage of radioactive elements from nuclear reactors, careless use of radioactive elements as fuel and careless handling of radioactive isotopes must be prevented.
- ✰ Safety measure against accidental release of radioactive elements must be ensured in nuclear plants.
- ✰ Unless absolutely necessary, one should not frequently go for diagnosis by x-rays.
- ✰ Regular monitoring of the presence of radioactive substance in high risk area should be ensured.
- ✰ Laboratory generated nuclear wastes should be disposed of safely and scientifically.
- ✰ Nuclear power plants should be located in areas after careful study of the geology of the area, tectonic activity and meeting other established conditions.
- ✰ Appropriate protection against occupational exposure
- ✰ Leakage of radioactive elements from nuclear reactors, careless use of radioactive elements as fuel and careless handling of radioactive isotopes must be prevented.
- ✰ Safety measure against accidental release of radioactive elements must be ensured in nuclear plants.
- ✰ Unless absolutely necessary, one should not frequently go for diagnosis by x-rays.
- ✰ Regular monitoring of the presence of radioactive substance in high risk area should be ensured.

5.8 Solid Waste Management

Solid waste refers to the range of garbage materials–rising from animal and human activities–that are cast-off as unwanted and useless. Solid waste is produced from industrial, residential, and profitable activities in a given area, and may be handled in a variety of ways. As such, landfills are characteristically confidential as sanitary, municipal, construction and demolition, or manufacturing waste sites.

Waste can be branded based on material, such as plastic, paper, glass, metal, and organic waste. Categorization may also be based on hazard potential, including dangerous, flammable, infectious, toxic, or non-toxic wastes. Categories may also relate to the origin of the waste, whether industrial, domestic, commercial, institutional, or construction and demolition.

Solid waste management is a term that is used to refer to the process of collecting and treating solid wastes. It also offers solutions for recycling items that do not belong to garbage or trash. As long as people have been living in payments and residential areas, garbage or solid waste has been an issue. Waste management is all about how solid waste can be changed and used as a valuable resource. Solid waste management should be embraced by each and every domestic including the business owners across the world. Industrialization has brought a lot of good things and bad things as well. One of the negative effects of development is the creation of solid waste.

Definition

According to Britannica, "Solid-waste management, the collecting, treating, and disposing of solid material that is discarded because it has served its drive or is no longer useful. Improper disposal of civic solid waste can create unhygienic conditions, and these conditions in turn can lead to pollution of the environment and to outbreaks of vector-borne disease–that is, illnesses spread by rodents and insects."

Solid waste management refers to the collecting, treating, and positioning of solid material that is discarded or is no longer useful. Solid waste management is an important aspect of urban area management. Improper removal of municipal solid waste can create unsanitary conditions, which can lead to ecological pollution and the outbreak of vector-borne disease.

Methods of Solid Waste Disposal and Management

Here are the methods of solid waste disposal and management:

1. *Solid Waste Open Burning*: Solid waste open burning is not the perfect method in the present scenario.
2. *Sea dumping process*: This sea dumping process can be carried out only in coastal cities. This is very costly procedure and not environment friendly.
3. *Solid wastes sanitary landfills*: Solid wastes sanitary landfills process is simple, clean and effective. In this procedure, layers are compressed

with some mechanical equipment and covered with earth, leveled, and compacted. A deep trench of 3 to 5 m is excavated and micro-organisms act on the organic matter and degrade them.

4. *Incineration method*: Incineration method is suitable for combustible refuse. High operation costs and construction are involved in this procedure. This method would be suited in crowded cities where sites for land filling are not available.
5. *Composting process*: Composting process is similar to sanitary land-filling and it is popular in developing countries. Decomposable organic matter is separated and composted in this procedure. Yields are stable end products and good soil conditioners. They can be used as a base for fertilizers.
6. *Disposal by Ploughing into the fields*: Disposal by ploughing into the fields are not commonly used. These disposals are not environment friendly in general.
7. *Disposal by hog feeding*: Disposal by hog feeding is not general procedure in India. Garbage disposal into sewers including BOD and TSS increases by 20-30 per cent. Refuse is ground well in grinders and then fed into sewers.
8. *Salvaging procedure*: Materials such as metal, paper, glass, rags, certain types of plastic and so on can be salvaged, recycled, and reused.
9. *Fermentation/biological digestion*: Biodegradable wastes are converted to compost and recycling can be done whenever possible. Hazardous wastes can be disposed using suitable methods.

5.8.1 Causes of Urban and Industrial Waste

Every day, tonnes of solid waste is disposed of at numerous landfill sites. This left-over comes from homes, offices, industries and various other agricultural related activities. These landfill sites produce foul smell if waste is not stowed and treated properly. It can pollute the surrounding air and can seriously affect the health of humans, wildlife and our environment. The following are major sources of solid waste:

Source of Waste Generation	*Action to be taken*
Household	• Not to throw any solid waste in the neighbourhood, on the streets, open spaces, and vacant lands, into the drains or water bodies • Keep food waste/biodegradable waste in a non-corrosive container with a cover (lid) Keep dry, recyclable waste in a bin or bag or a sack • Keep domestic hazardous waste if and when generated separately for disposal at specially notified locations

Source of Waste Generation	*Action to be taken*
Vegetable and Fruit Markets	• Provide large containers, which match with transportation system of the local body. • Shop keepers not to dispose of the waste in front of their shops or open spaces. Deposit the waste as and when generated into the large container placed in the market.
Meat and fish markets	• Not to throw any waste in front of their shops or open spaces around. Keep non-corrosive container/containers not exceeding 100-litre capacity with lid handle and the rim at the bottom and deposit the waste in the said containers as and when generated. • Transfer the contents of this container into a large container provided by the association
Street food vendors	• Not to throw any waste on the street, pavement or open spaces. Keep bin or bag for the storage of waste that generates during street vending activity. • Preferably have arrangements to affix the bin or bag with the hand-cart used for vending.
Marriage halls, community halls, kalyanamandapas	• Not to throw any solid waste in their neighbourhood, on the streets, open spaces, and vacant lands, into the drains or water bodies. • Provide a large container with lid which may match with the transportation system of the local body and deposit all the waste generated in the premises in such containers.
Hospitals, Nursing homes, *etc.*	• Not to throw any solid waste in their neighbourhood, on the streets,open spaces, and vacant lands, into the drains or water bodies. • Not to dispose off the biomedical waste in the municipal dust bins or other waste collection or storage site meant for municipal solid waste. • Store the waste as per the directions contained in the government of India, Ministry of Environment Biomedical Waste (Management and Handling) Rules, 1998.
Construction/ demolition waste	• Not to deposit construction waste or debris on the streets, footpaths, pavements, open spaces, water bodies *etc.* Store the waste within the premises or with permission of the authorities just outside the premises without obstructing the traffic preferably in a container if available through the local body or private contractors
Garden waste	• Compost the waste within the garden, if possible Trim the garden waste once in a week on the days notified by the local body. • Store the waste into large bags or bins for handing over to the municipal authorities appointed for the purpose on the day of collection notified.

5.8.2 Effects of Urban and Industrial Waste

Due to improper waste disposal systems chiefly by municipal waste management teams, wastes heap up and become a problem. People clean their homes and places of work and litter their environments which affects the environment and the community.

This type of dumping of waste materials forces decomposable materials to rot and decompose under improper, unhygienic and uncontrolled conditions. After a few days of decomposition, a foul smell is produced and it develops a breeding ground for different types of disease causing insects as well as communicable organisms. On top of that, it also spoils the artistic value of the area.

Solid wastes from industries are a basis of toxic metals, hazardous wastes, and chemicals. When released to the environment, the solid wastes can cause biological and physico-chemical **problems to the environment** and may affect or alter the productivity of the soils in that particular area.

Toxic materials and chemicals may seep into the soil and pollute the groundwater. During the procedure of collecting solid waste, the dangerous wastes usually mix with ordinary garbage and other combustible wastes making the disposal process even harder and risky.

When dangerous wastes like pesticides, batteries containing lead, mercury or zinc, cleaning solvents, radioactive materials, e-waste and plastics are mixed up with paper and other scraps are burned they produce dioxins and gasses. These toxic gases have a possible of causing various illnesses including cancer.

5.8.3 Control Measures of Urban and Industrial Waste

Two important steps involved in solid waste management are:

Reduce, Reuse and Recycle of Raw Materials

Recycling or recovery of capitals is the process of taking valuable but cast-off items for next use. Traditionally, these items are treated and cleaned before they are recycled. The process aims at plummeting energy loss, ingesting of new material and reduction of landfills.

Discarding Wastes

Disasters such as floods, earthquakes and hurricanes (cyclones) can produce large quantities of rubble. This will be a danger to people, block access roads, and hide stuck persons and chunk drainage channels. It will also delay the access of other spare services.

Role of NGOs

During the recent years, NGOs (non-governmental organizations) have taken up initiatives to work with local residents to improve sanitation. They have been playing an active role in organizing surveys and studies in specified disciplines of social and technological sciences. In the field of garbage management, such studies are useful in identifying areas of commercial potentials to attract private entrepreneurs. They can play a significant role in separation of waste, its collection and handling over to local authorities.

They are all positively creating consciousness among the countries about their rights and responsibilities towards solid waste and the cleanliness of their city. These governments promote environmental teaching and consciousness in schools and involve groups in the organisation of solid waste.

5.9 Role of an Individual in Prevention of Pollution

If man has to endure, he has to combat and overcome this enormous problem before it overcome him and smears out his being. So it is very vital to decrease and the feast of pollution. Each and every individual should be conscious on his role in the deterrence of pollution.

Individually man can participate in solving the problem of pollution mainly in two ways:

1. Direct role
2. Indirect role

Direct Role of an Individual

Direct Role of an Individual includes those purposes which are done by separately. Since nature cannot solve the problematic of contamination on its own basis, it is the responsibility of man to discovery means to resolve this problem.

Since the reason of effluence is mostly human beings, only they can minimalize it by stopping its upsurge and spread. One can recall the old proverb that 'prevention is better than cure'. A man should control and decrease the reasons of effluence at every step. Some simple principles to reduce the pollution are as follows:

Preventive Measures for Air Pollution

1. Establish or transfer factories or industries away from residential areas.
2. Reduce the use of automobiles and make use of public transport system.
3. Keep vehicle filters clean and use only quality fuels.
4. Make vehicle's engine off while taking or standing at any crossing.
5. Use CNG (Compressed Natural Gas) as fuel for auto engines instead of liquid petroleum.
6. Use self or low noise horns and avoid blowing them unnecessarily.
7. Use efficient silencer in vehicles.
8. Use of crackers should be avoided.
9. Use LPG for cooking instead of wood, coal.
10. Do not burn plastics.

11. Quit smoking.
12. Reduce the use of air conditioners, refrigerators *etc.*
13. Support companies that are committed to sustainable manufacturing practices and reducing pollution in the air.
14. Plant trees around our home and in our community that can help to reduce air pollutants.

Preventive Measures for Water Pollution

1. Sewage disposal is the main cause of water pollution in big cities. One should take care of proper disposal of sewage.
2. Sewage should be released to water only after treatment.
3. Industrial effluents should be released into water bodies only after proper treatment.
4. Use of chemical pesticides, weedicides, insecticides *etc.* should be minimized. Promote the use of biopesticides, fertilizers *etc.*
5. Proper checking of super tankers in the sea in order to prevent oil spillage.
6. Use the minimum amount of detergent.
7. Use only phosphate free soaps and detergents.

Preventive Measures for Solid Waste Pollution

1. Divide the domestic wastes into biodegradable and non biodegradable components before discharging
2. Biodegradable wastes can be used as fertilizers.
3. Non bio degradable waste should be collected and disposed for recycling processes
4. One should never allow scattering of domestic wastes
5. Recharge batteries rather than using disposable.
6. Give items to charity or sell in a yard sale rather than throwing them away.
7. Use of cloth bags rather than plastics.
8. Buy environmentally friendly electronics.
9. Recycle and reuse items rather than sending them to the landfill.
10. Ensure the proper disposal of solid wastes.

5.10 Role of a Community in Prevention of Pollution

The general community is involved in environmental law and choice making through a number of diverse ways. There can be straight contribution in the form of native consultations on individual planning or effluence control applications, or through seeking legal review medicines. Besides this a secondary method of contribution is through the mechanism of democracy, wherein local communities elect politicians who make environmental policy and decisions.

1. Support new developments that promote pedestrian traffic through the inclusion of sidewalks, pathways, and crosswalks.
2. Encourage alternative transportation by developing regional or local bus programs and bike lanes.
3. Consider passing a local ordinance to restrict or prohibit open burning. Smoke from open burning pollutes the air with fine particles and gases that may cause health problems or reduce visibility.
4. Use fewer materials.
5. Print double-sided.
6. Remove unnecessary light bulbs, such as those near windows.
7. Buy products with minimal packaging.
8. Use a different technology, process, or product that produces less waste or uses less energy.
9. Retrofit lighting. Replace incandescent bulbs with fluorescent ones.
10. Purchase energy-efficient equipment such as copiers and computers that power down when not in use.
11. Purchase wisely. Hazardous wastes, such as unused cleaning supplies, fertilizers, and paint, require extra care. When purchasing these types of products, look for less toxic varieties, such as citrus cleaners.
12. Reuse wastes.
13. Use shredded paper (that you would have shredded anyway) as packing instead of purchasing packing "peanuts."
14. Compost grass clippings or use composting mowers.
15. Reuse and recycle solvents, oils, and antifreeze from maintenance shops.
16. Recycle wastes that cannot be reused.

5.11 Pollution Case Studies

India Suffers Most Pollution-Linked Deaths in World: Case Study

India leads the world in pollution-linked deaths– followed by China and Nigeria – according to a report published on Wednesday that estimated the global impact of contaminants in the air, water and workplace.

The report by the Global Alliance on Health and Pollution (GAHP) found pollution to be the largest environmental cause of premature death on the planet, causing 15 per cent of all deaths – some 8.3 million people.

Among the 10 countries with the most pollution deaths in 2017, the latest year for which data was available, were some of the world's largest and wealthiest nations, as well as some poorer ones.

India and China led in the number of pollution deaths, with about 2.3 million and 1.8 million respectively, followed by Nigeria, Indonesia and Pakistan.

"The report reminds us all that pollution is a global crisis," said Rachael Kupka, acting executive director of GAHP. "It does not matter where you live. Pollution will find you."

Pollution-linked deaths rates were highest in some of the world's most impoverished countries where poor water sanitation and contaminated indoor air are major killers.

Chad, Central African Republic and North Korea saw the highest number of deaths per 100,000 people (287, 251 and 202 respectively), with India entering the per capita list at number 10 with 174 deaths.

"India has seen increasing industrial and vehicular pollution from urban growth while poor sanitation and contaminated indoor air persist in low-income communities," the report said.

On the other end of the scale, five nations in the Arabian Peninsula ranked among the 10 countries in the world with the lowest death rates from pollution, with Qatar reporting the lowest.

Drawing its data from the Institute of Health Metrics Evaluation, founded by the Bill and Melinda Gates Foundation, the report broke risk factors into four categories: air, water, occupational, and lead.

Air pollution represents a combination of household and outdoor contaminants as well as ozone, while water pollution included unsafe water and poor sanitation.

Occupational risk encompassed deaths from carcinogens, second-hand smoke, particulates, gases, and fumes, while lead pollution deaths were those associated with exposure to legacy emissions from leaded gasoline. This refers to the lead that was deposited, and still remains, in the soil from car exhausts.

The report also named ambient air pollution responsible for 40 per cent of all pollution-related deaths, led by China, India and Pakistan (1.2 million, 1.2 million and 130,000 respectively).

The number of global deaths linked to pollution just exceeded those from tobacco use, which stands at about 8 million, but greatly eclipsed deaths from alcohol and drugs, high-sodium diets, HIV, malaria, TB, and war, it said.

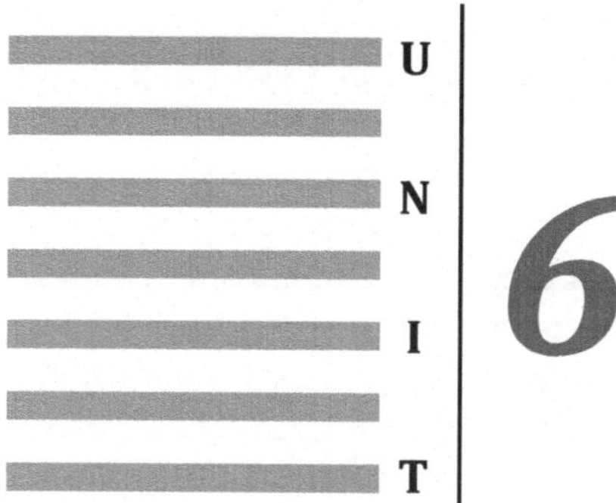

Social Issues and the Environment

6.1 From Unsustainable to Sustainable Development

Human beings co-operate together with the communal world and nature. Both, financial growth and steady environment are obligatory for the repeated improvement of lifestyle and living values. But until now, the growth was human concerned with and incomplete to rich nations. The development was attained by harmful the environment and over misuse of natural resources which were non-renewable. That caused unpredictability of environment and traversed the verge limit of environmental damage. Sustainable development can be defined as "meeting the needs of the present without compromising the ability of future generations to meet their own needs." It is real use of resources for economic development while conserving the environment and ecosystem so that not only the needs of presents are fulfilled but also for the future generations. Sustainable development also interlinks the development and carrying capacity of environment and ecosystem.

If we care for the ease of the current coevals just and do not believe of the stresses of the future coevals, and we injury the location by mixt development activities, these doings will be called as unmaintainable. In taking every action, little or large, the likely compensations to the environment must be given full deliberation and the action must not go forth behind a wicked environment. Strictly, sustainable development is distinct as a way of development in which no permanent and permanent harm is done to the environment and the resources are kept essential for the upcoming coevals. The Earth has the whole thing for each coevals, but it depends on the proper usage. The current coevals can last really well on the resources available, but they must besides go forth behind passable capitals for the future coevals. It is essential that a sustainable development way do not hold any undesirable factor that is responsible for doing unpromising impacts on the environment. A maintainable development programme is welcoming to the ecosystem in all respects and has the capacity to absorb disconnected alterations of the present and the henceforth. Sustainable development has also a strong constituent of socio-political development. Thus, sustainable development agenda must hold equal anxiety for all subdivisions of the society with a balanced economic development and environmental protection. The programme should hold a log-term position of future effects of any action taken today. In short, sustainable development has become the basis of development planning today and has besides become a chief tool of dialogue in international assistance bundles to the states.

The sustainable development can be broadly classified into three different types, *viz.*, environmental sustainability (no long term damage to the environment), economic sustainability (economy remains stable with equitable sharing of resources) and socio-political sustainability (maintaining social harmony and political stability).

There are few measures by adopting we can convert a unsustainable development to sustainable development. They are as follows:

1. Using suitable technology: It is one which is locally flexible, ecofriendly well-organized and socially suitable. It comprises local labours, less resources and produces minimum waste.
2. 3-R Approach: Reduce, Reuse and Recycle approach: Reduce the practise and also reduce the expenditure of reserve by creation things that last lengthier and are easier to recycle, reuse and repair.
3. Encouraging environmental education awareness: Environmental education will help in changing the thinking and attitude of people towards environment.
4. The demand should be reduced up to the system's carrying capacity that a system can sustain on a long term basis.
5. Consumption should not exceed regeneration and changes should not be allowed to occur beyond the tolerance capacity of the system.
6. Components of carrying capacity are supporting capacity, *i.e.*, the capacity to regenerate and Assimilative capacity, *i.e.*, the capacity to tolerate different stresses.
7. For sustainable development, human population growth should be controlled. Otherwise, no invention and development in science and technology will be able to prevent irreversible degradation of the environment and also to alleviate continued poverty in large elements of the globe.
8. The natural and social sciences will be crucial in developing new options for limiting population growth, protecting the environment, and improving the grade of human life. These should be the perspectives in

every short and long term planning for all governments so the challenges of today's and the future can be adequately addressed.

9. A judicious balance between developmental activities and environmental protection should be assured. It is possible only trough sustainable development.

6.2 Urban Problems Related to Energy

The world is progressively flattering urban with more and more people travelling to the cities in search of food, work and shelter. While some are being forced to move as chances dry up in their old-style lands, some are itinerant in search of a healthier life while still others are forced by climatic conditions that make their homes habitable.

With huge urbanization happening on a global scale, the state of the environs and human health in the cities of the world has turned into a prime concern.

Poverty and Health

Despite the unparalleled creation of wealth universal before two periods, the amount of people exist in in total poverty keeps growing steadily. Lack leftovers the main killer, with the indigent bearing a uneven share of the global burden of ill health. The poor live in unsafe and overfull housing, often in semi-urban and urban slums, with practically no access to safe water or to sewerage. These folks are also bare to pollution, traffic and industrial and other risks at home, at the job or in their communities, much more than the rich people surviving in the same cities. They may have inadequate food that too of low quality from the idea of nourishment.

Energy Ingestion

Compared to the rural countries, people in municipalities and city consume big amount of energy. For design, the old kind of Indian houses were complete of wood, clay and unburnt bricks which obligatory really small temperature housings and so really minor energy was required for warming and chilling. The current organisations in towns and metropolis are largely made of concrete, cement, steel, aluminium, marble, well-burnt bricks and glass. These stuffs are energy focussed and the houses made with them require a batch of power to uphold them comfortable during the winter and the summer and also for illuming and ornament. The process of industry of these things besides devours tremendous sum of energy in different signifiers.

The social and economic outcome of such an energy disaster are sure to be disastrous and the discussion about the ladders wanted to speech and alleviate have been raging for long. The answer is a huge drive to yoke non-conventional energies. However, these sources usually have high cost of technology as real entry barriers and may not be as plentifully satisfactory as fossil fuels are. The result too is foreseeable.

As urban communities eat more energy per capita than their rural complements, such an arrival is sure to have an opposite effect on the demand for energy. With usual systems that are deeply reliant on fossil fuels as the main source of energy, such rising demands are creating energy blocks which threaten to blow out of proportions if not addressed immediately. The global request for energy far outscores the supply for the same and with increased industrialisation and urbanization the gap is becoming wider by the hour. Add to it the fact that energy production cannot be scaled up the way previous generations had considering the fast depleting resource base and the concerns of sustainability and climate change. The per unit cost of cohort will also become a warning factor as the cost of energy, subsequent natural laws of demand and supply go up to levels unimaginable.

6.3 Water Conservation, Rainwater Harvesting, and Watershed Management

6.3.1 Water Conservation

The belief that water is plentiful makes people use lots of water for drinking, cooking and washing; much more water is used for producing things such as food, paper, cotton clothes, *etc.* All living organisms need water, plants put it to use in photosynthesis, persons and other animals drink it, and aquatic plants and animals are in it. Water also plays an significant part in many natural and human processes and is also a critical constituent of many physical and chemical reactions. In adding, it maintenances many economic activities. The quantity of water in the world has remained constant. In fact, water hasn't changed in amount or nature for an unbelievable number of years. It just keeps cycling and recycling from atmosphere to earth and again. Freshwater is a precious resource as it makes up less than 3 per cent of the earth's entire water resources. Because freshwater is so partial and plays such a key role in world health, economies, and environmental constancy, it must be conserved and found in a sustainable manner.

"The virtual-water content of a product (a commodity, good or service) is the volume of freshwater used to produce the product, measured at the place where the product was actually produced (production-site definition). It refers to the sum of the water use in the various steps of the production chain. The virtual-water content of a product can also be defined as the volume of water that would have been required to produce the product at the place where the product is consumed (consumption-site definition)".

The most intuitive approaches to conservation include:

- ☆ **Limiting the consumption**: Application of policies, economics, and technologies help regulate the demand. Policies may include restrictions and bans on certain types of water use, standards for fixtures and appliances, enforcement on recycling, *etc.* Economic measures, such as setting prices for water consumption, typically make people more diligent about their water-related activities. Technological advances may help use less water for the same function (water-economy washers, toilets, *etc.*)

- **Reuse and recycling**: Designing systems for reuse and recirculation of water in both domestic and industrial applications are involved in this approach. For example, treated wastewater can be used for irrigation; or water can be recycled in certain types of car wash systems.
- **Elimination of losses**: Regulation, metering, water-sensitive design, and smart technology are key factors contributing to this strategy. Leak identification systems, smart controllers for water use are some of such conservation technologies.
- **Pollution prevention**: Less pollution creates more opportunity for water reuse and conserves natural water reserves. Regulatory policies are

Case Study - Pani Panchayat scheme in Pune

Pani Panchayats are councils set up for conservation, management and distribution of water resources on equitable basis in Maharashtra. The average annual rainfall ranges in between 250mm and 500mm, but, the rainfall pattern is very erratic and uncertain. The area experiences drought almost once in every three years on an average.

The technological and social innovations in this council help repair and restore the degraded watersheds. This in turn increases water harvests and guarantee each family an equal share of the water harvested, meeting the goals of equity. Every member of the family is typically allowed water for cultivation at the rate of half an acre/head, subject to a maximum of 2.5 acres per family. In other words, 1000 m^3 of water per head/year subject to maximum 5000 m^3/family per year is supplied.

9.60 hectares were brought under protective irrigation, 2.40 hectares under afforestation and the remaining 4 hectares came under the percolation tank, well, field bunds, tract and other infrastructures. Initially the farmers of showed little interest, but were amazed to see the production of 200 quintals of food grains. 40 acres of their own land could hardly produce 10 quintals. They grew interest.

The number of lift irrigation schemes has gone up to more than 100 in about 10 years and most of them are functioning well. The villagers are able to produce two crops a year with an irrigation provision of eight months. Another implication of the Pani Panchayat has been the increase in employment opportunities, which also induced reverse migration. The reason behind the success lies in the unity between the villagers and strict obedience to the rules and regulations laid down by the Pani-panchayat which is a unanimously elected village body.

introduced on water discharging facilities. Also, numerous technologies exist to treat the effluents from domestic and industrial facilities.

Water conservation and demand management reduce the request for aquatic and energy resources. Reducing per capita usage and expenditure and leakage from the delivery system all contribute to reducing the pumping and treatment volumes. Any discount in the volumes of fresh and wastewater transported and preserved within the water cycle will have a direct and proportional reduction in energy consumption.

Surface water organisation needs real procedures to switch the excesses of floods and droughts, while upholding a trusted water supply to meet the basic needs of human life and the stresses of financial development. Because of the tall amount of erraticism of precipitation, water storage is a must to provide a year-round supply of water. The storage in the form of reservoirs, dams, dugouts, and natural lakes has been a shared practice. A significant challenge to water managers is locating surface water supplies relative to that of water users. Water users (domestic, municipal, agricultural, and industrial) usually need a range of resources and services and are not always located nearby the water sources. Providing a reliable supply of water to users often requires distribution networks comprising canals and pipelines.

6.3.2 Rainwater Harvesting

The code of collecting and using precipitation from a catchments surface. An old skill is gaining popularity in a new way. Rain water harvesting is enjoying a rebirth of sorts in the world, but it traces its history to biblical times. Wide rain water harvesting apparatus existed 4000 years ago in the Palestine and Greece. In antique Rome, residences were built with separate reservoirs and cemented patios to capture rain water to augment water from city's aqueducts. As early as the third millennium BC, farming communities in Baluchistan and Kutch confiscated rain water and used it for irrigation dams.

Some of the benefits of Rain Water Harvesting are mentioned below:

- Rainwater is a comparatively clean and unconditionally free source of water.
- It is informally suitable and environmentally responsible.
- It indorses self-sufficiency and helps conserve water.
- Rainwater is healthier for landscape plants and gardens since it is not chlorinated.
- It decreases storm water runoff from homes and businesses.
- It can resolve the drainage problems on your stuff while providing you with free water.
- It uses humble technologies that are cheap and easy to uphold.
- It can be used as a chief source of water or as a hold-up basis to wells and municipal water.

Case Study – Gansu Province in China

Gansu is a province in Central China with highly variable annual rainfall of 300 mm. The months July to September receives 60 per cent of the annual rainfall and this factor attribute the people of that region with poverty. People depend on rainwater and frequent storage of surface runoff in 20 cu meters of clay lined underground cisterns is seen. But this isn't enough for the dry periods and they either have to walk long miles for water of have to depend on government supply. The government provided each family with one clay tiled roof catchments area, two water cellars made of cement and plastic sheeting for directing rainwater runoff on one field. Traditional clay lined 'shuijiao' (water cellars) were lined with cement or concrete and upgraded and fitted with small metal pumps. The bare earth catchments were replaced by proper tiled roof catchments and cemented courtyards. A trench was dug around this was to collect rainwater and used for watering the gardens. This simple, operative, inexpensive, approach has helped over 200,000 families and warranted with sufficient water and good yield to around one million people was provided. By the year 2000, a total of 2,183,000 rainwater tanks had been constructed with a total capacity of 73.1 million cubic meters in Gansu Province that can supply drinking water to about 1.97 million people and supplement irrigation service for 236,400 ha of land.

- ☆ The scheme can be easily retrofitted to a current construction or built during new home construction.
- ☆ Schemes are very supple and can be modular in nature, allowing growth, reconfiguration, or relocation, if necessary.
- ☆ It can deliver an outstanding back-up source of water for emergencies

Rainwater harvesting is a likely source for meeting water plea in a few areas, but is frequently ignored by planners, engineers and builders. In many areas, where RWH has been presented as part of a broader drinking water supply programme, it was originally disliked, due to the fact little was known about the technology by the recipients. In most of these cases, the skill has quickly increased popularity as an individual realizes the advantages of a clean, reliable water source at the home, when the supply is unreliable or inadequate, or where local water sources dry up for an integral part of the year.

This water conservation method can be easily skilful in individual homes, apartments, parks, offices and temples too, crossways the world. Farmers have invigorated their dry bore wells, created water banks in drought areas, greened their farms, augmented sustainability of their water resources and even shaped a river. Technical know-how for the rooftop RWH with direct storage can be availed

Case Study – RWH in Frankfurt Airport, Germany

On an average the rainfall in Germany ranges between 563- 855 mm. It had cost about US $ 63,000 in the year 1963. It was installed with and expectation of handling 13 million people/year and saves about 1, 00,000 cu m of water annually. The roof with an area of 26,800 sq. meters provides as a catchment. The water is transported to six storage tanks of 100 cubic meter capacity in the basement. The water is used for washing clothes, toilet flushing, gardening and cleaning of air conditioners, *etc.*

for better application. RWH A real method in water scarce times, it is also an easily achievable practice.

Setting up a rain harvesting is not problematic but needs some sort of understanding of hydrology and building and as a result most people find it too complex to do it themselves. In order to make it simple and expedient for everyone to set up a rainwater harvesting system suitable for their needs, we have prepared a set of guidelines which will help you to set up your own rainwater harvesting system quickly and efficiently.

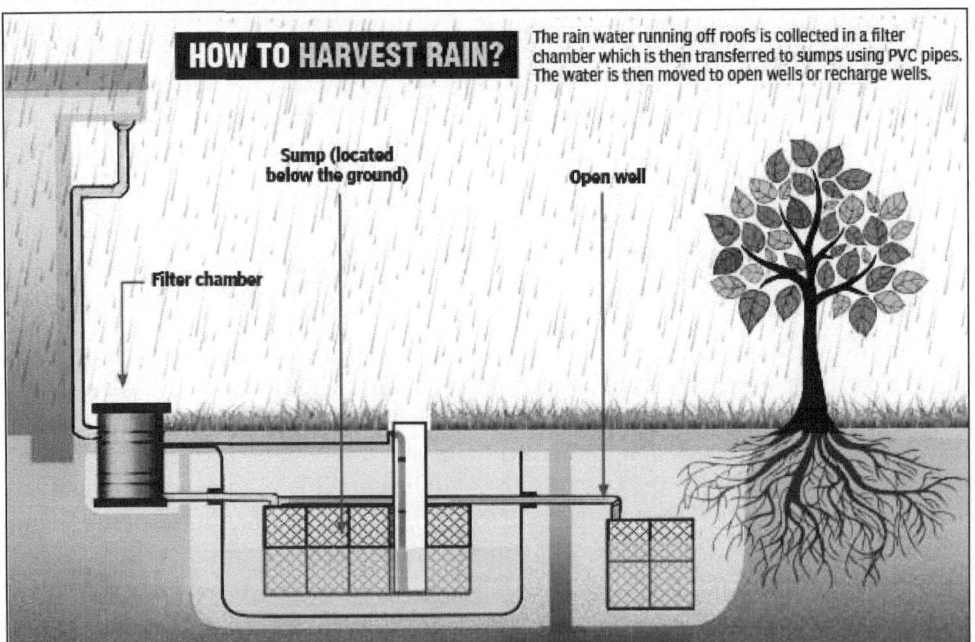

6.3.3 Watershed Management

A *watershed* is merely the physical area finished which water flows across the land and sanitations into a shared body of water, whether a stream, river, lake, or ocean. The disaster boundary will more or less follow the highest ridge-line around the watercourse channels and meet at the lowermost or lowest point of the land where water flows out of the watershed, the mouth of the waterway. Watershed is clear as a geo-hydrological unit draining to a common point by a system of drains. All lands on earth are part of one watershed or other. Watershed is therefore the land and water area, which donates runoff to a common point.

Watershed management is an attempt to halt land degradation and a holistic process for getting maximum production out of land. Watershed management implies rational utilisation of land and water resources for optimum and sustained production, with the minimum of hazard to natural resources. With a bit of acumen and proper judgement, the rain water can be made to move down the slopes slowly, ensuring optimum infiltration and percolation, automatically solving the problem of soil erosion. Reducing the impact of the rain on the soil, checking its speed at various intervals, taking up all operations on the contour and diverting the excess to prevent the, pressure, are some of the procedures one could adopt; a process that starts at the highest point of the watershed and ends right down in the lower parts.

Watershed management provides a outline for integrated decision-making to help: measure the nature and rank of the watershed; identify watershed issues; define and re-evaluate short and long-term objectives, movements and goals; measure welfares and costs; and implement and evaluate actions.

Steps in Watershed Planning

1. Groundwork of base maps for carrying out surveys.
2. Investigation survey of the watershed for overall development.
3. Measuring rainfall characteristics.
4. Groundwork of soil maps and classification of lands for different uses according to capability classification for agriculture, forestry, pasture, horticulture, *etc.*
5. Preparation of list of current land uses and farm sizes.
6. Assessment of agricultural production patterns and potentials, present and potential markets and possible group action arrangements.
7. Carrying out topographic and hydrologic surveys for engineering works.
8. Geo-hydrological survey to delineate areas suitable for groundwater development.
9. Formulation of an integrated time-bound plan for land and moisture conservation, groundwater. Recharge, development of productive afforestation, agriculture production, grasslands and horticulture.
10. Assigning of priorities for implementation of the project.
11. Assessing social costs and benefits.

Following can be considered as some of the benefits of Watershed Management:

Ecological Health

A healthy watershed purposes as a whole ecological system endorsing the health of all living creatures and landscapes within the watershed. A healthy, complete watershed reduces the influences of flooding and erosion and helps to filter residues and pollutants so they do not reach our streams, lakes, and groundwater.

Economic Health

A plentiful supply of clean water is vital for a lively economy. Homes, farms, municipalities and trades all need a plenty supply of clean water to operate effectively. Clean water allows municipalities, businesses, agricultural producers, and businesses to operate more cost effectively, saving money for taxpayers and consumers. Healthy rivers, lakes, wetlands and natural spaces are foundations for regeneration and tourism.

Human Health

Life needs a safe daily supply of water. But water is far more than that: clean superficial and groundwater is vital to sustenance our high quality of life and the social features of our communities. Clean rivers, lakes and streams deliver many healthy recreational opportunities including swimming, boating, and fishing.

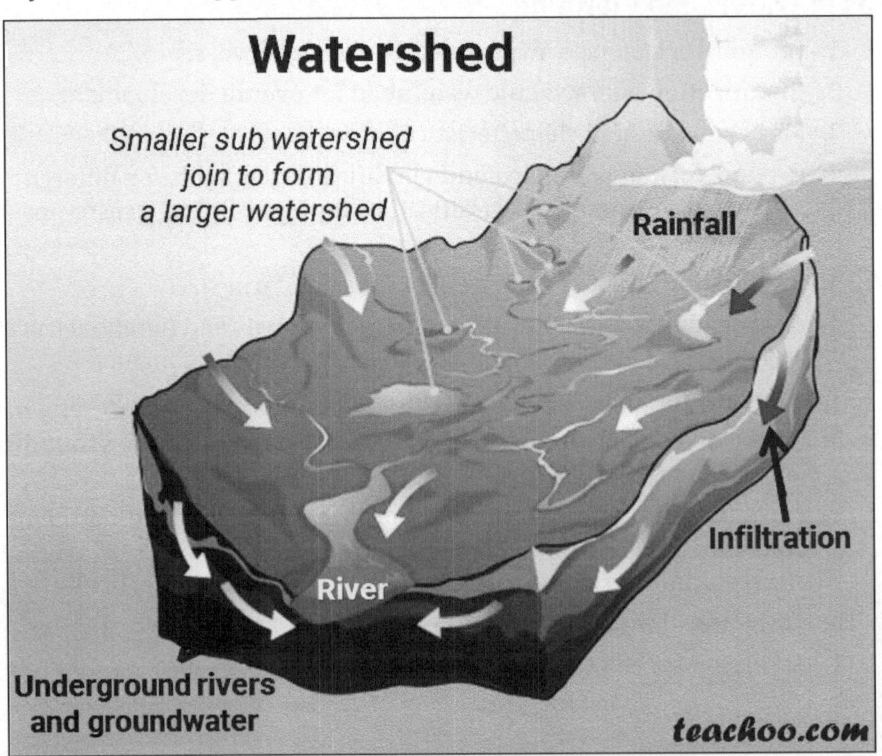

6.4 Environmental Ethics: Issues and Possible Solutions

Morals is clear as a set of rules or values that are followed by a approximately documented race or group (lawyer, writers *etc.*). So when we talk about ecological morals, we hope to define certain set rubrics and values that should be followed in thought with the environment. All other life forms are assistants and important only to the extent to which they affect the human beings. In this ethic, nature is only considered as because protection and squalor can benefit or harm the humanoid beings. It is just human bigotry and holds human to be the amount of all things. Western thinking is largely anthropocentrism.

Environmental morals is a branch of morals that teachings the comparative of human beings and the environment and how morals play a part in this. Environmental ethics trust that humans are a share of civilisation as well as other living beings, which includes florae and animals. These items are a very significant part of the world and are careful to be a functional part of human life. Thus, it is vital that every human being respect and honour this and use morals and ethics when dealing with these creatures.

According to Wikipedia, "*Environmental ethics is the part of environmental philosophy which considers extending the traditional boundaries of ethics from solely including humans to including the non-human world. It exerts influence on a large range of disciplines including environmental law, environmental sociology, Eco theology, ecological economics, ecology and environmental geography.*"

In contrast Deep ecology, the living setting is taken as a whole where all have the similar right to living and embellishment. It highlights on the inherent parity of species counting humankind.

6.4.1 Resource Consumption Patterns and the Need for their Equitable Utilisation

A resource is anything that can be used by humans. Resources that happen freely in countryside are natural resources while capitals developed my man for his own consumption are human resources like schools and hospitals are human resources. Resources can be further classified as renewable – that can be refilled and non-renewable, which once used cannot be reused like fossil fuels.

If an all-inclusive view is taken then it will be sufficiently clear that the resource consumption by the countries to the North of the divide is much higher than those in the South. This consumerist use of scarce capitals is giving rises to anxieties about ethicality, which when viewed in the light of Malthus' predictions take menacing proportions.

To put it bluntly, countries in the North are environmental debtors (whose ecological footprints are bigger than their bio capacities – ability of the environment to carry the people and consumption patterns), while those in the South are normally ecological creditors, their absence of development leaving resources underutilised.

And it is exactly here that the questions of equitable utilization comes in. Not only will resource use have to be maintainable in the future for a just and reasonable world order to emerge, the consumption designs will have to change radically to safeguard that the scarce resources are first used to meet the basic needs of the world's poor before they are deployed to feed the fancies of the rich.

6.4.2 Equity – Disparity in the Northern and Southern Countries

The truth, obviously fights somewhere in between, pulled as it is by these two diverse forces. The fact however, is that the more we are able to shrink this gap, the so-called north-south divide, and the faster we do it is not only good for the persons but also the world at large.

The division between the North and South is essentially on the basis of socio economic and political differences. The North comprises the West, first world and the richer and developed nations in contrast to the south having the poorer and less developed countries. All the G8 countries belong to the North Block and comprise North America, Western Europe, Australia and Japan. As states become more and more frugally developed, they gain influence and so become a part of North, irrespective of their geographical location. The South covers Africa, Asia and Latin America. 95 per cent of the people here lack adequate food and shelter. They are rich in capitals but lacks proper teaching and knowledge to utilise it. They do not have suitable technology and are governmentally unstable. A good measure to find out which block a country is located is the Human Development Index (HDI). Countries with value nearer to 1.0, indicates better level of country's development. A visual representation of the North – South Divide is the Brandt Line, proposed in 1980 by Willy Brandt, The West German Chancellor. It moves around the world at approximately 30oN latitude but dips down to include Japan, Australia and New Zealand into the North.

6.4.3 Urban – Rural Equity Issues

Rural lifestyles are more sustainable – cleaner, greener and healthier. However, the lure of the cities, with their tinsel tomes and seductive opportunities have been a attraction since the dawn of mankind with a steady trickle of migrating people flattering townsfolk. With the rural economies coming under increasing pressure, the urban- rural divide is flattering wider with the traditional trickle of refugees becoming a steady flow. This migrating hordes are not only putting the civic amenities of the cities under severe pressure but are also creating in their wake weights on the environment as well as life in the city is generally less sustainable than in the villages. More and more power is being consumed; water has to be provided for drinking and sanitation as the other basic amenities. All this ultimately lead to the burning of fossil fuels and it has been seen that marginal increases in the urban population leads to a disproportionately bigger carbon footprint.

6.4.4 The Need for Gender Equity

Women manage natural resources all the time in their roles as farmers and household felicitators. Characteristically, they are responsible for growing crops, and often have unique knowledge of local cropping patterns. To meet family needs, rural womenfolk walk long distances to collect wood and water. Despite their reliance on natural resources, women have historically had lesser access to and control over resources than men. Usually it is the men who put land, water, plants and animals to commercial use, which is often more valued than women's domestic uses – just like the lionesses of the pride doing the actual hunting while the lions have the manes.

Gender inequality is most evident in access to land. Custom prohibits women from owning land in many cultures and countries. Mostly, women have only use rights, that too, mediated by men, and those rights are highly precarious to put things mildly. The developing world's 1 billion plus rural poor make up the world's largest group of natural resources users and managers. Understanding their roles and responsibilities - including the gender dynamics of natural resources utilization and organisation - is the starting point for reversing the process of environmental squalor.

Improved water management, especially irrigation and fertilizers, is critical to higher agricultural productivity and conservation of the resource. Women farmers have limited access to irrigation networks or, when they do, to irrigation management decisions: membership of water users' associations is often linked to land ownership which naturally goes against them. Women's limited water entitlements force them to use subsistence agricultural practices that may lead to soil erosion, a major source of instability in watersheds apart from huge opportunity losses.

Over generations, small-scale farmers have shaped a wide diversity of crop species and animal breeds. Commercialization of agriculture, driven partly by global trade in high-yielding crops and animals, is responsible for a rapid decline in agro-biodiversity, which threatens not only local production but, ultimately, global food security, apart from pushing many a marginal farmer over the threshold.

To protect their natural resources, rural women (and men, to a lesser amount) must be empowered to participate in decisions that affect their needs and vulnerabilities. Addressing the gender dimensions of natural resources management will help policy makers formulate more effective interventions for their conservation and sustainable use of the earth's resources, which can and should be put on use for the greatest common good.

6.4.5 Preserving Resources for Future Generations

The very core of man' response to climate change and global warming is sustainability. Sustainability, put simply, is a practice that leads to rational use of capitals in the present so that enough resources are left for the use of the future generations. It is opposed to wanton consumerism and stresses on optimal and rational consumption in the present.

6.4.6 The Rights of Animals

Being the most evolved of all the creatures on Earth, and sitting bang at the top of the food chain, it is for the Humans to be compassionate about the rights of the animals. But what exactly are Animal rights? At its simplest, it is the idea that some or all non-human animals are entitled to the possession of their own lives. That, animals have a right to their most basic needs and interests – such as an interest in not suffering – and should be afforded the same consideration as to the similar interests of human beings. Supporters face the assignment of moral value and fundamental protections on the basis of species membership alone (an idea that gained currency in the 1970's as speciesism, when the term was coined by Richard D. Ryder). They agree for the most part that animals should no longer be viewed as property, or used as food, clothing, research subjects, entertainment (as in a circus), or beasts of burden. Every animal is a cohabitant on Earth and its rights to food, life and breeding should be honoured. All animals should be left to live the life their instinct leads them to live and should not be subjected to human intervention, leave alone use them as staples of our consumption, be that for physical needs like food and clothing or of entertainment.

A charitable organization, "People for the Ethical Treatment of Animals" (PETA), founded in 1980 by Ingrid Newkirk and Alex Pacheco, based in Norfolk, Virginia in US is perhaps one of the world's largest animal rights organization. It is based on the principle, "animals are not ours to eat, wear, experiment on, or use for entertainment."

6.4.7 The Ethical Basis of Environment Education and Awareness

Environmental education includes a code that safeguards better performance and arrogances among the groups towards the environment. It enables us to achieve social and economic goals based on ecologically sound principles to strive equilibrium between resource management and nature. Environmental Education (EE) lies within the notion of Sustainable Development. That is, "meeting the needs of the present without compromising the ability of the future generations to meet their own needs". (Brundtland, 1987). The needs can be social, political and economic. The needs of future generation can be met by minimizing the use of non-renewable resources and reducing the generation of wastes and also keeping in mind to keep the limit of was absorptive capacity of local and global sinks. The concept of environmental education for sustainable development has been defined by Fien in1993 as: "The development of human capacity and creativity to participate in determining the future, encourage technical progress as well as fostering the cultural conditions favouring social and economic change to improve the quality of life and more equitable economic growth while living within the carrying capacity of supporting ecosystems to maintain life indefinitely"

6.4.8 The Conservation Ethic and Traditional Value Systems of India

For 5000 years, the Indian civilization has changed a highly developed way of life that taught successive generations how to live in sync with nature. Our traditional value systems taught us the lessons of taking from nature only that part that was needed and to ensure that by conscious acts we ensured replacement and replenishment of the resources used.

Today, as we stand face to face with a prospect of India keeping her assignation with destiny, in a world order that is facing the resource crunch, we are flattering increasing aware of our legacy. On the one hand is the agenda of feeding the teeming millions and helping them rise above the poverty line, while on the other is the acute need to optimally use our resources so that we may morph into a strong and vibrant economy with a solid industrial base. These two extremes have to be addressed using sustainable means so that we may achieve our development without causing harm to fragile eco system in which we operate in.

And it is here that out traditional knowledge about the forces of nature and the means to harness them in the most sustainable and eco-friendly way is coming in. Yes, we will have to use the raw power of science but with the ethically correct mind-set that is enriched by traditional ways so that we may win, win and win again.

6.5 Climate Change, Global Warming, Acid Rain, Ozone Layer Depletion, Nuclear Accidents and Holocaust

6.5.1 Climate Change

Climate Change is the important issue of our time and we are at a defining moment. From instable weather designs that loom food production, to rising sea levels that increase the risk of disastrous flooding, the influences of climate change are world-wide in scope and unparalleled in scale. Without drastic action today, familiarising to these impacts in the future will be more problematic and expensive. The Earth's average temperature is about 15C but has been much higher and lower in the past. There are natural variations in the climate but scientists say temperatures are now rising earlier than at many other times.

Earth's climate is now altering earlier than at any point in the history of modern civilization, primarily as a result of human activities. Global climate change has already resulted in a wide range of influences across every region of the country and many sectors of the economy that are predictable to grow in the coming decades. This is linked to the greenhouse effect, which describes how the Earth's atmosphere traps some of the Sun's energy.

Solar energy radiating back to space from the Earth's surface is absorbed by greenhouse gases and re-emitted in all directions. This heats both the lower atmosphere and the surface of the planet. Without this effect, the Earth would be about 30C colder and hostile to life.

The atmosphere is prejudiced by and linked to other features of Earth, including oceans, ice masses (glaciers and sea ice), land surfaces, and vegetation. Together, they make up an integrated Earth system, in which all mechanisms interact with and influence one another in often complex ways. For instance, climate effects the delivery of flora on Earth's surface (*e.g.*, deserts exist in arid regions, forests in humid regions), but flora in turn influences climate by reflecting radiant energy back into the atmosphere, transferring water (and latent heat) from soil to the atmosphere, and swaying the horizontal movement of air across the land surface.

Causes of Climate Change

It is much easier to text the evidence of weather erraticism and past climate alteration than it is to determine their underlying mechanisms. Climate is prejudiced by a multitude of factors that operate at time-scales ranging from hours to hundreds of millions of years. The most important mechanisms are described in this section.

Solar Variability

The luminosity, or brightness, of the Sun has been cumulative steadily since its formation. This marvel is important to Earth's climate, because the Sun delivers the energy to drive atmospheric circulation and establishes the input for Earth's heat budget. Low solar luminosity during Precambrian time motivates the faint young Sun paradox, labelled in the section Climates of early Earth.

Volcanic Activity

Volcanic activity can effect climate in a number of ways at dissimilar time-scales. Individual volcanic flare-ups can release large amounts of sulphur dioxide and other aerosols into the stratosphere, plummeting atmospheric transparency and thus the amount of solar radiation reaching Earth's surface and troposphere. A recent example is the 1991 eruption in the Philippines of Mount Pinatubo, which had quantifiable influences on impressive circulation and heat budgets.

Tectonic Activity

Tectonic movements of Earth's crust have had profound effects on climate at time-scales of millions to tens of millions of years. These actions have changed the shape, size, position, and elevation of the mainland masses as well as the bathymetry of the oceans. Topographic and bathymetric changes in turn have had strong effects on the flow of both the atmosphere and the oceans.

Greenhouse Gases

In general, greenhouse gas attentions have been chiefly high during warm periods and low during cold phases. A number of procedures effect greenhouse gas attentions. Some, such as tectonic activities, operate at periods of millions of years, whereas others, such as vegetation, soil, wetland, and ocean sources and sinks, operate at periods of hundreds to thousands of years. Human activities–specially fossil-fuel combustion since the Industrial Revolution–are responsible for steady increases in atmospheric concentrations of various greenhouse gases, especially carbon dioxide, methane, ozone, and chlorofluorocarbons (CFCs).

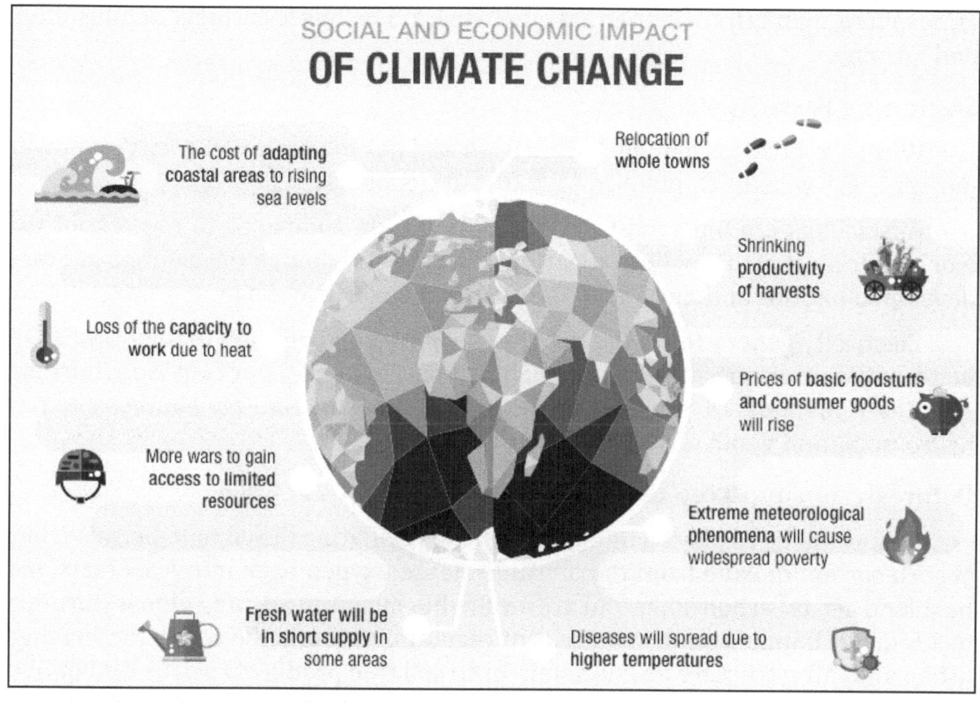

6.5.2 Global Warming

Global warming is chiefly a problem of too much carbon dioxide (CO_2) in the atmosphere–which turns as a blanket, deceiving heat and warming the planet. As we burn fossil fuels like coal, oil and natural gas for energy or cut down and burn forests to create pastures and plantations, carbon accrues and surpluses our atmosphere. Global warming was once a rare term used by a few scientists who were rising worried over the effects of decades of pollution on long-term weather patterns. Today, the idea of global warming is well known, if not well understood. It is not unusual to hear somebody complaining about a hot day or a freak storm and remark, "It's global warming." Certain waste management and agricultural practices aggravate the problem by releasing other potent global warming gases, such as methane and nitrous oxide.

Global warming causes climate change, which postures a serious threat to life on earth in the forms of widespread flooding and extreme weather. Scientists continue to study global warming and its impact on Earth. With the rapid increase in world's population, the consumption of natural resources has increased several times. This has degraded our planet's ability to provide the services we humans need. The consumption of resources is going at a faster rate than they can naturally replenish. It is caused by increased concentrations of greenhouse gases in the atmosphere, mainly from human activities such as burning fossil fuels, deforestation and farming.

Sweltering Fossil Fuels

When we burn fossil fuels like coal, oil and gas to create electricity or power our cars, we release CO_2 pollution into the atmosphere.

Australians are big producers of CO_2 pollution compared to the rest of the world. Our level of CO_2 pollution per person is nearly double the average of other developed nations and more than four times the world average.

Electricity generation is the main cause of carbon pollution in Australia as 73 per cent of our electricity comes from burning coal and 13 per cent from burning gas. The remaining 14 per cent comes from renewable energy sources such as hydro, solar and wind, which do not emit carbon.

Deforestation and Tree-Clearing

Plants and trees play an important role in regulating the climate because they absorb carbon dioxide from the air and release oxygen back into it. Forests and bushland act as carbon sinks and are a valuable means of keeping global warming to 1.5°C. But humans clear vast areas of vegetation around the world for farming, urban and infrastructure development or to sell tree products such as timber and palm oil. When vegetation is removed or burnt, the stored carbon is released back into the atmosphere as CO_2, contributing to global warming.

Agriculture and Farming

Animals, particularly livestock like sheep and cattle, produce methane, a greenhouse gas. When livestock are grazed at a large scale, as in Australia, the amount of methane produced is a big contributor to global warming.

Some fertilisers that farmers use also release nitrous oxide, which is another greenhouse gas.

Global Warming is the result of the temperature of the Earth increasing. Some of the Sun's rays bounce back from the atmosphere, whereas some are trapped by the greenhouse gases like Carbon Di Oxide and Methane which keeps the Earth warm. This process is known as the greenhouse gas effect and contrary to popular beliefs; it is as old as life and is actually beneficial. These greenhouse gases are both naturally created and man-made (anthropogenic) and comprise of the following: Carbon dioxide, Hexafluoride (SF6), Hydrofluorocarbons (HFCs), Methane, Nitrous Oxide, Ozone, Perfluorocarbons (PFCs) and Water Vapour.

6.5.3 Acid Rain

Acid rain, or acid deposition, is a broad term that includes any shape of precipitation that incorporates acidic additives, consisting of sulphuric acid or nitric acid, in line with the Environmental Protection Agency (EPA). Acid rain is a broad time period that is frequently used to describe several types of acid deposition.

Wet deposition is while rain, snow, fog, or mist incorporates excessive amounts of sulphuric and nitric acid. When sulphur dioxide and nitrogen oxides are emitted into the atmosphere, they dissolve in water and fall as precipitation. Dry deposition takes place while dirt and smoke that contain excessive amounts of sulphur dioxide and nitrogen oxides settle to the floor, or onto buildings, cars and vegetation. These gases are converted to acids after they touch water. The acidity of acid rain can vary. Pure water has a pH of seven and everyday rainwater has a pH round 5.6. In 2000, the maximum acidic rain that fell within the United States had a pH of 4.3.

Acid Rain is any form of precipitation with acidic components that falls to the floor from the ecosystem. It can be moist or dry. When nitrogen oxides (NOx) and sulphur dioxide (SO_2) are launched into the ecosystem, they react with water and create sulphuric and nitric acids. These compounds may be carried away via air currents for loads of miles. They fall returned to earth within the form of rain, sleet, hail, snow and dirt.

Causes of Acid Rains

Acid rain is related to both natural and man-made sources. Nitrogen oxides are formed through the acute heating of air whilst a thunderstorm produces lightning. Also, sulphurous gases are discharged from erupted volcanoes and rotting vegetation.

Volcanic eruptions, earthquakes, herbal fires, lightning and some microbial strategies launch sulphur dioxide and nitrogen oxides into the environment. However, it's miles human movement that causes most sulphur dioxide emissions because of the burning of fuels in industry and strength stations, in addition to half of nitrogen oxide emissions due to gases produced with the aid of motor vehicles. Likewise, even though to a lesser extent, extensive livestock farming produces ammonia from the decomposition of natural matter.

These 3 pollutants, which can be transported over long distances from in which they originate, oxidise in contact with the atmosphere and result in the formation of sulphuric acid and nitric acid. These acids dissolve in water droplets in the clouds and fall at the earth's floor as acid rain, which can also occur inside the form of snow or fog.

Effects of Acid Rains

The pH of rain adjustments when combined with sulphuric acid and nitric acid, so when it falls onto the floor or water it alters their chemical characteristics and endangers the balance of the ecosystems. This is referred to as acidification of the environment, a phenomenon that has extreme effects:

Oceans can lose biodiversity and productivity. The reducing of the pH of marine waters harms phytoplankton, a food supply for one-of-a-kind organisms and animals that can regulate the food chain and result in the extinction of various marine species.

Inland waters also are acidifying at a totally fast rate, which is especially worrying as even though handiest 1 per cent of the planet's water is fresh, 40 per cent of fish stay in it. This acidification increases the awareness of metal ions – especially aluminium ions – which could result in the demise of many fish, amphibians and aquatic flora in acidified lakes. In addition, heavy metals are transported to under-groundwaters, which come to be fallacious for consumption.

In forests, the low pH stage of the soil and the attention of metals which includes aluminium save you flora from properly soaking up the water and vitamins it needs. This damages roots, slows boom and makes vegetation weaker and extra vulnerable to diseases and pests.

Acid rain also affects artistic, ancient and cultural heritage. In addition to corroding metallic elements of homes and infrastructures, it deteriorates the external look of monuments. The greatest harm happens to calcareous constructions, along with marble, which regularly dissolve due to the impact of acids and water.

Solutions

There are several solutions to stopping artificial acid rain. Regulating the emissions coming from automobiles and buildings is an important step, in line with the EPA. This can be finished through restricting the usage of fossil fuels and focusing on greater sustainable strength resources inclusive of solar and wind electricity.

Also, everyone can do their component with the aid of reducing their vehicle use. Using public transportation, walking, riding a motorbike or carpooling is a good start, in line with the EPA. People also can lessen their use of electricity, which is widely created with fossil fuels, or switch to a sun plan. Many electricity corporations offer solar packages to their clients that require no installation and occasional costs.

Acid Rain is any form of precipitation with acidic additives that falls to the floor from the ecosystem. It may be wet or dry. When nitrogen oxides (NOx) and sulphur dioxide (SO_2) are released into the surroundings, they react with water and create sulphuric and nitric acids. These compounds may be carried away with the aid of air currents for loads of miles. They fall back to earth within the shape of rain, sleet, hail, snow and dust.

Causes of Acid Rains

Acid rain is connected to both herbal and man-made resources. Nitrogen oxides are fashioned through the intense heating of air when a thunderstorm produces lightning. Also, sulphurous gases are discharged from erupted volcanoes and rotting plants.

Volcanic eruptions, earthquakes, herbal fires, lightning and some microbial procedures release sulphur dioxide and nitrogen oxides into the atmosphere.

However, it's far human action that causes most sulphur dioxide emissions because of the burning of fuels in enterprise and power stations, in addition to half of of nitrogen oxide emissions due to gases produced with the aid of motor automobiles. Likewise, despite the fact that to a lesser extent, in depth cattle farming produces ammonia from the decomposition of natural matter.

These three pollutants, which can be transported over lengthy distances from in which they originate, oxidise in contact with the surroundings and bring about the formation of sulphuric acid and nitric acid. These acids dissolve in water droplets within the clouds and fall at the earth's surface as acid rain, which can also occur in the form of snow or fog.

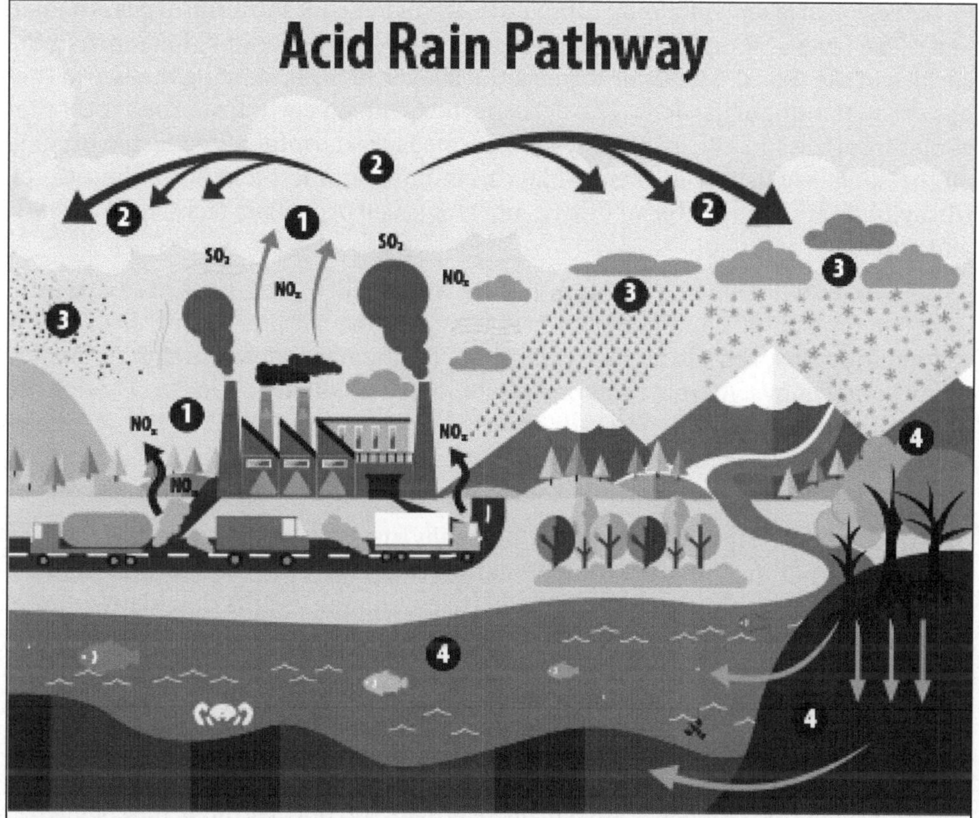

This image illustrates the pathway for acid rain in our environment: (1) Emissions of SO_2 and NO_x are released into the air, where (2) the pollutants are transformed into acid particles that may be transported long distances. (3) These acid particles then fall to the earth as wet and dry deposition (dust, rain, snow, etc.) and (4) may cause harmful effects on soil, forests, streams and lakes.

6.5.4 Ozone Layer Depletion

The earth's atmosphere consists of the many layers, every taking part in a big role. the primary layer stretching around ten kilometres upwards from the earth's surface is thought because the layer. a great deal of human activities like gas balloons, mountaineering, and tiny craft flights present itself inside this region.

The layer is that the next layer higher than the layer stretching around fifteen to sixty kilometres. The layer sits within the lower region of the layer from regarding 20-30 kilometres higher than the surface of the world. The thickness of the layer is regarding three to five millimetre, however it just about fluctuates reckoning on the season and earth science.

Ozone layer may be a deep layer in earth's atmosphere that contain gas that may be a present molecule containing 3 gas atoms. These gas molecules type a volatilised layer within the Earth's higher atmosphere referred to as layer. This lower region of layer containing comparatively higher concentration of gas is named layer. The layer is found 15-35 kilometre (9 to twenty two miles) higher than the surface of the world. The layer forms a thick layer in layer, skirting the world that has great amount of gas in it. The layer protects life on earth from robust UV that comes from the sun. Ultraviolet rays square measure harmful rays which will come on the danger of deadly disorders like carcinoma, cataracts and harm the system. Ultraviolet rays are capable of destroying single cell organism, terrestrial flowers, and aquatic ecosystems.

An essential property of gas molecule is its ability to dam star radiations of wavelengths but 290 nanometres from reaching Earth's surface. During this method, it conjointly absorbs ultraviolet radiations that square measure dangerous for many living beings. Actinic radiation might injure or kill life on Earth. Though the absorption of actinic radiation radiations warms the layer however it's necessary forever to flourish on planet Earth. Analysis scientists have anticipated disruption of prone terrestrial and aquatic ecosystems thanks to depletion of layer.

Ozone depletion is that the term normally accustomed describe the dilution of the layer within the layer. Gas depletion happens once the natural balance between the assembly and destruction of gas within the layer is tipped in favour of destruction. Human action is that the major think about tipping that natural balance, largely from emotional artificial chemicals, called ozone-depleting substances (ODS), to the atmosphere. These square measure stable substances that don't break down within the lower atmosphere and contain either/both halogen and/or Br.

When halogen and Br atoms get contact with gas within the layer, they destroy gas molecules. One halogen atom will destroy over 100,000 gas molecules before it's off from the layer. Gas will be destroyed a lot of quickly than it's naturally created.

Since regarding 1975, scientists have detected a severe call gas concentration within the layer over the Antarctic continent every spring. things then reached associate degree dismaying scale in 1987 once a world expedition found that 1/2 the Antarctica's gas have disappeared over a district double the dimensions of the us, making a massive "hole" within the layer. Concentrations of gas fell by the maximum amount as five hundredth of the norm at altitude of eighteen kilometre.

At mid-latitudes within the hemisphere, up to three decrease in gas concentration was conjointly determined.

The chief ozone-depleting substances embrace chlorofluorocarbons (CFCs), perchloromethane, hydrochlorofluorocarbons (HCFCs) and alkyl group chloroform. Halons, generally called brominated fluorocarbons, conjointly contribute mightily to gas depletion. However, their application is greatly restricted since they're utilised in specific fireplace extinguishers. The draw back to halons is that they square measure thus potent that they're able to drop the layer ten times quite ozone-depleting substances.

There are two regions in which the ozone layer has depleted. In the mid-latitude, for example, over Australia, ozone layer is thinned. This has led to an increase in the UV radiation reaching the earth. It is estimated that about 5-9 per cent thickness of the ozone layer has decreased, increasing the risk of humans to over-exposure to UV radiation owing to outdoor lifestyle. In atmospheric regions over Antarctica, ozone layer is significantly thinned, especially in spring season. This has led to the formation of what is called 'ozone hole'. Ozone holes refer to the regions of severely reduced ozone layers. Usually ozone holes form over the Poles during the onset of spring seasons. One of the largest such hole appears annually over Antarctica between September and November.

Natural causes of depletion of ozone layer: Ozone layer has been found to be suffering from sure natural phenomena which include Sun-spots and stratospheric winds. But this has been determined to cause no longer greater than 1-2 pletion of the ozone layer and the outcomes are also thought to be best temporary. It is also believed that the predominant volcanic eruptions (in particular El Chichon in 1983 and Mt. Pinatubo in 1991) has additionally contributed toward ozone depletion.

Man-made reasons of depletion of ozone layer: The main cause for the depletion of ozone is decided as excessive launch of chlorine and bromine from man-made compounds along with chlorofluorocarbons (CFCs). CFCs (chlorofluorocarbons), halons, CH_3CCl_3 (Methyl chloroform), CCl_4 (Carbon tetrachloride), HCFCs (hydro-chlorofluorocarbons), hydrobromofluorocarbons and methyl bromide are observed to have direct effect at the depletion of the ozone layer. These are labelled as ozone-depleting substances (ODS).

Solutions to Ozone Layer Depletion

Less use of Fuel: by decreasing the usage of the fuels and petroleum utilized in vehicles nowadays we are able to assist in lowering the ozone layer depletion

Less use of pesticides: pesticides facilitates in growing your farms and plants but cause damage to the ozone layer and make contributions to ozone layer depletion.

Limited use of harmful chemical substances for cleaning: the chemical compounds used for making cleaning products results in depletion of the ozone layer.

6.5.6 Nuclear Accidents and Nuclear Holocaust

Nuclear accidents and screw ups have led to extremely huge results, and what has been addressed to save you reoccurrences of such catastrophic activities in the future. The preferred population, on average, is unaware of the actions, or lack thereof, that led up to the events of more than one nuclear injuries, together with Chernobyl inside the former Soviet Union, Three Mile Island in PA, United States, and Fukushima-Daichi in Japan. Because of each of those disasters, protection precautions have advanced and been carried out as prevention measures toward future nuclear catastrophe potentials. This overview of activities aims to look at the aftermath and effects of nuclear disasters on populations, the environment, and the requirements for protection and prevention of destiny injuries, as well as the price involved on a big scale. The severity of devastation that nuclear injuries or disasters are capable of creating is regularly overlooked.

CHERNOBYL ACCIDENT 1986

Chernobyl is taken into account the world's worst nuclear disaster thus far. It occurred on Apr twenty six, 1986, once a unexpected surge in power throughout a reactor systems take a look at resulted in associate explosion and hearth that destroyed Unit four. huge amounts of radiation loose and unfold across the western land and Europe. As a results of the disaster, some 220,000 individuals had to be resettled from their homes.

Nearing a routine close-down, on Apr twenty five, 1986, a crew of staff ready to check the turbines spin period and power provide, once losing main wattage, to current pumps at metropolis reactor four. Multiple actions by operators, like disabling the mechanisms for automatic close-down, occurred before the take a look at once the operator finally took action to close up reactor four, it had been already considerably unstable. A style error within the management rods resulted in an exceedingly huge surge of power wherever the rods entered the reactor. Extraordinarily hot fuel interacted with cooling water and caused multiple reactions leading to pressure increase. Characteristics of the Soviet designed reactor determined that harm to fuel assemblies may, and did, destroy the reactor in entireness. Excessive pressure detached the 1000-ton covering for reactor four, that caused the fuel channels to rupture and also the management rods to jam. fast generation of steam crammed the core, inflicting associate explosion and inflicting fission merchandise to be discharged into the atmosphere. a couple of seconds once, another explosion occurred, spewing fuel channel and hot plumbago fragments. The fuel and plumbago started multiple fires, that caused emission to be discharged into the setting. over half the emission that was discharged was of biologically-inert noble gases. Nearly 5 thousand a lot of dolomite, boron, clay, lead, and sand were born by chopper onto the reactor's burning core in an effort to extinguish the hearth and limit the radioactive particles being discharged into the setting. For 10 days, the hearth burned.

The actions taken by metropolis plant staff were in direct violation of the technical specifications that were *in situ*. The plant was operated at extraordinarily low power with inadequate safety precaution, additionally as poor communication with safety personnel. At low power, the RBMK reactors were extraordinarily unstable. All of those factors LED to the ability surge that inevitably destroyed metropolis reactor range four. A unexpected heat increase caused a surge of power that caused fuel containing pressure tubes to rupture. Most atomic power plants, in alternative areas of the globe, unremarkably had containment structures. Chernobyl, however, didn't have one in all these structures, and while not it, the setting was left unprotected from the material that loose the reactor core.

Fukushima Nuclear Tragedy

The Fukushima Daiichi accident was much different than the incidents at Chernobyl and Three Mile Island. On March 11, 2011, three reactors at the Fukushima Daiichi nuclear power plant were disabled after the power supply and cooling were disabled, following a fifteen-meter tsunami that resulted from a major earthquake. In three days, immediately following the events, all three reactor cores melted. Units 1-3 instantly shut down operation automatically. Reactor units 4-6 were not in operation during the event, but were still affected. During days four, five, and six, radioactive release totaled roughly 940 PBq. Because of damage from the accident, four other reactors were inevitably written off. Two weeks later, reactor units 1-3 stabilized after water addition, and they were able to be cooled with water that was recycled from the new plant by July. In mid-December, it was officially announced that the plant was in 'cold shutdown condition'.

Three Mile Island

Middletown, Pennsylvania, USA,

March 28, 1978

The partial meltdown at Three Mile Island Unit 2 is considered the most serious nuclear accident in U.S. history, although it resulted in only small radioactive releases.

How did it happen?

The accident began with failures in the non-nuclear secondary system, followed by a human-operated relief valve in the primary system that stuck open, which allowed large amounts of nuclear reactor coolant to escape. Plant operators' initial failure to correctly identify the problem compounded it. In particular, a hidden indicator light led to an operator manually overriding the automatic emergency cooling system because he mistakenly believed that too much coolant water in the reactor had caused the steam pressure release. Eventually the reactor was brought under control, although the full extent of the accident was not understood until later.

Nuclear Holocaust

Nuclear holocaust refers to a possible nearly entire annihilation of human civilization by means of nuclear warfare. Under the sort of scenario, all or most of the Earth is made uninhabitable by using nuclear weapons in future world wars. Nuclear physicists and others have speculated that nuclear holocaust could bring about an end to human life, or at the least to trendy civilization on Earth due to the immediate effects of nuclear fallout, the transient loss of much current technology because of electromagnetic pulses, or nuclear iciness and resulting extinctions. Importantly however, despite contemporary high civilization being at risk, assuming weapons stockpiles at the previous cold warfare heights, analysts and physicists have determined that billions of human beings would despite the fact that continue to exist a worldwide thermonuclear war. The risk of a nuclear holocaust performs an critical role inside the popular belief of nuclear weapons. It functions in the security idea of mutually assured destruction and is a common state of affairs in survivalism. Nuclear holocaust is a common feature in literature, especially in speculative genres such as science fiction, dystopian and post-apocalyptic fiction.

Nuclear Holocaust in Hiroshima, Japan

On August 6, 1945, at about 8:15am Japanese time, the US aircraft Enola Gay dropped an untested uranium-235 gun-assembly bomb nicknamed "Little Boy" over Hiroshima.

The devastation was unlike anything in the history of warfare, ushering in the era of weapons of mass destruction.

Hiroshima was immediately flattened. The resulting explosion killed 70,000 people instantly; by December 1945, the death toll had risen to some 140,000.

The radius of total destruction was reportedly 1.6km.

"The impact of the bomb was so terrific that practically all living things - human and animal - were literally seared to death by the tremendous heat and pressure set up by the blast," Tokyo radio said in the aftermath of the explosion, according to a report by The Guardian in August 1945.

"All the dead and injured were burned beyond recognition. Those outdoors were burned to death, while those indoors were killed by the indescribable pressure and heat.

"But the damage did not end there. The radiation released from the explosion caused further suffering.

Thousands more died from their injuries, radiation sickness and cancer in the years that followed, bringing the toll closer to 200,000, according to the Department of Energy's history of the Manhattan Project.

Effects of Nuclear Holocaust

Nuclear bombardment will purpose combustion of wood, plastics, forests *etc.* Large quantity of soot will be performed into the atmosphere. Black soot absorb all UV radiation and will no longer permit the radiation to reach the earth. Therefore cooling will result. This reduces evaporation of water. In stratosphere there won't be sizeable moisture to rain out the black soot. Due to nuclear explosion a manner contrary to worldwide warming will occur. This is known as Nuclear wintry weather.

A nuclear war would begin fires in cities and industrial areas and pump a number of smoke into the stratosphere above wherein we live. In the upper atmosphere, there's no climate or rain to clean it out, and the smoke enveloping the earth would final for years, sending temperatures plummeting.

6.6 Wasteland Reclamation

Environment keeps changing over time obviously and it is also agreeable to vicissitudes by human beings. Thanks to scientific and technological developments, our ability to alter the environment has increased extremely, whereas the capacity of environment to cope with those alterations is limited. Nature's bounty and abundance are vanishing at a rapid rate now in many regions of the world including India due to the human alterations of the environment. All this has brought to the fore the need for protection and preservation of environment and the earnestness of developing sound environmental policies and programmes. Without them, development would not only be unsustainable but would be equal to decline. The challenge of creating and upholding a sustainable environment is probably the single most pressing issue challenging us today and will remain so in the foreseeable future.

Wasteland" being a broader term has been defined by authors in different ways. According to Dudley Stamp (1954), "Wasteland is that land which has been previously used, but which has been abandoned, and for which no further use has been found." Wasteland Survey and Reclamation Committee, Ministry of food and Agriculture, government, of India (1961) has defined wastelands as those lands which are either not available for cultivation or left out without being cultivated for some reason or other. The former includes uncultivable barren lands which are either waterlogged, saline or badly dissected and the later the cultural wastelands. Bhumla and Khare (1987) gives a working definition of wastelands. According to them "Those lands are wastelands (a) which are ecologically unstable, (b) whose top soil has nearly been completely lost, and (c) which have developed toxicity in the root zones or growth of most plants, both annual crops and trees"

Distribution of Wasteland

Wastelands are lands which are economically unproductive, ecologically unsuitable and subject to environmental deterioration. The official estimate of wasteland in India in 2004 was nearly 63.85 million ha – more than 20 per cent of the country's geographical area.

According to the state wise break up, the state with the greatest percentage of its area that is classified is Rajasthan with 45 per cent, followed by the hill states Jammu and Kashmir, Tripura, Himachal Pradesh, Manipur, Sikkim, Mizoram, Meghalaya and Nagaland. This predominance of the hill states at the top of the list is because of the extent of snow-covered and glacial land in these states.

- ✰ Cultivable Wasteland- The land which is has potential for the development of vegetative cover and is not being used due to different constraints of varying degrees, such as erosion, water logging, salinity *etc.*
- ✰ Uncultivable Wasteland– The land that cannot be developed for vegetative cover, for instance the barren rocky areas and snow covered glacier areas.

The non-technical definition of wasteland from the Cambridge Advanced Learner's Dictionary is 'An empty area of land, especially in or near a city, which is not used to grow crops or built on, or used in any way and/or a place, time or situation containing nothing positive or productive, or completely without a particular quality or activity'.

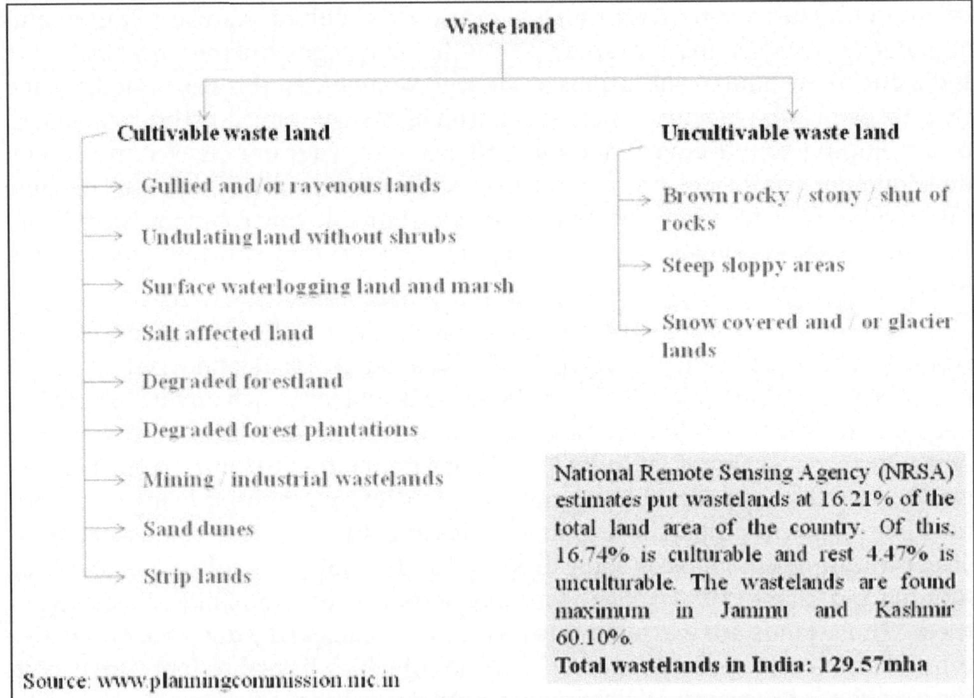

What is Wasteland Reclamation?

Reclaiming lands that have been laid waste in an extraction or industrial system is "desolate tract reclamation." Strip-mining coal produces wastelands. Using chemical substances in an industrial method, then dumping the used chemical substances either at the land or into a move creates wastelands or releasing chemicals into the air in an industrial process can create waste lands. When there's no regulation of wastes disposal with the aid of the enterprise can create wastelands. And finally, nuclear injuries can create wastelands.

It is predicted that in India in 1994, approximately 188 million ha of land, which is 57 in step with cent of the country's total geographical vicinity of about 329 million ha, turned into degraded. Of the 188 million ha of degraded land, approximately 149 million ha changed into stricken by water erosion, 13.five million ha via wind erosion, approximately 14 million ha via chemical deterioration and 11.6 million ha by using water-logging (Sehgal and Abrol, 1994). A recent survey with the aid of the National Bureau of Soil Survey and Land Use Planning found out that 66 in line with cent of India's overall geographical location (round 192 m ha) became at various stages of degradation (quoted in Haque, 1997).

The method used for improvement for barren region will depend on the form of wilderness. Some traditional strategies may be the *in situ* conservation methods for improving the moisture content of soil (bunding, terracing *etc.*); to sell natural regeneration; water harvesting; promoting agro- forestry.

Methods of Wasteland Reclamation

The different strategies used for the reclamation of wastelands are:

1. These lands may be brought under cultivation by using excess water and fertilizers.
2. Afforestation and agronomical methods are used to conserve the soil, which help us to apply it for agriculture.
3. Contour bunds are built which afford safe disposal of water of the catchment areas.
4. These lands can be used for settling the landless agricultural labourers.

A Degraded Wasteland

National Wasteland Development Board become hooked up in 1985 beneath the Ministry of Forests and Environment specially to address the trouble of decay of lands, recuperation of ecology and to meet the growing demands of fuel wood and fodder at the countrywide level. During the Seventh Five Year Plan, the strategy adopted by using the National Wasteland Development Board emphasised greater on tree planting activities instead of Community Participation for desolate tract improvement, In the year 1992, the new Department under the Ministry Of Rural Development (now Ministry of Rural Areas and Employment) turned into created and the National Wasteland Development Board was positioned beneath it. The Board became reconstituted in August 1992 and turned into made liable for particularly development of wastelands in non-forest regions in totality through involving local people at each degree of development. It aims at growing a state of affairs wherein the Government acts as a facilitator and the human beings on the grass root level end up the real executioner of the programme. Major programme carried out for enhancing the productiveness of waste and degraded lands maintaining in view the poverty, backwardness, gender and equity is Integrated Wasteland Development Programme.

Mining Wasteland

Surface mining (sometimes known as quarrying or opencast mining) calls for the elimination of topsoil (the fertile layer of soil and natural matter this is mainly valuable for agriculture) to get on the precious rocks below. Most metals, for example, occur in rocky mixtures referred to as ores, from which the treasured factors should be extracted through chemical, electrical, or different processes. That leaves at the back of waste merchandise and the chemical substances used to technique them, which historically were sincerely dumped back at the land. Since all of the waste became left in one place, the concentration of pollution often became dangerously high. Industrial interest generate waste which incorporates any material this is rendered useless all through a manufacturing process along with chemical solvents, paints, sandpaper, paper products, industrial by means of-products, metals, and radioactive wastes.

Reclamation of Wasteland

With the gradual upward thrust of land useful resource demand, reclamation of wilderness look like one of the major responsibilities of the country.

The reclamation and improvement of wilderness has 4 primary ecological objectives:

Case Study: The Sukhomajri - Watershed Management Project

Sukhomajri, a small hamlet of about one hundred families with average land holding of 0.57 ha, is located in the foothills of Shivaliks in Panchkula district of Haryana. It is at a distance of about thirty kilometers by road to the north-east of Chandigarh. A successful experience of participatory natural resource management, which has been proved to be conducive for bringing about socio-economic and cultural transformation of the village community, has been initiated by Central Soil and Water Conservation Research and Training Institute, Chandigarh. Until 1975, Sukhomajra had no source of regular irrigation. The entire agricultural land (52 hectares) was under rain-fed single cropping. Small land holdings (less than one hectare per family) coupled with frequent crop failures due to erratic distribution of rainfall, made agriculture least dependable as a means of adequate livelihood. Consequently, the people of Sukhomajri were forced to keep a large number of sheep, goats and cows to eke out a living. But, once the domestic animals, especially the goats and cows, were allowed to graze freely in the nearby hills, followed by indiscriminate felling of trees for fuel and other domestic consumption, the hill slopes, once covered with lush green vegetation, soon became bare and not even a blade of grass was to be seen.

1. To improve the physical shape and the pleasant of the soil,
2. To improve the availability and pleasant of water,
3. To save you the transferring of soil, landslides and flooding, and
4. To conserve the biological sources of the land for sustainable use.

There are some of methods and means by way of which desert may be reclaimed for productive purpose in extra green manner. Among them watershed management, barren region afforestation and social forestry programmes are noteworthy. In addition for development of land condition, manipulate of soil erosion and enhancement of soil fertility an amazing wide variety of measures has to be taken up.

Need for Wasteland Development

- Wasteland development provides a source of income for the rural poor.
- Trees help in holding moisture and reduce surface run-off rates.
- It ensures a constant supply of fuel, fodder and timber for local use.
- Makes the soil fertile by preventing soil erosion and conserving moisture.
- Increases the forest cover maintains the local climatic conditions.
- Helps maintain an ecological balance in the area.
- There is need to develop policies, which would result in the best use and sustainable management of land and water resources so as to prevent land becoming degraded and waste in the light of country's food and livelihood security.

Wasteland Development Schemes

Till the sixth five year plan, no specific programme of wasteland development was taken up. It is only in1985 with establishment of NWDB that the problem of wasteland development received a new thrust. With the setting up of NWDB, a number of new schemes were initiated to secure people's participation, besides continuation of ongoing afforestation schemes.

These are: Grants-in-aid to voluntary agencies:

- Decentralized People's nurseries
- Silvipasture farms
- Seed development
- Area oriented fuel wood and fodder projects
- Ariel seeding programme
- Plantation of minor forest produce
- Margin money schemes
- Rural employment scheme

Ninth Five Year Plan

The applications for the improvement of degraded and wastelands got a first-rate boost during the Ninth Plan. A Two centrally backed schemes for soil conservation and integrated watershed management inside the catchments of flood-inclined regions brought in 1961 and 1982 aimed at enhancing productiveness of degraded lands, minimizing siltation of reservoirs and probabilities of floods. During the Ninth Plan both those schemes were merged into a brand new one- soil conservation for reinforcing productivity of degraded lands within the catchments of river valley tasks and flood susceptible rivers. The scheme is being applied in 53 catchments having a total vicinity of 113.40 million hectare spread over in 27 States. The scheme for reclamation of alkali soils introduced in 1985-86 become prolonged to all States in the Ninth Plan. It attempts to improve land and crop productiveness by means of taking over production of crops, including horticulture, fuel wood plantation and fodder species appropriate to the soil conditions. Integrated wasteland development venture being applied, because 1989-90, is based totally on village/micro watershed plans, which might be organized after deliberating the land capability, site situations and local wishes of the humans. The scheme additionally targets at rural employment except enhancing the contents of people's participation inside the development method at all stages that is ensured by using presenting modalities for equitable and sustainable sharing of benefits and usufructs bobbing up from such tasks.

6.7 Consumerism and Waste Products

A customer is someone who buys and uses goods and offerings. Consumers are not best located in human society however additionally in different societies. Consumerism additionally encompasses the evolving set of activities of government business employer and independent customer agency which are designed to protect the rights and pursuits of consumers. Hence consumerism is a international problem affecting each segment of the society.

Consumerism is regularly harassed with capitalism but the latter is an financial device, whilst the former is a pervasive cultural attitude. A model combining the 2 is sometimes referred to as consumer capitalism, a device in which client demand for items is deliberately increased through manipulation as a way of increasing sales. The model is based on stimulating purchaser preference for goods a ways in excess of satisfying needs. Mechanisms to do so consist of promoting of luxurious items, new technologies and new fashions of existing technology.

Financial mechanisms have additionally advocated consumerism. The advent of credit cards allowed humans to spend money that they did now not have; debit playing cards gave human beings quick get admission to to extra cash than they were currently carrying and regularly the choice of overdraft, as properly.

Although purchaser tradition remains prevalent, critics accept as true with that environmental issues render it unsustainable. Alternative fashions that have arisen, which consist of collaborative intake and the circular financial system, emphasize maximizing the green use of substances and having access to products and services over private ownership.

Efficient consumerism is the maximum crucial detail to minimize waste and to sell the financial system of a nation. Inefficient consumerism results inside the following troubles:

- Uncontrolled manufacture of foods main to inferior satisfactory
- Rampant adulteration main to health and hygiene troubles
- Improper offerings resulting in dissatisfaction and stress
- Production of masses of waste leads to depletion of herbal assets and environmental imbalance.

Evolution of Consumerism

The concept of 'consumerism' changed into first conceived in USA around the start of this century. Its evolution may be studied in 3 awesome stages as follows:

1. Around 1900

The business firms dealing in meat packing had been least involved with the clients. Meat used to be sold inside the unhealthiest way. This affected the health of the consumers. Even many different companies used to supply dangerous, unwanted merchandise and drugs and sell them to the clients with the aid of adopting manipulative devices. Conscious and realistic became disgusted with this type of affairs and started campaign to preserve the hobbies of the consuming public.

2. Around 1930

Consumerism assumed extra significance because people in preferred have become more enlightened and concerned with the requirements of good satisfactory products. This become viable as a consequence of education, cognizance and political consciousness. Although consumerism did not come to be a severe public motion at some point of this time, the Government came up with a legislation known as Miller-Tiching Act 1936 so that you can regulate certain advertising and marketing mal-practices.

3. 1960's

It is within the sixties of this century consumerism became a very forceful social movement. Late President Kennedy, within the year 1962, handed a rules to shield the clients' rights particularly in regards to false marketing and unhealthy packaging of food and other articles. The consumerism motion reached its peak when a severe grievance was levelled in opposition to the 'un-safety' of automobile businesses which induced dying to many human beings.

Problems due to Consumerism in India

- Consumerism is still in its infancy and now not properly organised. Majority of Indian clients aren't consciously aware about their rights.
- Shortage of critical commodities occur very frequently in India. Such imbalances cause hoarding and black-advertising, profiteering and corruptions.
- Many customers are ignorant and uneducated and in such situations, the marketer take advantage of the client. There are many such instances in India.
- Producers advertise their merchandise, no longer that allows you to serve the public, but on the way to get rid of their dead products at an excellent profit.
- Consumers emerge as clean victims, inside the absence of information, and buy sub-preferred and defective products.
- The court technique in India is a time eating and tiresome technique. Thus, purchasers avoid prison actions. People are blind to the simple procedures underneath the Consumer Protection Act.
- Supplier, and no longer the purchaser, will become the king within the market.
- A cycle of waste, disparities and poverty
- Poverty, land manipulate and ownership, pollutants and so on, are largely components of financial and ideological structures too. As exemplified by means of the Lawrence Summers quote above, a cost is positioned at the environment, on life, on one-of-a-kind cultures and so on.
- This is so ingrained into the cultures of the wealthy nations, that the thought of big adjustment of existence and economic structures to an extra sustainable consumption seems too much to consider. Instead the system is continued and maintained. Built into the device itself are mechanisms that inspire this, without knowing the costs.

Consequence of Consumerism: The Waste Problem

As modern-day electronics enterprise encourages stop customers to update their old client merchandise with brilliant new topics from the shelf, the natural question arises in which clients take away their vintage stuff. Used electronics consists of hazardous materials like mercury, lead, silver and flame retardants. On the other hand, many factors and components contain precious raw materials like gold, copper, titanium and even platinum, albeit via and big in very small amounts. Officially declared as used, working customer electronics to be offered on 2d markets in the Third World, most used electronics become non-functioning and vain and consequently come to be as virtual waste. While it isn't always genuinely immoral to take gain of those international locations and use them as cheap waste dumps, it additionally bears dangers for people and the environment within the affected countries.

Electronic waste consists of many poisonous substances which need to be well disposed of. This way that decomposing of the waste has to be accomplished in appropriately ready recycling centers in which the character aspect substances of the waste may be separated from each unique in a safe way such that personnel are never uncovered to toxic fumes or liquids.

Waste is a huge worldwide issue. Increasing volumes of waste are being generated because the global population and residing standards rise.

The environmental effect is tremendous, with massive volumes of waste generated yearly with best simple or little remedy to minimise its effect. People are increasingly more worried approximately the manufacturing of waste and its effect, and are looking for methods to address the trouble. They are also asking governments to take steps to minimise the environmental effect of waste.

While recycling and specific simple strategies to deal with waste had been brought, the conversion of waste lower back to its constituent components remains in its infancy. Another hassle is the fact that getting rid of the junk is hazardous. For example, mercury will leach when sure electronic devices, together with circuit breakers are destroyed.

Batteries are an environmental hazard. The acid leaches no longer simplest into the soil but moreover goes into the floor water. Disposing of them also creates their own problems because the lead is likely to remain in the ash and be launched in the air.

Solution

Zero waste is not any single technology, program, or policy. Zero waste is a goal, a method, and a vision that shifts how we reflect on consideration on and use resources: it is a whole-system approach that targets a number one change inside the manner substances float through our economic system. It is the other of a quit-of-pipe solution. Instead, 0 waste is a bold method to waste management that looks at each the both the front cease (production and design) and the again quit (reuse and reprocessing) of fabric glide and solutions to connect the

Zero waste centres round lowering needless intake, minimizing waste, maximizing recycling, and incentivizing the manufacturing of products that may be deliberately reused, repaired, or recycled returned into the marketplace.

Working toward zero waste method the opportunity to advance economic fairness and right historic wrongs. Studies show marginalized businesses are through some distance the most affected by way of manner of poisonous waste. In a take a look at through the Nuclear Regulatory Commission, they found corporations with or greater centres or one of the nation's five largest landfills, the average minority per cent of the network was extra than 3 times that of corporations without these facilities. By moving far from those harmful techniques, those groups in the long run have an equal opportunity to thrive. Our zero waste initiative stands for secure jobs for enterprise employees, the advantages of a 0-waste economy shared equitably, and a waste gadget that is healthy for all. Waste Disposal Solutions

Eco-duty – "Reduce, Re-use, Recycle"

Eco-responsibility pertains to the three R's mantra of Re-use, Reduce, and Recycle. Local groups, government and states need to put greater efforts in the direction of the training of waste control. Essentially, the slogan can help lessen the levels of unsustainable waste that prove complicated in numerous environments across the globe. With the implementation and consistent exercise of the three R's, groups and nearby authorities in addition to states will no longer only be capable of manipulate waste however also move within the course of achieving 0 waste.

More emphasis must be placed on responsible useful resource use with an goal of avoidance, maximizing recycling and waste discount methods. Avoidance and waste reduction involves strategies such as restore of damaged matters instead of buying new, buying and re-using second-hand items, and designing reusable and recyclable products.

Effective Waste Disposal and Control

An effective approach for municipal waste disposal and control can provide stepped forward answers for the diverse problems associated with waste materials. It ensures there's gradual improvement of recent and cost-powerful facilities which intention to encourage higher environmental protection requirements. An powerful control approach may also see to it that landfills are purposefully located to ease waste collection, transfer, and monitoring or recycling. This can be achieved via the implementation of waste disposal plan which ought to include proper monitoring and regulation of municipal stable and meals waste, cattle waste, sewage sludge, medical waste, and construction waste.

Polluter Pays Principle and Eco-product duty

Polluter pay principle is wherein the regulation calls for the polluter to pay for the impact prompted to the environment. When it comes to waste management, the principle will require folks who generate waste to pay for the correct disposal of non-reclaimable materials. For the effectiveness of the pay precept, it need to include charging schemes on all waste disposal components including construction waste and home waste thru public fill reception centres.

Control and Monitoring of Land Filling and Fly-Tipping Activities

Thousands of heaps of production and demolition substances are generated by way of various local construction industries. In most cases, a huge portion of those waste substances may be re-used, reclaimed or recycled. With the manage and tracking of land filling and fly-tipping activities in the vicinity of public works, creation and demolition substances can be resourcefully reclaimed, reused or recycled in different projects including landscaping, village houses, recreation centers or automobile parks, or roads. By making use of those strategies and tracking fly-tipping activities, the development and demolition materials that sometime go into landfills which in addition worsen the control of solid waste can without difficulty be managed.6.8 The Environment (Protection) Act 1986

In India, environmental protection was not much of an issue till the mid-1980. That is, environmental protection as a whole was never seriously considered. The government did enact various legislation pertaining to the environment in the 1970's and 80's; most notably the Water Act, 1974, Air Act, 1981, Indian Forest Policy, 1988. But it took two major pushes, one from the international community and one from the national community, to create the Environment Protection Act.

India experienced huge decline in the environment quality, evidenced by increasing global warming, effects on flora and fauna, disturbance in the food chains. The protection of environment become a growing concern. The Government of India participated in the United Nations Conference on the Human Environment held at Stockholm in June, 1972. Before the conference various measures was taken but after the conference it further leads to implementation of the decision taken in the conference. Thus, the Environment Protection Act, 1986 was passed and was enacted under Article-253 of Indian Constitution.

Objectives of the Environmental Protection Act

The Environmental Protection Act, 1986 (EPA) was passed with the following objects:

1. It was enacted to implement the decisions which were made at the United Nation Conference on the Human Environment held at Stockholm in June 1972.
2. Creation of authority for government protection.
3. Coordinating the activities of various regulating agencies which is done under the existing law.
4. The main task is to enact general laws for environmental protection, which could be unfolded in areas of severe environmental hazards.
5. Providing deterrent punishment to those who inculcate in endangering the human environment, safety and health.
6. The main goal for the environment should be sustainable development and it can be regarded as one of the goals for Environment Protection Act, 1986.
7. Sustainable development includes achieving the object and the purpose of the act as well as the protection of life under Article 21 of the Indian Constitution.

Some of the special features of this Act include:

1. The Act covers all forms of pollution; air, water, soil and noise.

2. It provides the safe standards for the presence of various pollutants in the environment.
3. It prohibits the use of hazardous material unless prior permission is taken from the Central Government.
4. It allows the central government to assign authorities in various jurisdictions to carry out the laws of this
5. Purpose of Environment Protection Act
6. The purpose of the *Environmental Protection Act* (*EPA*) is to support and promote the management, protection, enhancement and wise use of the environment, while recognizing the following:
7. Preventing, mitigating and remediating environmental impacts is important in making decisions and taking actions.
8. Where there are threats of serious or irreparable harm to the ecological integrity, lack of complete certainty is not to be a reason for postponing reasonable environmental protection measures.
9. All persons are responsible, financially and otherwise, for impacts on the environment as a result of their actions or inaction.
10. Administrative, management and regulatory processes need to be adaptive, responsive, fair, effective and timely.
11. Scope and commencement of the Act
12. The Environment Protection Act, 1986 extends to whole India and it came into force on 19th November.
13. Section 2 of the Environmental protection Act, 1986 (EPA) deals with some of the information about the definition of the Act and these definitions are as follows:
14. "Environment" the word environment includes water, air, land and also the inter-relation between their existences. It also includes human beings and other living creatures such as plants, micro-organisms and property.
15. "Environmental Pollutants" means any substance in solid, liquid or gaseous form which in consideration is injurious to the health of living beings.
16. "Handling" means any substance which is in the relation of being manufactured, processed, collected, used, offered for sale or like of such substance.
17. "Environmental Pollution" includes the presence of environmental pollutants in the environment.
18. "Hazardous substance" includes the substance or the preparation by which the physical-chemical property is liable to harm the human beings or other living creatures such as plants, microorganisms and the property.

19. "Occupier" is in the relation of factory or any other premises which means a person who has control over the affairs of it.
20. From the above definitions given the Environmental protection Act tends to cover a wide range of matters related to the environment protection.

6.9 The Air (Prevention and Control of Pollution) Act

India donated in the United Nations Conference on the Human Environment held in Stockholm in June 1972 to take suitable stages for the conservation of the natural incomes of the earth which, among other things, include the conservation of the excellence of air and control of air pollution. Based on the final rules of this conference, the Air Act was expressed by the government of India in 1981.

The Air Act (1981)

This is an Act to deliver for the deterrence and resistor and reduction of air pollution through numerous organisation rules and limitations. One of the major purposes of this performance is loud out the drives mentioned above by transmission a set of errands, powers, and functions to the Boards for the deterrence and control of air pollution.

The act was presented and combined into the Constitution of India in 1981 laying down guidelines for the protection of the excellence of air and control of air pollution. In difference to the water act of 1974 which covers only 13 states and union territories of India, the air act spreads to the whole of India.

The Air (Prevention and Control of Pollution) Act, 1981 is a central Act of Parliament, which delivers for the deterrence and control of air pollution and preserving the virtuous value of air. It has also provided for the founding of the Central and State Boards for the Prevention and Control of Air Pollution under sections 3 and 4.

According to the Act "Air Pollution means the presence in the atmosphere of any air pollutant", and "air pollutant" means "any solid, liquid or gaseous substance present in the atmosphere in such a concentration as may be or tend to be injurious to human beings or other living creatures or plant or property or environment."

Powers and Functions of Boards

Functions

Section-16 of this act suggests the capacities of the Central Board and comprises giving exhortation to the Central Government, direction and specialized offer assistance to the State Sheets, planning the exercises of the state, preparing, mindfulness and dispersal of data and arranging a across the nation program for the avoidance and control of discuss contamination. The guidelines for discuss quality have moreover to be endorsed by the Central Board. The Central Board may too set up a research facility or research facilities to help it in performing its capacities. It may moreover appoint any of its capacities to any committee shaped by it for the purpose. Section17 of the act endorses capacities of the State Sheets. It incorporates preparing, mindfulness, arranging and executing programs for discuss contamination control, prompting the state government in things relating to discuss contamination, collecting and spreading data, reviewing premises and giving reasonable bearings for discuss contamination control, surveying the encompassing discuss quality, laying down emission standards in consultation with the Central Pollution Control Board (Standards prescribed by the Central Government or the Central Pollution Control Board cannot be relaxed by the State Boards, they can only be made more stringent),advising the state government on the suitability of any location for an air polluting industry and setting up or recognizing a laboratory or laboratories to enable the state board to perform its functions.

Functions of the Central Board

The most functions of the Central Board shall be to enhance the standard of air and to stop, control or abate pollution within the country. The Board may additionally establish or recognize a laboratory or laboratories to enable the Central Board to perform its functions under this section efficiently.

- Highlight functions of the Central Pollution electrical device (CPCB) are as follows:
- Advise the Central Government on any matter concerning the development of the standard of air and therefore the prevention, control or abatement of pollution;
- Plan and cause to be executed a nation-wide programme for the prevention, control or abatement of pollution;
- Co-ordinate the activities of the State and resolve disputes among them;
- Provide technical assistance and guidance to the State Boards, perform and sponsor investigations and research regarding problems of pollution and prevention, control or abatement of pollution.
- Plan and organize the training of persons engaged or to be engaged in programmes for the prevention, control or abatement of pollution on such terms and conditions because the Central Board may specify;
- Organize through mass media a comprehensive programme regarding the prevention, control or abatement of pollution;
- Collect, compile and publish technical and statistical data regarding pollution and therefore the measures devised for its effective prevention, control or abatement and prepare manuals, codes or guides regarding prevention, control or abatement of air pollution.
- Lay down standards for the standard of air.
- Collect and disseminate information in respect of matters regarding pollution.

- Functions of the State Boards: The functions of the State Board, as laid out in Section 17, shall be:
- To plan a comprehensive programme for the prevention, control or abatement of pollution and to secure the execution thereof.
- To advise the regime on any matter concerning the prevention, control or abatement of air pollution.
- To collect and disseminate information regarding pollution.
- To collaborate with the Central Board in organising the training of persons engaged or to be engaged in programmes regarding prevention, control or abatement of pollution and to organise mass-education programmer relating thereto
- To inspect, in any respect reasonable times, any control equipment, works or manufacturing process and to provide by order, such directions to such persons because it may consider necessary to require steps for the prevention, control or abatement of air pollution.

The Objectives of the Air (Prevention and Control of Pollution) Act of 1981

The objective of this Act is to offer for the prevention, manage and abatement of air pollution, for the establishment, on the way to sporting out the aforesaid purposes, of Boards, for conferring on and assigning to such Boards powers and functions touching on thereto and for subjects linked therewith.

Decisions have been taken on the United Nations Conference at the Human Environment held in Stockholm in June 1972, wherein India participated, to take suitable steps for the preservation of the natural sources of the earth which, among different things, includes the renovation of the high-quality of air and manage of air pollution.

Therefore its miles requisite to implement the choices aforesaid insofar as they relate to the renovation of the pleasant of air and manage of air pollutants.

Conclusion: It is a citizen's responsibility to document to the local government inclusive of the Collector or the Pollution Control Board, and the press approximately offences made with the aid of a polluter so that action may be taken against the offender. It is equally essential to save you and record to the government on cutting down of trees, as this reduces nature's capacity to maintain the carbon dioxide and oxygen tiers. Preventing air pollution and maintaining the satisfactory of our air is a responsibility that each man or woman need to help so that we can breathe air that will now not ruin our health.

6.10 The Water (Prevention and Control of Pollution) Act, 1974

Water (Prevention and Control of Pollution) Act, 1974 is a comprehensive regulation that regulates organizations liable for checking on water pollutants and ambit of pollution manipulate boards each on the centre and states. The Water (Prevention and Control of Pollution) Act, 1974 become adopted with the aid of the Indian parliament with the aim of prevention and manipulate of Water Pollution in India.

The government formulated this act in 1974 to prevent the pollution of water by using industrial, agricultural and family wastewater which can contaminate our water sources. Wastewaters with high degrees of pollution that enter wetlands, rivers, lakes, wells as properly as the sea are serious fitness hazards.

The essential targets of the Water Act are to offer for prevention, manipulate and abatement of water pollutants and the upkeep or recuperation of the wholesomeness of water. It is designed to evaluate pollutants ranges and punish polluters. The Central Government and State Government have set up PCBs to screen water pollution.

The Water Act, 1974 with positive amendments in 1978 is an in depth legislation with extra than sixty sections for the prevention and manipulate of water pollutants. Among other things, the Act gives for charter of critical and State Boards for preventing water pollutants, strength to take water samples and their analysis, discharge of sewage or exchange effluents, appeals, revision, minimal and maximum penalties, booklet of names of offenders, offences with the aid of businesses and Government departments, consciousness of offences, water laboratories, analysis *etc.*

Functions of the State Board

Section 17 of the Water (Prevention and Control of Pollution) Act, 1974 in reality lists all functions of the respective kingdom boards for countering water pollution. The country board of respective states is empowered to plan a complete software for the prevention, control or abatement of pollution of streams and wells, collect and disseminate information referring to water pollutants and encourage, conduct and participate in investigations and studies relating to troubles of water pollution and prevention.

The state water forums also have the right to inspect sewage or exchange effluents, works and flora for the remedy of sewage and change effluents and to study all water purification vegetation. The Board may also establish or understand a laboratory or laboratories to permit the Board to carry out its features beneath this segment efficiently, which includes the analysis of samples of water from any circulation or properly or of samples of any sewage or alternate effluents.

Consent of the State Board Is important to Discharge Sewage

Section 25 of the Water (Prevention and Control of Pollution) Act, 1974 states that Prior Consent of the State Board beneath section 25 is important to set up any industry, plant or manner which is in all likelihood to discharge sewage or trade effluent right into a circulate or nicely or sewer or on land or convey into use any new or altered shops for the release of sewage or begin to make any new discharge of sewage. The segment similarly states that every State Board is susceptible to keep a register containing details or situations imposed underneath the section associated with any outlet, or to any effluent, from any land or premises which ought to be open to inspection through the nation board.

Power to take Emergency Measures

Section 32 of the Water (Prevention and Control of Pollution) Act, 1974 describes the strength to take emergency measures in case of pollutants of flow or properly. Under the act, State Board may trouble orders to re pass the be counted, which is, or might also purpose pollution; or remedy or mitigate the pollution, or problem prohibition orders to the involved persons from discharging any toxic or noxious or polluting count.

Section 24 and forty three of the Water (Prevention and Control of Pollution) Act, 1974 relate to prohibition on use of stream or properly for disposal of polluting be counted and penalty for contravention thereof Under the scope of the provision, no man or woman shall knowingly motive or allow any toxic, noxious or polluting mater as decided by means of the State Board to go into any flow or sewer or on land. Anyone failing to abide by using the legal guidelines of beneath is liable for imprisonment underneath Section 24 and Section 43 ranging from now not much less than one year and six months to 6 years along with economic fines.

Penalties and fines

Section forty two of the of the Water (Prevention and Control of Pollution) Act, 1974 states penalties and fines for positive acts which include flattening pillars, Obstructs any person acting underneath the orders or path of the Board, Damages any works or assets belonging to the Board and Failure to furnish any officer other worker of the Board any data required. The exceptional and penalty consists of Imprisonment for a term which may also expand up to a few months or with quality to Rs. 10,000/- or both.

Functions of Central Board (Section 16)

Advise the Central Government on any depend regarding the prevention and manage of water pollutants.

Coordinate the activities of the State Boards and resolve disputes among them.

Provide technical help and steering to the State Boards, perform and sponsor investigations and research regarding troubles of water pollution and prevention, manage or abatement of water pollutants.

Plan and organize the training of persons engaged or to be engaged in for the prevention, manipulate or abatement of water pollution on such terms and situations as the Central Board might also specify.

Organize through mass media a complete application regarding the prevention and manage of water pollution.

Conclusion

Water pollution is a big trouble in India and controlling and preventing it is another massive issue, till now we aren't able to create awareness among the people concerning the importance of keeping water bodies, this Act honestly provides numerous agencies so one can look to save you and manage water pollution, the Act lays down diverse methods for submitting a grievance and the powers of every and every board. However more wishes to be done and the Act ought to be made more comprehensive, more participation have to be given to the locals and punishments must be made stricter so that it Acts as a strong deterrence. Above all these more emphasis have to be given at the implementation factor as simply through making laws you cannot manage pollutants, proper implementation is also required.

6.11 The Wildlife Protection Act

This Act provides for the protection of the country's wild animals, birds and plant species, in order to ensure environmental and ecological security. Among other things, the Act lays down restrictions on hunting many animal species. The Act was last amended in the year 2006. An Amendment bill was introduced in the Rajya Sabha in 2013 and referred to a Standing Committee, but it was withdrawn in 2015.

The Wild Life Act provides for:

- ☆ State wildlife advisory boards,
- ☆ Regulations for hunting wild animals and birds,
- ☆ Establishment of sanctuaries and national parks,
- ☆ Regulations for trade in wild animals, animal products and trophies, and
- ☆ Judicially imposed penalties for violating the Act.
- ☆ Harming endangered species listed in Schedule I of the Act is prohibited throughout India.
- ☆ Hunting species, like those requiring special protection (Schedule II), big game (Schedule III), and small game (Schedule IV), is regulated through licensing.
- ☆ A few species classified as vermin (Schedule V), may be hunted without restrictions.
- ☆ Wildlife wardens and their staff administer the act.
- ☆ An amendment to the Act in 1982, introduced a provision permitting the capture and transportation of wild animals for the scientific management of animal population.

Wildlife Protection Amendment Act 2006

The act was amended in the year 2006 and its purpose is to strengthen the conservation of tigers and other endangered species by combating crimes against them through the special Crime Control Bureau.

Wildlife Protection Amendment Bill 2013

- The Wildlife Protection Amendment Bill, 2013, was introduced by Environment Minister Jayanthi Natarajan in the Rajya Sabha on August 5, 2013
- The bill seeks to further amend the Wild Life Protection Act, 1972
- It has a provision of imprisonment up to seven years for poaching or hunting of wild animals
- The Bill also protects the hunting rights of Scheduled Tribes in the Andaman and Nicobar Islands
- Withdrawal of the amendment bill
- In the year 2015, Prakash Javadekar asked to withdraw the amendment bill to further amend the Wild Life (Protection) Act, 1972, and announced a lot of changes which needs to be made before actually implementing it. These changes were announced in Rajya Sabha.

Protected Areas under the Wildlife Protection Act

There are five types of protected areas as provided under the Act. They are described below.

1. Sanctuaries

"Sanctuary is a place of refuge where injured, abandoned and abused wildlife is allowed to live in peace in their natural environment without any human intervention."

- They are naturally-occurring areas where endangered species are protected from poaching, hunting and predation.
- Here, animals are not bred for commercial exploitation.
- The species are protected from any sort of disturbance.
- Animals are not allowed to be captured or killed inside the sanctuaries.
- A wildlife sanctuary is declared by the State government by a Notification. Boundaries can be altered by a Resolution of the State Legislature.
- Human activities such as timber harvesting, collecting minor forest products and private ownership rights are permitted as long as they do not interfere with the animals' well-being. Limited human activity is permitted.
- They are open to the general public. But people are not allowed unescorted. There are restrictions as to who can enter and/or reside within the limits of the sanctuary. Only public servants (and his/her family), persons who own immovable property inside, *etc.* are allowed. People using the highways which pass through sanctuaries are also allowed inside.
- Boundaries of sanctuaries are not generally fixed and defined.

- ☆ Biologists and researchers are permitted inside so that they can study the area and its inhabitants.
- ☆ The Chief Wildlife Warden (who is the authority to control, manage and maintain all sanctuaries) may grant permission to persons for entry or residence in the sanctuary for the study of wildlife, scientific research, photography, the transaction of any lawful business with persons residing inside, and tourism.
- ☆ Sanctuaries can be upgraded to the status of a 'National Park'.
- ☆ Examples: Indian Wild Ass Sanctuary (Rann of Kutch, Gujarat); Vedanthangal Bird Sanctuary in Tamil Nadu (oldest bird sanctuary in India); Dandeli Wildlife Sanctuary (Karnataka).

2. National Parks

"National Parks are the areas that are set by the government to conserve the natural environment."

- ☆ A national park has more restrictions as compared to a wildlife sanctuary.
- ☆ National parks can be declared by the State government by Notification. No alteration of the boundaries of a national park shall be made except on a resolution passed by the State Legislature.
- ☆ The main objective of a national park is to protect the natural environment of the area and biodiversity conservation.
- ☆ The landscape, fauna and flora are present in their natural state in national parks.
- ☆ Their boundaries are fixed and defined.
- ☆ Here, no human activity is allowed.
- ☆ Grazing of livestock and private tenurial rights are not permitted here.
- ☆ Species mentioned in the Schedules of the Wildlife Act are not allowed to be hunted or captured.
- ☆ No person shall destroy, remove or exploit any wildlife from a National Park or destroy or damage the habitat of any wild animal or deprive any wild animal of its habitat within a national park.
- ☆ They cannot be downgraded to the status of a 'sanctuary'.
- ☆ Examples: Bandipur National Park in Karnataka; Hemis National Park in Jammu and Kashmir; Kaziranga National Park in Assam.

3. Conservation Reserves

The State government may declare an area (particularly those adjacent to sanctuaries or parks) as conservation reserves after consulting with local communities.

4. Community Reserves

The State government may declare any private or community land as a community reserve after consultation with the local community or an individual who has volunteered to conserve the wildlife.

5. Tiger Reserves

These areas are reserved for the protection and conservation of tigers in India. They are declared on the recommendations of the National Tiger Conservation Authority.

6.12 Forest Conservation Act

Forest Conservation Act launched 1980, which become amended in 1988.

The Indian Forest Act of 1927consolidated all of the previous laws regarding forests that had been passed earlier than the 1920s. The act gave the government and Forest Department the power to create Reserved Forest, and the right to apply of resources through nearby humans turned into controlled. It offers priority to preserving environmental stability and ecological balance. It expressly states that the network of Protected Areas have to be bolstered and extended.

The woodland conservation act of 1980 was enacted to manipulate deforestation. It ensured that forestlands couldn't be de-reserved without previous approval of the Central authorities.

The wooded area conservation Act of 1980, which became amended in 1988 it, is essential to apprehend its historic background. The Indian woodland Act of 1927 consolidated all of the previous laws regarding forests that had been passed earlier than the 1920s. The Act gave the government and Forest Department the energy to create Reserved Forests, and the strength to create reserved forests, and the right to use Reserved Forests for the authorities use alone.

The Act remained in force fill the Eighties while it was found out that the protective forests for timber manufacturing alone was not accceptable.

India's, those states had regularized encroachments and resettled 'challenge Affected people' from development projects consisting of dams in those de-reserved areas.

Forest Act might additionally come within the purview of the Forest Conservation Act 1980. The Supreme Court has also held that "wooded area" as understood in the dictionary feel could also be included underneath "wooded area land". The term "wooded area" shall no longer be applicable to the plantation raised on personal land except notified non-public woodland. Tree falling in such plantation would but be governed through nation acts and guidelines. The term "tree" will have the same which means as defined in section 2 of the Indian Forest Act 1927.

Objects and Reasons of the Forest Conservation Act

Deforestation causes ecological imbalance and ends in environmental deterioration. Deforestation were taking area on a massive scale inside the United States of America and it had caused sizeable concern. The act seeks to test upon deforestation and de-reservation of forests.

Restriction at the de-reservation of forests or use of woodland-land for non-forest functions:

Section 2 of the Act offers with a restriction on the de-reservation of forests or the use of forest-land for non-forest functions. It offers that whatever contained in any other law for the time being in force in a kingdom, no kingdom government or another authority shall make, except with earlier approval of the central authorities, any order directing:

1. That any reserved woodland declared below any law in the meanwhile in force in that nation or any component, shall quit to be reserved.
2. That any wooded area land or any portion thereof may be used for any non-woodland cause.
3. That any wooded area land or any element may be cleared of timber that have grown certainly in that land or element, for the motive of the use of it for re-afforestation.

The time period 'non-woodland purposes' implies the breaking up or cleaning of any woodland-land or part of forest land for-

1. The cultivation of tea, coffee, spices, rubber, palms, oil-bearing, plants, or medicinal plants,
2. Any functions apart from re-afforestation,

But does no longer encompass any work related to conservation, improvement, and management of forests and wildlife.

Penalties

For offences in reserved forested: no man or woman is authorized to make clearings or set fire to reserved forest. Cattle aren't authorized to trespass into the reserved woodland. Felling, amassing of timber, bark or leaves, quarrying, or amassing any woodland product is punishable with imprisonment for a term of six months, or which may amplify to Rs.500 or both.

Penalties for offences in protected forests: someone who commits any of the following offences like felling trees, stripping the bark or leaves of timber, setting fireplace to such forests, kindling a hearth without taking precaution to save you its spreading, dragging wooden, or permitting livestock to damage any tree, shall be punishable with imprisonment for a time period which may additionally make bigger to 6 months or with a quality which may exceed to Rs. 500, or both.

What can an individual do to assist the act?

- ☆ Acquaint yourself with the legal guidelines, detailed policies and orders issued via the government.
- ☆ Be in touch with worried local NGOs and associations. Organize one with different like-minded people if none exist on your area.
- ☆ Create focus approximately the lifestyles and price of country wide parks and sanctuaries and increase a public opinion towards illegal activities in the wooded area or disturbance to wildlife.
- ☆ Pressurize the authorities to enforce the wooded area and wild life laws and rules to protect inexperienced areas.
- ☆ Take criminal motion if essential and if viable thru a public interest litigation (PIL) against the offending party. Use the assist of NGOs who can adopt felony motion.
- ☆ Help to create public pressure to change regulations, legal guidelines and processes while vital.
- ☆ Use better, ecologically touchy public shipping and bicycle tracks, do now not muddle in a woodland area.
- ☆ Participate in maintenance of greenery, by means of planting, watering being concerned for plants.

6.13 Issues Involved in Enforcement of Environmental Legislation

In India, greater than three-quarters its population depends at once for their livelihoods on sports primarily based on herbal sources and the the rest of the populace relies on these assets not directly for food, fuel, industrial output and recreation. Their financial well-being is inextricably tied to the productivity of herbal resources and first-class of surroundings. Sadly, most of the herbal resources including environment in India are in a serious state of degradation. For example, agricultural lands be afflicted by soil erosion, water-logging, salinity and fashionable loss of fertility, making them much less productive, and water for drinking and irrigation is an increasing number of getting scarce and polluted. Fishery yields are falling, and air best is deteriorating. Increasing levels of air, water and land pollutants pose a extreme risk to human health and longevity. Good control of herbal sources and surroundings is vital to obtain and sustain economic increase and development. It isn't always, as is from time to time mistakenly asserted, just a luxurious for wealthy countries concerned with aesthetics.

The capacity hole between the formal law and its enforcement is visible in lots of fields of regulation, however it raises unique challenges within the discipline of environmental protection. In areas of regulation together with competition, social security, or consumer protection there are honestly described sufferers with felony interests who can and will make sure that the law is enforced. In contrast, the environment is often unowned in legal terms - with the consequence that the surroundings dies in silence, it's been said. The obligation for its criminal safety lies largely on public government - the police, nearby authorities, or specialized regulatory agencies - often underneath competing policy priorities and severe aid constraints.

Yet, as this Thematic Issue demonstrates, in latest years some distance greater attention is being paid to the question of enforcement of environmental law - how it have to maximum correctly be implemented, how pleasant to ensure compliance, and how first-class to cope with breaches of environmental regulation in which they occur.

These troubles can raise sensitive political issues at both countrywide and regional stages. Deciding how to appoint assets and reply to breaches of environmental law regularly entails massive discretion amongst enforcement authorities, and countrywide and local administrations have their personal traditions and culture in which they operate. Imposing over-elaborate, pinnacle-down solutions might also consequently be inappropriate. Within the European Union, environmental regulation has commonly left the query of enforcement to the discretion of Member States, and it is rare for EU Regulations or Directives to specify the kind of sanction that ought to be employed. The Court of Justice of the European Union has been equally reticent to trespass at the discretion of countrywide authorities in this context, and in reality relied upon the general precept that any sanctions employed need to be effective, proportionate, and dissuasive.

Three troubles which are especially important for environmental law are:

1. The precautionary precept This principle has developed to deal with risks and uncertainties faced by using environmental control. The precept implies that of prevention is worth a pound of cure– it does not prevent troubles but may additionally lessen their incidence and helps make certain contingency plans are made.
2. The polluter-will pay principle In addition to-the obvious–the polluter will pay for the damaged caused by a development–this principle also implies that a polluter can pay for monitoring and policing. A hassle with this approach is that fines may additionally bankrupt small businesses, yet be low sufficient for a massive business enterprise to write them off as an occasional overhead, which little for pollution manipulates.
3. Freedom of information: Environmental making plans and control is hindered if the public, NGOs or maybe official bodies are not able to get information.

Drawbacks of Natural World Protection Act, 1972

- ✰ Since this act has been enacted simply as a fallout of Stockholm convention held in 1972, it has not blanketed any regionally advanced conservation measures.

- The ownership certificate for a few animal articles (Ex: Leopard and Tiger skins) frequently function a device for unlawful buying and selling.
- Jammu and Kashmir have their own flora and fauna acts, consequently, searching and trading of many endangered species, prohibited in different states are allowed in Jammu and Kashmir.
- The offenders of this act are not problem to any harsh fines. The exceptional is handiest Rs. 25000 or imprisonment for up-to a few years.

Drawbacks of the Forest (Conservation) Act, 1980

- This act best transfers the powers from state to centre to determine the conversion of reserve forests to non-forest regions.
- The power has been centralized at the pinnacle and nearby groups were completely ignored from the decision making process regarding the nature of forest areas.
- Tribal humans living in forests are absolutely dependent on forest resources. If they are stopped from exploiting forests for his or her livelihood, they hotel to criminal sports like smuggling, killing, *etc.*
- This law is focused on shielding trees, birds and animals however no longer on defensive poor humans
- The forest dwelling tribal communities have a rich expertise approximately forest assets, their importance and conservation. However, their role and contribution isn't always acknowledged.

Drawbacks of Pollutants Associated Acts

- The strength and authority has been given handiest to the central authorities with little energy to the kingdom government. This hinders powerful implementation of the act within the states.
- The consequences imposed through this act are very small when as compared to the harm as a result of big industries due to pollution.
- A person cannot directly report a petition in the court.
- Litigation, associated with the environment is high priced, because it involves technical understanding.
- For small industries, it's far very highly-priced to install a person custom-made effluent remedy plant
- The role of chairman of the board of most industries is occupied by a political appointee. Hence it becomes difficult to implement the act without political interference.
- Litigation is outwardly an expensive affair. Environmental litigation is greater costly than different forms of disputes, since it involves professional testimony, technical evidences.

Conclusion

Although several laws relations to environment have been enacted from time to time (including country wide Green tribunal Act, 2010), but the overall influences of those Environmental legislations isn't encouraging due to numerous reasons widely recognized to the pollution control authorities, polluting industrial establishment,N G O,s and those in well-known.

Environmental legal guidelines have little threat of acquiring real targets unless observed by a composite set of promotional measures, starting from direct financial subsidies to value sharing, *e.g.* In installation of ETP's by way of Polluting Industries. However, the maximum extensive step will be to create environmental cognizance among humans as a whole regarding their obligations and rights in regards to the protection of the Environment

6.14 Public Awareness

Since our environment is getting degraded due to human sports, we want to do something about it to sustain the excellent. We frequently experience that authorities should take proper measuring steps. But all people are equally responsible to defend our environment. Hence, public attention needs to be created. Both print media and electronic media can strongly influence public opinion. Politicians have to reply undoubtedly to a strong publicly supported activity. NGOs can take active role in growing consciousness from grass root stages to the top-most policy decision makers.

Environment is an integration of both living and non-residing organisms. Water, air, soil, minerals, wild life, grass lands, forests, oceans, agriculture are all life helping systems. Since these herbal assets are limited, and human sports are the causative factors for environmental degradation, each certainly one of us want to sense accountable to defend the environment.

It is crucial to make the public aware of the formidable consequences of the Environmental Degradation. If these aren't addresses and reformative measures undertaken, the extinction of existence is inevitable. We are facing numerous environmental demanding situations as mentioned below:

Growing Population

A population of over heaps of tens of millions is growing. Over 17 million humans are added every 12 months. It puts significant strain on its natural resources and reduces the gains of development. Hence, the greatest project earlier than us is to restrict the populace growth. Although populace control does robotically lead to development, yet the development leads to a lower in population growth rates.

Poverty

India has frequently been described a wealthy land with negative humans. Poverty and environmental degradation have a nexus among them. The good sized majority of our human beings are immediately depending on the natural assets of the us of a for his or her basic needs of food, fuel, refuge and fodder.

Agricultural Growth

People should be acquainted with the methods to sustain and growth agricultural growth with out unfavourable the surroundings. High yielding varieties have reason soil salinity and harm to bodily shape of soil.

Need for Groundwater

It is important to rationalize the usage of groundwater. Factors like community wastes, commercial effluents and chemical fertilizers and pesticides have polluted our surface water and affected exceptional of the groundwater. It is important to repair the water fine of our rivers and different water bodies as lakes. Deciding on suitable technologies for restoring the high-quality of groundwater aquifers is essential.

Development and Forests

Forests serve catchments for the rivers. The growing call for of water, led to using rivers for irrigation projects. These precipitated forests to submerge and displacement of neighbourhood human beings apart from destructive the neighbourhood plant life and fauna.

Forests in India had been shrinking for numerous centuries owing to pressures of agriculture and other uses. Vast areas that were as soon as green, stand nowadays as wastelands. These areas are to be brought back under vegetative cover. The tribal communities inhabiting forests respects the timber and birds and animal that offers them sustenance. We have to understand the position of these human beings in restoring and keeping forests.

The present day information and skills of the wooded area branch need to be incorporated with the conventional know-how and revel in of the neighbourhood groups. The strategies for the joint control of forests should be evolved in a well-planned way.

Degradation of Land

A small portion of land possesses ability for production. Agricultural land suffers from varying stages of soil degradation. Land degradation mainly occurs due to overgrazing and soil erosion because of wind and water.

Reduction of Genetic Diversity

Proper measures to conserve genetic diversity want to be taken. At present most wild genetic stocks were disappearing from nature. Wild animals consisting of the Asiatic Lion are facing hassle of lack of genetic diversity. The protected areas network like sanctuaries, national parks, biosphere reserves are keeping apart populations. So, they're lowering chances of one organization breeding with another. Remedial steps are to be taken to test reducing genetic diversity.

Evil Consequences of Urbanisation

Nearly 27 according to cent Indians stay in urban areas. Urbanisation and industrialisation has given start to a top notch variety of environmental trouble that need pressing attention. Over 30 percentage of urban Indians live in slums. Out of India's 3,245 cities and towns, best 21 have partial or complete sewerage and treatment facilities. Hence, handling fast urbanization is a major mission.

Air and Water Pollution

Majority of our industrial flowers are the use of old and population technologies and makeshift facilities without any provision of treating their wastes. A high-quality range of cities and commercial areas that had been diagnosed as the worst in phrases of air and water pollution. Acts are enforced within the use of a, however their implementation isn't always easy. The cause is their implementation needs high-quality resources, technical expertise, political and social will. Again the people are to be made aware about these rules. Their support is critical to put into effect these rules.

Examples of Environmental Issues that Need Fixing

Here are numerous purpose-and-effect troubles that harm our surroundings:

Oil Drilling

This difficulty is one that causes a first-rate deal of environmental destruction. Our dependence on fossil gas is a international addiction that influences everything of the global. Oil spills and offshore drilling poison marine existence, oil drilling (on land) suffocates the earth, and the combustion of fossil fuels add to the increased atmospheric CO, which in turns reasons the development of global warming and ocean acidification. This is a multifaceted difficulty and is a great motive to get worried with as it covers one of these broad spectrum of problems.

Deforestation

Millions of acres of forest are cut down for industrial benefit, such as large scale farming, oil mining, and the manufacturing of paper goods. Deforestation reasons flora and fauna and biodiversity extinction. The International Union for Conservation of Nature (IUCN) has a Red List of environmentally threatened species with up-to-date information. Oftentimes, the reason for their threatened lifestyles is indexed as loss of habitat as it is for lots Amazonian species.

Production of Plastic Goods

Currently our society creates a incredible deal of waste and lots of that waste is comprised of plastic. According to the Environmental Protection Agency (EPA) in 2010 on my own 31 million tons of plastic waste changed into created. This waste finally ends up all around the globe in each land and water, an amazing instance is the Great Pacific Garbage Patch. Not handiest is plastic waste an difficulty, but the manufacturing of plastic is likewise depending on fossil gasoline combustion. According to the U.S. Energy Information Administration (EIA) in 2010 191 million barrels of liquid petroleum gases (LPG) and herbal gasoline liquids (NGL) have been used in the U.S. by myself to supply plastic goods.

Actions

During the next 3 years, the Government will continue to increase its Public Outreach and Education Programme by running with NGOs and the National Environmental Education Committee. In particular, Government will guide efforts to create stakeholder associations and promote alternative monetary activities which might be environmentally sound amongst charcoal burners and other similar groups. The NRCA will also continue to work to see environmental issues included in the school's curriculum.

The NRCA has prepared a Directory of Environmental Education Activities in both the formal and casual sectors, country wide co-ordination of all environmental training sports had been reinforced and might be carried forward in the direction of the development of a National Environmental Education Programme.

The NRCA will increase and implement a National Environmental Education Policy and act as the focus for a National Environmental Education Network.

Government through the NRCA will keep to preserve local linkages thru the Caribbean Conservation Association and the UNESCO-UNEP Caribbean Environmental Education Programme, with a view to maximise the usage of available sources and analyze from as well as contribute to the enjoy of others on this subject. It may even observe international environmental treaty responsibilities and use Public Education and Awareness programmes to obtain compliance.

The authorities plays an essential function in growing public focus. The 'Environmental Education, Awareness and Training' programme of the Government of India objectives at growing public expertise and abilities to shield the surroundings.

As part of the programme, during the year 2007-2008, 83372 eco-clubs had been created in India. Between 2007 and 2008, 9938 associations have been included within the recognition programme and monetary assistance turned into provided. National TV and radio had been used as a way of spreading attention and 'Vatavaran', an environmental and wildlife festival became held in 2007.

How to Promote Public Awareness

When learning about the environment's declining health it is straightforward to experience discouraged, however what keeps us preventing for a healthful global is the future of our children. They have to now not ought to inherit our environmental issues and so as to hold their destiny bright, spreading focus is imperative.

Make a Difference

After you've got end up environmentally aware you could start coaching the ones around you. The symposium acts a catalyst to be able to start your position as an environmental steward and offers online guides and different sources to hold you on track.A good route of movement that ensures your persisted participation as a international citizen is to pick out an environmental issue that moves you as maximum urgent. The amount of environmental trouble seems endless and while they're all vital it is great to to begin with attention on one cause.

Share Your Knowledge

Once you have got chosen your motive and figured out some beneficial activities, percentage it with the world! Get your community, friends, and family worried. Use your political voice in your local city council and advise environmentally sustainable and useful projects. Start a community garden, buy from neighbourhood shops and farmers markets, make your public transportation more efficient and/or start enforcing clean strength strategies in your network centres.

Public Awareness thru Adult Education

The intention of grownup schooling need to be to create a collective action in solving the environmental troubles. The new styles of complicated environmental troubles require knowledge and for this reason a new approach to adult training. A new approach have to be relevant in regards to being able to fulfil set up environmental objectives. For effective grownup education, functionaries should construct rapport with the aid organizations.

Establishment of Eco-Clubs

National Green Corps, a movement of green children, has been spreading attention approximately environmental safety amongst children. Under this programme, more than 70,000 eco-clubs are setup all over India covering about 150 colleges district wise. Young students not handiest learn about environment but take part in subject activities in and around their colleges.

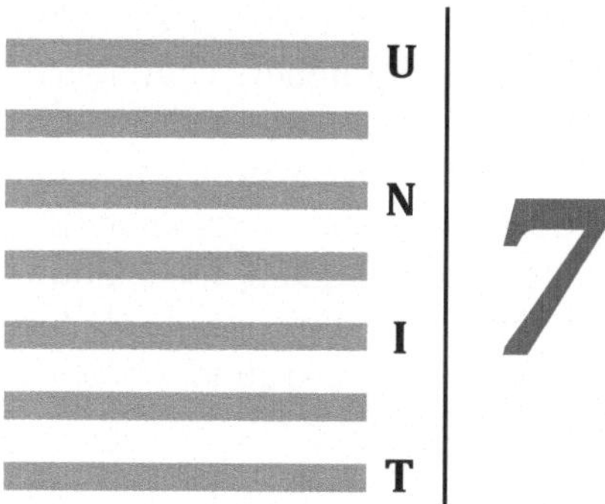

Human Population and the Environment

7.1 Population Growth, Variation among Nations

Population increase, inequality and monetary improvement are amongst the most pressing social troubles confronting us today. This research argues that those national problems are embedded inside the context of increasingly complicated multi-dimensional global networks, commonly called globalization.

Our global human populace, 6 billion at present, will cross the 7 billion mark by using 2015. The desires of this massive wide variety of human beings cannot be supported through the Earth's herbal resources, without degrading the first-rate of human life. In the close to future, fossil fuel from oil fields will run dry. It may be impossible to satisfy the needs for food from current agro systems. Pastures will be overgrazed by way of home animals and industrial growth will create ever-greater troubles because of pollutants of soil, water and air. Seas will now not have enough fish. Larger ozone holes will develop due to the discharge of business chemical substances into the atmosphere, with a view to affect human health. Global warming because of commercial gases will lead to a rise in sea ranges and flood all low-lying areas, submerging coastal agriculture as properly as towns and cities. Water 'famines' due to the depletion of sparkling water, will create unrest and sooner or later make countries go to war. The manage over nearby biological diversity, which is crucial for generating new medicinal and business products, will cause grave economic conflicts among biotechnologically superior international locations and the bio-rich countries. Degradation of ecosystems will result in extinction of lots of species, de-stabilizing natural ecosystems of wonderful value.

After centuries of very gradual and uneven boom, the world populace reached one billion in 1800. The cutting-edge expansion of human numbers commenced then, growing at a gradual however steadier pace over the subsequent 150 years to 2.five billion in 1950. During the second half of the 20th century, however, boom rates accelerated to traditionally unprecedented tiers. As a result, world populace greater than doubled to 6.five billion in 2005 (United Nations 1962, 1973, 2007). This populace growth is expected to preserve for several greater decades before peaking near 10 billion later inside the twenty-first century. Around 2070, the world's populace will be 100 instances larger than in 1800.

Population Boom

The fee of population growth is the charge of herbal boom mixed with the effects of migration. Thus a high charge of herbal growth can be offset by means of a big net out-migration, and a low rate of herbal boom can be countered by a high degree of net in-migration. Generally speaking, however, those migration results on population growth fees are far smaller than the effects of modifications in fertility and mortality.

Population "Momentum"

A critical and regularly misunderstood function of human populations is the tendency of a tremendously fertile population that has been growing swiftly in size to hold to do so for decades after the onset of even a tremendous decline in fertility. This consequences from the younger age shape of any such populace, as mentioned below. These populations contain large numbers of children who have still to develop into adulthood and the years of reproduction. Thus even a dramatic decline in fertility, which affects best the numbers at age 0, cannot prevent the continuing boom of the quantity of adults of childbearing age for at the least two or three decades.

At the cease of the demographic transition herbal population increase reaches zero once three conditions are met:

Fertility levels-off at the substitute level of approximately 2.1 bpw (more precisely, the internet reproduction price need to be 1). If fertility remains above substitute, population increase continues.

Mortality stops declining. In practice, this isn't possibly to happen due to the fact improvements in medical generation and healthcare as well as modifications in lifestyles, *etc.* Will probably ensure continued will increase in LE.

The age structure has adjusted to the post-transitional stages of fertility and mortality.

Global human populace growth is around 75 million annually, or 1.1 per cent in line with year. The worldwide populace has grown from 1 billion in 1800 to 7 billion in 2012. It is expected to maintain growing, even though predictions fluctuate as to while and if this increase will plateau. The "populace boom fee" is the price at which the variety of individuals in a populace will increase in a given term as a fragment of the initial population. Specifically, populace increase fee refers to the change in populace over a time period, frequently expressed as a per cent of the quantity of people inside the populace at the beginning of that period. This may be written because the formula:

$$\text{Population growth rate} = \frac{P(t_2) - P(t_1)}{P(t_1)}$$

Globally, the boom charge of the human population has been declining since 1962 and 1963, while it became 2.20 per cent in line with annum. In 2009, the anticipated annual growth charge changed into 1.1 per cent. The CIA World Fact ebook offers the arena annual birth-charge, mortality charge, and increase rate as 1.89 per cent, 0.79 per cent, and 1.096 per cent respectively. The ultimate a hundred years have seen a rapid increase in population due to medical advances and massive boom in agricultural productivity.

Each vicinity of the globe has seen discounts in increase charge in current decades, though increase fees stay above 2 per cent in a few international locations of the Middle East and Sub-Saharan Africa, and also in South Asia, Southeast Asia, and Latin America. This does not mean that the populace is declining; rather, it means the populace is growing extra slowly. However, a few countries do revel in negative population growth, mainly because of low fertility prices, high death costs and emigration.

According to the UN's 2010 revision to its populace projections, world population will height at 10.1 billion in 2100 as compared to 7 billion in 2011. However, a few experts dispute the UN's forecast and have argued that birth-charges will fall underneath substitute fees (the number of births wanted to hold a stable population) within the 2020s. According to those forecasters, population increase will be only sustained until the 2040s through rising longevity, however will peak below nine billion by means of 2050, accompanied via a long decline.

Maternal and Child Health

High populace growth – through high fertility charges – may have an adverse impact at the fitness of ladies and youngsters, especially in growing countries that lack trained health personnel, commodities, and equipment/infrastructure. In 2015, an envisioned 303 000 women died in pregnancy or childbirth globally, with most of these deaths (201 000) concentrated in SSA where fertility remains excessive (WHO, 2015). The risk of death from a pregnancy-related cause is positively related to women's lifetime fertility. Young girls, especially those below age 18, are more likely to die in being pregnant due to physical immaturity. Unintended pregnancy also influences maternal fitness outcomes and is focused among girls with unmet want for birth control. In growing international locations, women with unmet want for circle of relatives making plans account for 81 per cent of overall unintended pregnancies (Singh *et al.*, 2014). If unmet want become fulfilled, 90 per cent of abortion deaths and 23.6 per cent of obstetric deaths consistent with year could be prevented. This would bring about a reduction of maternal deaths through 28–32ch year (Cleland *et al.*, 2012).

"Birth control" become superior as alternative to the then-elegant terms "own family limitation" and "voluntary motherhood." Family limitation cited planned attempts by using couples to give up childbearing after the favoured range of youngsters have been born. Voluntary motherhood have been coined by using feminists in the 1870s as a political critique of "involuntary motherhood" and expressing a choice for ladies' emancipation. Advocates for voluntary motherhood disapproved of birth control, arguing that women have to only have interaction in intercourse for the reason of procreation and recommended for periodic or permanent abstinence. In contrast the birth control movement advocated for contraception with a view to allow sexual sex as preferred without the risk of pregnancy. By emphasising "manage" the contraception movement argued that ladies ought to have manage over their replica and the motion had near ties to the feminist motion. Slogans which includes "manipulate over our own bodies" criticised male domination and demanded girls' liberation, a connotation this is absent from own family making plans, population manage and eugenics. Though within the Eighties birth manipulate and population manipulate firms co-operated in annoying rights to birth control and abortion, with a growing emphasis on "choice."

Long-term Consequences of Population Growth

The "population explosion" seen inside the ultimate century has led to dire predictions. In 1968, biologist Paul Ehrlich wrote, "The conflict to feed all of humanity is over. In the 1970s, loads of hundreds of thousands of people will starve to death in spite of any crash applications embarked upon now. At this late date, nothing can save you a great growth inside the world dying fee. " Although many critics view Ehrlich's view as an exaggeration, the human populace continues to develop exponentially. The legal guidelines of nature dictate that exponential increase can't preserve indefinitely.

Despite efforts to curb population growth, which include the "one-baby policy" in China (delivered in 1979 however relaxed within the early 2000s), the human populace keeps to develop. A primary concern regarding this growth is that the demand for ever-extra meals will lead to extensive shortages, as forecast through Ehrlich.

In addition to the danger of food shortages, human population growth is destructive to the surroundings in probably everlasting ways. Most scientists agree that weather change caused by the emission of the greenhouse gasoline carbon dioxide (CO_2) is a widespread effect of human activities. In a chain of treaties in the overdue 20th century, many countries dedicated to reducing their CO_2 emissions to prevent continuous worldwide warming; however these treaties have now not been ratified with the aid of each country, largely due to financial and political concerns. The role of human interest in weather exchange is hotly debated in some circles. The destiny holds huge uncertainty for curbing human population growth and defensive the environment.

Overcoming Density-Dependent Regulation

Humans have surpassed density-dependent limits on population with the aid of enacting various environmental changes to deal with our desires for hygiene, shelter, and food.

Humans are uniquely capable of consciously modify their surroundings to growth its sporting capacity. This functionality is an underlying purpose for human populace growth as human beings are able to conquer density-established limits on population growth, in comparison with all other organisms.

Human intelligence, society, and communication have enabled this capacity. For instance, humans can assemble shelters to shield them from the elements; meals supply has increased due to agriculture and domestication of animals; and human beings use language to pass on era to new generations, permitting continual improvement upon preceding accomplishments. Migration has additionally contributed to human population boom. Originating from Africa, people have migrated to nearly each inhabitable location at the planet.

Developing international locations have also made advances in curtailing mortality from infectious disorder. For example, deaths from infectious and parasitic illnesses in Brazil fell from second region as the most critical causes of death in 1977 to fifth place in 1984. The improvement is attributed in component to increased get admission to important goods and services, reflecting the country's rising prosperity. Through changes in financial status, as in Brazil, as well as international sickness control efforts, human populace growth these days is much less limited by using infectious ailment than has been the case historically.

Age Structure, Population Growth, and Economic Development

A populace's boom is strongly influenced by the proportions of people in extraordinary age brackets, which in flip is influenced by way of economic development.

The variation of populations over time, additionally referred to as population dynamics, depends on biological and environmental procedures that decide populace adjustments. A population's boom charge is strongly influenced through the proportions of people of particular ages. With expertise of this age structure, population boom may be greater as it should be predicted. Age shape information allow the charge of boom (or decline) to be associated with a population's level of monetary development.

For example, the populace of a country with rapid growth has a triangle-shaped age shape with a extra percentage of younger folks that are at or close to reproductive age. This pattern typically takes place wherein fewer people stay to antique age because of sub-optimal dwelling standards, which include happens in underdeveloped nations.

Some evolved nations, along with the United States, have a slowly-growing populace. This outcomes in a column-shaped age structure diagram with steeper sides. In these cases, the population has fewer younger and reproductive-elderly people, with a greater share of older people. Some advanced international locations, consisting of Italy, have zero population increase. Countries with declining populations, consisting of Japan, have a bulge within the centre of their age shape diagram. The bulge shows relatively-few young people, and a better proportion of middle-elderly and older individuals.

Globally, less-economically developed countries in Africa and Asia have the highest growth quotes, main to populations consisting on the whole of younger humans. Improved fitness care, beginning within the 1960s, is one of the leading reasons of the increased boom fees that created the populace explosion. For example, in the Middle East and North Africa, around 65 per cent of the population is under the age of 30. These high growth prices lead to the so-called "youngsters bulge," which some specialists consider is a purpose of social unrest and financial problems along with high unemployment.

7.2 Variation among Nations

It goes without announcing that populations are scattered throughout space. The traditional measure of populace in relation to land area that of populace density, is usually a meaningless one, since one of a kind regions vary significantly in their cost for agricultural or different human purposes. Moreover, an excessive population density in an agrarian society, based upon agriculture for its sustenance, is probable to be a severer constraint upon human welfare than might the same density in a notably industrialized society, in which the majority of country wide product isn't of agricultural origin.

The distribution of global population densities show that even as the exquisite majority of the land surface is carefully or reasonably populated, but a few limited regions are densely populated. The densely populated areas include Western Europe, the Indian subcontinent, the plains and river valleys of China, and north-jap USA. High concentrations of humans are also discovered in a few distinctly smaller regions, for example–the Nile valley of Egypt, the Islands of Java in Indonesia and the Southern part of Japan.

In phrases of continents and countries, the world's population could be very ill-balanced. More than half of the world's humans live in Asia (approximately 3.7 billion), which accounts for handiest one-fifth of the international's land vicinity; even as North, Central and South America together occupying extra than 1/4 of the land surface, have best one-fifth of the populace (1.3 billion). The African continent additionally bills for 1/4 of the land surface but has simply over one-eighth (840 million) of the global populace. On the other hand, Europe whose vicinity is most effective one twenty-5th of the overall has about one-ninth (729 million) of the world's people.

The distribution in the continents is also uneven. In Asia, China alone, with approximately 1.29 billion people, debts for one-0.33 Asian and one-fifth of the world populace. The Indian subcontinent has a further 1.three billion people–India, 1.05 billion; Pakistan, 143.5 million; Bangladesh, 133.6 million; Nepal, 23.9 million; Sri Lanka, 18.nine million; Bhutan, 0.9 million; and Maldives, 0.three million. In Europe too, the population is a flippantly distributed. For much less humans stay in Northern European countries than in other European international locations. The most populous European international locations are Russia (143.five million), Germany (82.4 million), United Kingdom (60.2 million), France (59.five million); Italy (58.1 million), Ukraine (48.2 million), Spain (41.3 million) and Poland (38.6 million). In Africa and Americas humans are for the maximum component spread very thinly throughout the land, leaving big sections including Northern Canada, Southwest USA, the Sahara wilderness and the Amazon forest almost uninhabited.

Also of significance in phrases of geographical distribution is the division among rural and concrete areas. For many decades there was a almost frequent flow of populations from rural into urban regions. While definitions of urban regions differ from country to USA and place to area, the maximum enormously urbanized societies inside the global are the ones of western and northerly Europe, Australia, New Zealand, temperate South America, and North America; in all of these the fraction of the populace dwelling in urban regions exceeds seventy five percent, and it has reached 85 per cent in West Germany. An intermediate level of urbanization exists in the nations making up plenty of tropical Latin America, in which 50 to 65 per cent of the populace lives in cities. Finally, in a lot of the growing countries of Asia and Africa the urbanization manner has only recently begun, and it is not uncommon to find much less than one-0.33 of the population dwelling in urban regions.

The rapidity of urbanization in some countries is pretty astonishing. The populace of Mexico City in 1960 become around 5,000,000; it turned into estimated to be approximately 17,000,000 in 1985 and changed into projected to reach 26,000,000 to 31,000,000 by using 2000. A rule of thumb for a good deal of the growing world is that the price of boom of urban areas is two times that of the population as a whole. Thus in a populace developing three percentage annually (doubling in about 23.1 years), it's miles probably that the urban growth charge is at the least 6 per cent annually (doubling in approximately 11.6 years).

Trends in World Population

Before thinking about contemporary population tendencies one at a time for developing and industrialized nations, it's miles useful to provide an overview of older trends. It is typically agreed that handiest 5,000,000–10,000,000 humans (*i.e.*, one one-thousandth of the prevailing international populace) were supportable earlier than the agricultural revolution of approximately 10,000 years ago. By the beginning of the Christian era, 8,000 years later, the human populace approximated 300,000,000, and there has been apparently little boom inside the resulting millennium up to the 12 months 1000 CE. Subsequent populace increase become gradual and fitful, especially given the plague epidemics and other catastrophes of the middle Ages. By 1750, conventionally the beginning of the Industrial Revolution in Britain, world population may also have been as high as 800,000,000. This method that within the 750 years from one thousand to 1750, the annual population growth fee averaged best about one-tenth of 1 percent.

The motives for such gradual boom are well known. In the absence of what's now taken into consideration basic information of sanitation and fitness (the position of bacteria in disease, for example, become unknown until the nineteenth century), mortality costs were very high, especially for babies and children. Only approximately half of new-born infants survived to the age of 5 years. Fertility changed into additionally very excessive, as it needed to be to maintain the existence of any populace under such conditions of mortality. Modest populace increase might occur for a time in these circumstances, but ordinary famines, epidemics, and wars stored lengthy-term growth close to zero.

From 1750 onward populace boom elevated. In some measure this changed into a consequence of rising requirements of residing, coupled with stepped forward delivery and communication, which mitigated the results of localized crop failures that previously could have ended in catastrophic mortality. Occasional famines did occur, however, and it was now not until the nineteenth century that a sustained decline in mortality took place, inspired by the enhancing economic conditions of the Industrial Revolution and the growing know-how of the need for sanitation and public fitness measures.

The international populace, which did now not reach its first 1,000,000,000 until about 1800, introduced some other 1,000,000,000 persons via 1930. (To expect further discussion below, the 1/3 become introduced by using 1960, the fourth through 1974, and the 5th earlier than 1990.) The maximum rapid boom in the nineteenth century came about in Europe and North America, which skilled gradual but subsequently dramatic declines in mortality. Meanwhile, mortality and fertility remained high in Asia, Africa, and Latin America.

Beginning inside the Thirties and accelerating unexpectedly after World War II, mortality went into decline in a whole lot of Asia and Latin America, giving rise to a brand new spurt of populace growth that reached charges a long way higher than any formerly skilled in Europe. The rapidity of this increase, which a few described as the "populace explosion," changed into because of the sharpness in the falls in mortality that in turn have been the result of enhancements in public health, sanitation, and vitamins that have been basically imported from the evolved nations. The outside origins and the speed of the declines in mortality intended that there has been little danger that they could be followed via the onset of a decline in fertility. In addition, the marriage styles of Asia and Latin America were (and maintain to be) pretty different from the ones in Europe; marriage in Asia and Latin

America is early and almost conventional, while that during Europe is commonly past due and widespread probabilities of humans never marry.

These high boom fees befell in populations already of very huge size, that means that global populace boom have become very rapid both in absolute and in relative terms. The peak charge of increase changed into reached in the early 1960s, when every yr the world population grew with the aid of approximately 2 percent, or approximately 68,000,000 human beings. Since that time both mortality and fertility charges have decreased, and the annual boom rate has fallen reasonably, to about 1.7 percent. But even this lower price, as it applies to a larger populace base, way that the wide variety of humans brought every yr has risen from about 68,000,000 to 80,000,000.

Population Projections

Demographic exchange is inherently a long-time period phenomenon. Unlike populations of insects, human populations have not often been situation to "explosion" or "collapse" in numbers. Moreover, the effective lengthy-term momentum that is constructed into the human age shape approach that the effects of fertility changes emerge as apparent only inside the far destiny. For those and different reasons, it's far through now conventional exercise to employ the technology of populace projection as a means of better knowledge the consequences of tendencies.

Population projections represent certainly the playing out into the destiny of a hard and fast of assumptions about destiny fertility, mortality, and migration charges. It cannot be said too strongly that such projections aren't predictions, though they're misinterpreted as such regularly enough. A projection is a "what-if" exercise primarily based on explicit assumptions that may or won't themselves be correct. As long as the arithmetic of a projection is carried out correctly, its software is determined by using the plausibility of its critical assumptions. If the assumptions encompass doable destiny tendencies, then the projection's outputs can be doable and useful. If the assumptions are implausible, then so is the projection. Because the course of demographic traits is tough to count on very far into the destiny, maximum demographers calculate a hard and fast of alternative projections that, taken together, are predicted to define a variety of achievable futures, in place of to are expecting or forecast any single future. Because demographic trends sometimes trade in sudden ways, it is critical that every one demographic projections be updated on a everyday foundation to incorporate new trends and newly advanced data.

7.3 Population Explosion

The speedy boom of populace is possibly the maximum obvious element affecting the prevailing and destiny countrywide and regional development, however it's miles by no means the most effective population problem inside the international today. The main hassle is that of 'Population Explosion'. A populace explosion is a sudden boom in the range of people in a selected species. The term is generally used in connection with the world's human population, in particular the population increase that accompanied the give up of World War II due to a combination of an extended birth fee, lower in little one mortality, and an increased existence expectancy

Effects of Population Explosion

Following are the main effects of population explosion:

1. Difficulty in Investment Condition

Indian population is growing at a rate of 1.8 per cent per annum. In order to achieve a given rate of increase in per capita income, larger investment is needed. This adversely affects the growth rate of the economy. In India, annual growth rate of population is 1.8 per cent and capital output ratio is 4:1. It means that in order to stabilize the existing economic growth rate (4 X 1.8) = 7.2 per cent of national income must be invested.

2. Problem of Capital Development

Composition of population in India hampers the increase in capital formation. High birth rate and low expectancy of life means large number of dependents in the total population. In India 35 per cent of population is composed of persons less than 14 years of age. Most of these people depend on others for subsistence. They are unproductive consumers. The burden of dependents reduces the capacity of the people to save. So the rate of capital formation falls.

3. Consequence on per Capita Revenue

Large size of population in India and its rapid rate of growth results into low per capita availability of capital. From 1950-51 to 1980-81. India's national income grew at an average annual rate of 3.6 per cent per annum. But per capita income had risen around one percent. It is due the fact that population growth has increased by 2.5 percent.

4. Consequence on Food Problem

Rapid rate of growth of population has been the root cause of food problem.

5. Problem of Redundancy

Large size of population results in large army of labour force. But due to shortage of capital resources it becomes difficult to provide gainful employment to the entire working population. Disguised unemployment in rural areas and open unemployment in urban areas are the normal features of an under developed country like India.

6. Stumpy Standard of Existing

Rapid growth of population accounts for low standard of living in India. Even the bare necessities of life are not available adequately. The Human Population requires 121 lakh tonnes of food grains, 1.9 lakh metres of cloth and 2.6 lakh houses and 52 lakh additional jobs.

7. Scarcity

Rising population increases poverty in India. People have to spend a large portion of their resources for bringing up of their wards. It results into less saving and low rate of capital formation. Hence improvement in production technique becomes impossible. It means low productivity of labour.

8. Weight of Barren Customers

In India, a large number of children are dependent. Old persons above the age of 60 and many more in the age group of 15-59 do not find employment. In 2001, working population was 39.2 per cent while 60.8 per cent are unproductive workers. This high degree of dependency is due to high rate of dependent children. This dependency adversely affects effective saving.

9. Influence on Maternity Well-being

In India, population explosion is the result of high birth rate. High birth rate reduces health and welfare of women. Frequent pregnancy without having a gap is hazardous to the health of the mother and the child. This leads to high death rate among women in the reproductive age due to early marriage. Hence to improve the welfare and status of women in our society, we have to reduce the birth rate.

10. More Weight on Land

Rising rate of population growth exerts pressure on land. On the one hand, per capita availability of land goes on diminishing and on the other, the problem of sub-division and fragmentation of holdings goes on increasing. It adversely affects the economic development of the country.

Available Measures to Control Population

The government of India has been organizing several packages for restricting the population increase and has been spending thousands and thousands of bucks on controlling the beginning fee. Some of the applications have been successful, and the charge of boom has also reduced, but has nevertheless to attain the sustainable fee. The important factors affecting the populace boom of India are the rapidly increasing birth price and reducing death rates. We can comply with strict delivery manipulate measures like China to lower the birth price, however we can't go and decrease our technological advancements to decrease the dying charge. Thus, our important emphasis falls on reducing the beginning fee. Several authorities-funded companies like the Family Planning Association of India spend hundreds of hundreds of bucks on selling family making plans. These agencies intention to promote circle of relatives making plans as a fundamental human right and the norm of a two-baby own family on a voluntary basis, to gain a stability between the population length and assets, to put together younger humans for responsible attitudes in human sexuality, and to offer education and offerings to all. The circle of relatives making plans methods furnished by way of the circle of relatives planning application are vasectomy, tubectomy, IUD, conventional contraceptives (this is condoms, diaphragms, jelly/cream tubes, foam tables) and oral pills. In addition, precipitated abortion is to be had, free of charge, in institutions identified by means of the government for this purpose. However, the success of the own family planning software in India depends on numerous elements like literacy, faith and the place in which the couple live.

Problems with Implementing Measures to Control Population

The fulfilment of family planning greatly relies upon on girls and their status. Thus, it's miles imperative for the girls to be educated so that they can determine at the quantity of children they need and be privy to the to be had beginning manage measures. Also, in a country like India, it's far crucial for the women to have identical rights inside the decision approximately the variety of children to be produced. Women also want to get educated approximately the influences of getting so many children on their personal bodies and the impacts on their children. In addition, not handiest the younger ones, however also the older ladies want to be educated so that they can teach the right family making plans to their own daughters. However, in India, almost no importance is given to the education of ladies because of the monetary situations in some households and the non-secular and social norms in some. In such a case, educating girls approximately circle of relatives planning will become an even extra difficult task.

India, being a diverse country, has human beings from all religions and of different family backgrounds. As referred to earlier, in Islam, youngsters are taken into consideration to be a present of God, and as a result Muslims do no longer accept as true with in delivery manage measures. Also, a few of the Catholics, abortion is taken into consideration a sin. Thus, the people from such sects of the society do now not follow own family making plans.

In India, maximum of the populace live inside the rural areas. However, own family making plans isn't always widely marketed in rural areas. Also, in rural regions, social and spiritual norms are greater strictly followed. As a result, all of the above-noted problems are even more intense in rural regions with the addition of the lack of circle of relatives making plans facilities in those regions. In fact, from my own experience, own family making plans is taken into consideration a sin in maximum of the tribal and rural communities.

7.4 Family Welfare Programme

In the 12 months 1952, India launched a nation-wide own family planning programme making it the first country in the international to do so. Unfortunately, family making plans in India is associated with numerous misconceptions–one in all them is its sturdy association within the minds of human beings with sterilization, while others equate it with delivery manage. The popularity of its 'welfare idea' came simplest whilst the family planning programme was named as 'Family Welfare Programme' inside the year 1977. The idea of welfare could be very complete and is basically associated to exceptional of existence. The Family Welfare Programme targets at achieving a better end- that is, to improve the pleasant of lifestyles of the people.

Women's health is crucial during all stages of their lifestyles. When she offers beginning. She passes on the present of appropriate fitness to the next technology. A healthful child grows up into a healthful adolescent, true health at some stage in adolescent period results in desirable health for the duration of reproductive years, the cycle Continues into the next technology while a healthful pregnant is capable of deliver beginning to a wholesome toddler. RCH Program will cross a long manner closer to improving the overall health of girls and that of society as a whole.

In order to obtain reproductive health girls empowerment is essential. It is additionally necessary that required modifications be made in the social factors that hinder girls overall development and their rights. Reproductive rights means ladies' proper to determine whether, while and how to have children – regardless of nationality, class, ethnicity, race, age, religion, disability, sexuality or marital status in the social, economic and political conditions that make such decisions possible." The current technique in favour today is one in every of involvement and integration. The idea is to fee folks who stand to enjoy the programme and integrate the numerous attempts to propagate the identical. Family Welfare Programme with such an approach can lessen the populace boom to extra workable levels. Presently, the Family Welfare Programme seeks to promote on a voluntary basis, responsible and Planned Parenthood with one infant norm, male or female, through independent preference of circle of relatives welfare methods nice suitable to acceptors.

Objectives

The primary goal of the Family welfare programme is to stabilize the populace and to provide qualitative fitness services consisting of immunization to both-pregnant mother and kids. Since the last 40 years, the Family welfare programme is being applied in the State with economic help of the vital Govt.

The family planning programme isn't merely intended for populace stabilization now however it has a lot large angle in holistic approach toward Reproductive and toddler fitness. Services, circle of relatives planning, Immunization, training *etc.* include a comprehensive fitness care package. The Family welfare sports are guided as in step with the policies, tips and investment by way of Govt. Of India.

To ensure qualitative healthful transport and population manipulate the Reproductive and baby Health programme has been implemented in each country with community wishes assessment approach. To boost the health care transport, every State Govt. Has formulated fitness reforms and carried out the equal to give higher autonomy to the health institutions.

Problems of own Family Welfare Programme

The two primary hassle of Family Welfare Programme are:

1. Generally women are the fundamental goals of circle of relatives planning programmes.in keeping with National Family Health Survey, the maximum broadly used approach of circle of relatives planning in India is female sterilization. This shows that circle of relatives making plans has in large part remained a ladies-focused programme. Due to reluctance of fellows to apply everlasting methods, women are pressured to accept family making plans methods. Gender specificity or gender subordination must be eliminated within the method within the own family making plans programmes as a ways as possible.
2. The imbalance inside the intercourse ratio (female/one thousand male) throughout the nation, that is 933, is another worrisome factor. In states like Haryana (SR=861), Punjab (SR=874), U.P. (S.R=898), Delhi (SR=821), Sikkim (SR=875) and others, the girl-toddler is being discriminated against even before delivery. The example of girl infanticide in these and other states has brought down the sex ratio to an all-time low. Though there may be a law banning the willpower of the sex of the infant in the womb, unscrupulous scientific practitioners and short-sighted parents connive to prevent the birth of lady kids. There is, accordingly, an urgent want to save you the misuse of technology via education and awareness.

7.5 Demographic Patterns and Impact on Agriculture

In 2015 the world population is extra than 7.3 billion human beings. That's more than seven billion 3 hundred million our bodies that want to be fed, clothed, saved warm and ideally, nurtured and educated. More than 7.3 billion people who, whilst busy ingesting resources, also are producing widespread portions of waste, and our numbers hold to grow. The United Nations estimates that the world populace will reach 9.2 billion by means of 2050.

For maximum of our lifestyles the human populace has grown very slowly, stored in check via disease, weather fluctuations and other social factors. It took till 1804 for us to attain 1 billion humans. Since then, continuing enhancements in nutrition, medicine and generation have visible our population boom rapidly.

Population Length

It's no marvel that because the world population keeps to grow, the boundaries of critical global sources such as potable water, fertile land, forests and fisheries have become more obvious. You ought not to be a maths whizz to training session that, on the whole, extra human beings use greater resources and create extra waste. The maximum essential demographic parameter is the quantity of people within a population (Lebreton *et al.*, 1992). Population length is described because the range of people present in a subjectively designated geographic range. Despite the simplicity in its concept, locating all people for the duration of a census (a complete count of every individual) is almost impossible, so ecologists generally estimate populace length by counting individuals within a small sample location and extrapolating that sample to the larger populace. Regardless of the challenges in measuring population size, it's miles a crucial feature of a population with giant implications for the dynamics of the population as a whole (Lebreton *et al.,* 1992).

Population Distribution

The ways wherein populations are spread across Earth has an impact at the environment. Developing nations tend to have better birth charges because of poverty and decrease get entry to own family making plans and education, even as evolved international locations have lower birth rates. In 2015, 80 consistent with cent of the world's populace live in less-advanced nations. These faster-developing populations can add pressure to nearby environments.

Globally, in almost every country, people also are becoming extra urbanised. In 1960 much less than one 1/3 of the world's population lived in cities. By 2014, that figure became 54 consistent with cent, with a projected upward thrust to sixty six per cent through 2050.

Population Composition

Life expectancy has expanded via about 20 years given that 1960. While this is a triumph for mankind, and certainly a good aspect for the individual, from the planet's point of view it's miles just another frame that is persevering with to eat resources and bring waste for round 40 in line with cent longer than inside the past.

Ageing populations are another detail to the multi-faceted implications of demographic populace change, and pose challenges in their own. For example among 1970 and 2006, Japan's share of human beings over 65 grew from 7 per cent to more than 20 in step with cent of its populace. This has big implications on the workforce, as well as authorities spending on pensions and fitness care.

Impact of Population on Agriculture

Some theories hypothesized that increases in RPD will lead to more use of labour intensive inputs, inflicting a shift faraway from long-fallow periods in the direction of annual and multi-cropping practices.1 Driven in large part by means of adjustments in costs of land and labour, Boserup's hypothesis shows that population strain will growth demands for current inputs, ultimately increasing productiveness consistent with unit of land. Additional theories expand this argument to the cost of labour, consisting of Hayami and Ruttan (1970). Their theory utilizes the underlying concept that expenses circuitously inspire behavioural changes with a view to adapt and survive beneath converting conditions. Often known as the "caused innovation hypothesis", Hayami and Ruttan's theory indicates that the nice affect population has on agricultural productivity occurs due to the fact the rate of labour declines relative to the charge of land. This generates demand for labour and encourages elevated use of labour-in depth inputs, which include inorganic fertilizer and modern types of seed, which should in the end increase output per unit of land.

In order to examine the effect of growing populace on agricultural productivity and intensification and productivity, we first estimate how populace affects landholding and output fees directly. Subsequently, we estimate the effect of populace density on intensification, measured as fertilizer use per hectare, maize yield, and farm earnings in line with hectare in a machine of equations. We take a look at populace's direct effect on these measures of intensification and productivity, along with population's joint indirect effect on intensification as measured via its impact on landholding, wage charges and staple crop charges.

7.6 Environment and Human Health

The natural environment is the skinny layer of life and existence supports, referred to as the biosphere that carries the earth's air, soil, water, and living organisms. The connection among protective the herbal surroundings and safeguarding human health has been identified for some time. In current decades the point of interest of research and regulation has been identifying and regulating environmental toxics to reduce dangerous human exposures. The effect of numerous environmental exposures, including toxic chemical substances, air pollution, and biological retailers on the human body, is commonly perceived as the relevant problem in environmental health. However, keeping a healthful surroundings extends past controlling those hazards.

The interconnectedness of life on this planet approach that nothing we put into the ecosystem certainly disappears. In different words, what we do to the planet, we do to ourselves. Pollutants reach human beings through the food we consume, the water we drink, and the air we breathe and endanger our fitness in both immediately and long-lasting ways. Outdoor air pollutants on my own is related to greater than 1,000,000 deaths and countless illnesses every year throughout the globe. Children and the elderly – the most inclined members of our society – are especially liable to pollution like mercury and pesticides.

The herbal environment contributes considerably to human's fitness via the exceptional of air we breathe, the food we consume and the water we drink. On the only hand, it gives fitness enhancing economic and recreational opportunities, at the same time as at the different, it is threatened by using activities together with shipping, industrial processes, agricultural and waste control practices. Environmental pollutants and doubtlessly pathogenic organisms can harm human being's fitness through a sequence of complex shipping and exposure pathways. These are regions of developing public and authorities' hobby and this is reflected in UK and global priorities: for example inside the Environment Research Funders' Forum, the EU, the World Health Organisation (Millennium Ecosystem Assessment, 2005), and International Year of Planet Earth 2006 (Earth and Health theme).

Environmental health is the technology and practice of stopping human damage and contamination and promoting well-being by using

- Identifying and evaluating environmental sources and dangerous marketers and
- Prescribing exposures to unsafe physical, chemical, and organic retailers in air, water, soil, meals, and different environmental media or settings which could adversely have an effect on human health.

Importance of Environmental Health

Maintaining a healthy surroundings is important to increasing satisfactory of existence and years of healthy existence. Globally, 23 per cent of all deaths and 26 per cent of deaths among children beneath age 5 are due to preventable environmental elements. Environmental factors are numerous and a ways reaching. They include:

- Exposure to hazardous materials in the air, water, soil, and meals
- Natural and technological disasters
- Climate change
- Occupational hazards
- The built environment

Poor environmental exceptional has its greatest effect on people whose fitness repute is already at risk. Therefore, environmental fitness must address the societal and environmental factors that increase the likelihood of publicity and disease.

How Environment affects Human Health

Outdoor Air Quality

Poor air exceptional is connected to untimely death, cancer, and long-term harm to respiration and cardiovascular systems. Progress has been made to reduce bad air emissions, but in 2008, approximately 127 million human beings lived in U.S. Counties that exceeded countrywide air nice standards.2 Decreasing air pollution is an crucial step in creating a healthy surroundings.

Surface and Groundwater

Surface and floor water exceptional concerns practice to both ingesting water and leisure waters. Contamination through infectious sellers or chemicals can cause moderate to extreme contamination. Protecting water assets and minimizing publicity to contaminated water resources are essential components of environmental health.

Toxic Substances and Hazardous Wastes

The health outcomes of poisonous materials and dangerous wastes are not yet fully understood. Research to higher understand how those exposures may impact fitness is ongoing. Meanwhile, efforts to reduce exposures continue. Reducing exposure to toxic substances and risky wastes is fundamental to environmental health.

Homes and Communities

People spend most of their time at home, work, or school. Some of these environments may expose people to:

- ☆ Indoor air pollution
- ☆ Inadequate heating and sanitation
- ☆ Structural problems
- ☆ Electrical and fire hazards
- ☆ Lead-based paint hazards
- ☆ Environment-Related Illnesses
- ☆ Cancer
- ☆ Heart disease
- ☆ Diabetes
- ☆ Asthma
- ☆ Chronic obstructive pulmonary disease
- ☆ Obesity
- ☆ Occupational injuries
- ☆ Arthritis
- ☆ Parkinson's disease
- ☆ Malaria
- ☆ Dysentery
- ☆ Depression

Impact of Environment on Human Health

Physical Health

Though it's far widely established that wholesome eating and everyday exercising have major impacts on bodily fitness, within the beyond 30 years studies has additionally diagnosed that exposure to nature (*e.g.*, visual, multisensory, or by means of active engagement) is similarly powerful for regulating our diurnal body rhythms to ensure bodily vitality.

Mental Health

Mental fitness research in the context of connecting with nature have additionally generated a growing research base for the reason that emergence of the Basophilic concept in the mid-1980s. Much of its studies in the Evolutionary Psychology field examines the recuperative outcomes of nature on well-being and its useful houses following researcher's arguments of humanity's association for nature. Supporting studies has been properly documented in literature at some point of the last few decades. These include "Heraclitean motion" or natural movement, natural sounds, children's engagement activities within inexperienced settings as well as aesthetic alternatives for nature and natural forms.

Social Health

In the final two decades, the relationship between people and region inside the context of green areas has obtained tons attention in educational literature with reference to its importance for the vitality of communities and their surrounding environments. As research have shown, the presence of green space can promote social concord and group-based activities, factors that are critical for preserving social ties, developing communities, and increasing character's well-being (*e.g.*, horticulture and ecological restoration). Examples of findings consist of usage of out of doors area exponentially increases with wide variety and locality of trees, children's activities in inexperienced spaces improves social improvement as well as accessibility to inexperienced spaces complements social bonds in communities

Precautionary Measures

- ✰ You can limit your publicity to toxins as much as possible at the individual level by way of eating natural food, averting places and times of heavy air pollutants, and getting your water examined for dangerous metals and chemicals. But more importantly, we need advocates who will prevent pollutants at its source.
- ✰ Monitor the health of our local environment actively and continuously
- ✰ Create outreach packages for educating individuals approximately environmental health troubles
- ✰ Continue to address troubles related to pollution
- ✰ Base policy approximately the surroundings and fitness on sound science
- ✰ Strategies of enhancing environmental fitness want to encompass the unique circumstances of each locality

7.7 Human Rights

Human rights are rights inherent to all human beings, irrespective of race, sex, nationality, ethnicity, language, religion, or some other status. Human rights include the proper to lifestyles and liberty, freedom from slavery and torture, freedom of opinion and expression, the proper to paintings and education, and plenty of more. Everyone is entitled to those rights, without discrimination.

Human rights are rights inherent to all human beings, something our nationality, location of residence, sex, countrywide or ethnic origin, colour, religion, language, or any other status. We are all similarly entitled to our human rights without discrimination. These rights are all interrelated, interdependent and indivisible.

Universal human rights are often expressed and guaranteed via regulation, within the kinds of treaties, normal global law, trendy concepts and other assets of international law. International human rights regulation lays down responsibilities of Governments to act in positive ways or to chorus from certain acts, so as to sell and protect human rights and fundamental freedoms of individuals or groups.

Human rights are applicable to all of us, now not just folks who face repression or mistreatment.

The protection of human rights is honestly one of the most critical elements of development. Nevertheless, it receives plenty much less interest than other elements, possibly in element because it is so very hard to measure. If one is interested in empirically studying the protection of human rights, it's miles now not enough to rely countries that ratify human rights treaties; instead, the quantitative have a look at of human rights goals to decide whether or now not certain human rights are included in practice.

Since the early 1990s, the worldwide community has an increasing number of normal the inextricable link between human rights and the surroundings. For example, it has stated that procedural rights which include the rights to assembly, expression, or information, are vital to environmental protection. Several environmental problems, including pollutants and weather change, are diagnosed to have an effect on a host of rights. In addition, many constitutions and some of regional human rights frameworks encompass an impartial proper to an adequate and healthy surroundings.

Work in the direction of sustainable improvement is an increasing number of recognising the significance of a human rights technique. This should not be surprising: the safety of human lifestyles in relation to existence, fitness, way of life and dwelling standards is significant to any social, environmental or monetary programmes. The proper to life cannot be realised without the basic proper to clean, water, air and land. A human rights method permits the first-rate of lifestyles of human beings, especially the maximum vulnerable, to be integrated into environmental decision making.

There are main strategies to human rights and the surroundings:

- ✰ The use of existing human rights, and
- ✰ The want for new human rights for a secure and clean environment.

Such rights consist of the proper to existence, equality, political participation and association. They are couched maximum virtually within the Universal Declaration of Human Rights (1948) and International Covenant on Civil and Political Rights (1966). When realised civil and political rights are essential to making certain a political order supportive of sustainable improvement. They can guard civil mobilisation round environmental safety and equity. Economic, social and cultural rights are often referred to as 'second generation' rights. These provide substantive requirements for a character's properly-being.

The Need for Environmental Rights

It is nearly thirty years since the UN Stockholm Conference in 1972 placed environmental troubles at the international agenda. Since then there was a whole lot difficult paintings to guard and enhance the surroundings globally, nationally and locally. There has been an awful lot progress, but it's miles clear that in all components of the arena our surroundings is below threat and that many troubles are getting more serious. The Rio summit recognised those issues and sought to remedy them via Agenda 21 and the numerous UNCED Conventions.

Work on imposing these conventions, which are based on the principle of commonplace but differentiated responsibility, has gradually discovered that these much less than best deals have:

1. Failed to area constraints on country wide techniques that may cause unsustainable growth.
2. Failed to make sure the implementation of countrywide framework legal guidelines and enforcement strategies which include National Environmental Action Plans (NEAP).
3. Failed to control perverse state useful resource use and damaged intergenerational equity.
4. Increasingly discovered the inadequacies of investment by way of those companies which fund strictly environmental work that does now not do not forget socio-monetary factors. It is failures together with these which have led to require a new approach to sustainable development.

How does the UN Sell and Defend Human Rights?

High Commissioner for Human Rights

The Office of the UN High Commissioner for Human Rights (OHCHR) has lead responsibility within the UN machine for the promoting and safety of human rights. The office supports the human rights additives of peacekeeping missions in several countries, and has many country and nearby offices and centres. The High Commissioner for Human Rights often comments on human rights situations within the world and has the authority to research conditions and difficulty reports on them.

Human Rights Council

The Human Rights Council, hooked up in 2006, changed the 60-year-vintage UN Commission on Human Rights because the key unbiased UN intergovernmental body chargeable for human rights.

Human Rights Treaty Bodies

The human rights treaty bodies are committees of independent professionals that reveal implementation of the core international human rights treaties

Special Procedures

The special tactics of the Human Rights Council are prominent, independent experts operating on a voluntary basis, who examine, screen, publicly report and suggest on human rights from a thematic or use-precise perspective.

UNDG-HRM

The UN Development Group's Human Rights Mainstreaming Mechanism (UNDG-HRM) advances human rights mainstreaming efforts within the UN improvement device.

Special Advisers on the Prevention of Genocide and the Responsibility to Protect

The Special Adviser on the Prevention of Genocide acts as a catalyst to raise cognizance of the causes and dynamics of genocide, to alert relevant actors where there is a risk of genocide, and to advocate and mobilize for suitable action; the Special Adviser on the Responsibility to Protect leads the conceptual, political, institutional and operational improvement of the Responsibility to Protect.

7.8 Value Education

Environmental training or environmental literacy is something that every person must be well versed with. Values address one's own standards and requirements from which we judge what is right and wrong behaviour. Values in environment training must bring in several new concepts. Why and how can we use less resources and strength? Why do we want to keep our environment clean? Why must we use much less fertilizers and pesticides in farms? Why is it important for us to save water and preserve our water assets clean? Or separate our garbage into degradable and non-degradable types earlier than disposal? All these problems are related to the great of human life and go past simple financial growth. They cope with a love and admire for nature. These are the values that will result in a better humanity, one wherein we can stay healthy, productive and satisfied lives in harmony with nature.

Environmental Values

Every human being has a superb form of feelings for distinctive aspects of his or her surroundings. True environmental values go beyond valuing a river for its water, a wooded area for its timber and NTFPs, or the ocean for its fish. Environmental values are inherent in emotions that result in sensitivity for keeping nature as a whole. This is a greater spiritual, Eastern, conventional cost. There are several writings and sayings in Indian notion that help the concept of the oneness of all creation, of respecting and valuing all of the specific components of nature. Value system has been altered with time and circumstances. With massive numbers of people throwing away huge quantities of non-degradable waste, it is unfavourable to the surroundings and fee gadget need to prevent all this thru a strong environmental fee training gadget. Pro-environmental movements need to start to move from the domain of individuals to that of a network. Environmental values should pressure at the significance of keeping historical structures. The feature architecture, sculpture, works of art and crafts of historical cultures are worthwhile environmental belongings.

Valuing Nature

We ought to analyse to fee and respect numerous human cultures. We have an outstanding obligation to defend life in all its wonderful bureaucracy and must therefore admire the wasteland with all its dwelling creatures. On one hand, we need to guard herbal ecosystems; even as on the alternative hand, we have to guard the rights of neighbourhood people. We must additionally attempt to repair degraded areas to their former natural ecological state.

Valuing Cultures

Every subculture has a right to exist. Tribal humans are often linked intently with nature and we've got no proper to disturb and disrupt their lifestyles. We need to understand that many historic and tribal cultures have a expertise and know-how of their own environments that is based on a deep experience of respect for nature. Tribal have produced unique artwork forms, including painting, sculpture and crafts, which are lovely and may enrich residing stories for everyone. The world might be culturally impoverished if we lose this conventional understanding.

Human Heritage

The earth itself is a background left to us with the aid of our ancestors. Heritage upkeep is now a growing environmental concern, because we've undervalued a great deal of this history over the last several decades and is vanishing at an surprising pace. Though we appreciate and value the Ajanta and Ellora Caves, Taj Mahal and environmentally-pleasant colonial buildings, we've finished little to actively hold them. As environmentally-conscious individuals, we want to lobby for the protection of the desert and our wonderful architectural history.

Equitable Use of Sources

The equitable use of sources is seen as an vital factor of human properly-being and have to grow to be a shared factor of view amongst all socially and environmentally-aware people. In spite of the exceptional quantity of human beings inside the more populous developing nations, the smaller quantity of human beings in the advanced international locations makes use of extra sources and electricity than those inside the growing international. Similarly, the small range of rich human beings in bad nations who's in line with capita use of electricity and sources, and the generation of waste based at the one-time use of disposable merchandise leads to splendid pressures on the surroundings. We need to discourage this type of intake and need more sustainable lifestyles.

Common Assets Sources

There are numerous generally-owned sources that all people use as a community. The water that nature recycles, the air that all of us breathe, the forests and grasslands which hold our climate and soil, are all commonplace property resources. Managing local forests via village-degree FPCs has proven that if humans realize that they can enjoy the forests, they'll start to shield them. This essentially way sharing the power to govern forests among the Forest Department and the nearby human beings.

Ecological Degradation

In many situations, treasured ecological assets are grew to become into extreme environmental problems. The modifications in landuse from natural ecosystems to extra extensive utilization or marginal lands into extensive agricultural patterns or adjustments into city or industrial land convey an ecological price. Wetlands offer usable assets and lots of services, whilst destroyed to offer extra farmland, produce decrease returns. A natural forest provides treasured NTFPs, whose monetary returns are high that provided by means of felling the wooded area for wooden. These values need to form a part of a new conservation ethic.

Social Justice

We want to appreciate and value the diverse elements of the societies. If negative are not respected as they are lacking the first-rate matters in lifestyles, they'll rebel, and anarchy and terrorism will spread. The developing international will face a crisis than the advanced nations, unless we shield the rights of poor human beings.

Good best education is an important tool for achieving a greater sustainable world. This become emphasized at the UN World Summit in Johannesburg in 2002 wherein the reorientation of modern schooling systems became outlined as key to sustainable development. Education for sustainable development (ESD) promotes the development of the information, skills, understanding, values and actions required to create a sustainable international, which ensures environmental protection and conservation, promotes social equity and encourages economic sustainability. The idea of ESD developed largely from environmental training, which has sought to expand the expertise, skills, values, attitudes and behaviours in humans to care for their surroundings. The aim of ESD is to enable humans to make decisions and carry out moves to enhance our satisfactory of lifestyles without compromising the planet. It additionally ambitions to combine the values inherent in sustainable improvement into all factors and degrees of learning.

Value Education may be divided in two awesome categories:

- ☆ **Explicit Value Education:** Explicit price education directly relates to our teaching gadget in relation to price question like how the trainer or educators use different pedagogies, techniques or programmes to create a clean learning experience for the aspirants.

- **Implicit Value Education:** Implicit fee schooling covers that element of educational revel in which pertains to the idea of hidden curriculums of learning.

Objectives of Value Education

- The main goal of Value Education is to enhance the integral boom of human beings.
- It enables for the improvement in the sustainable lifestyle of the individual.
- Value Education enables to growth attention approximately our state like the history of the country, constitutional rights, country wide integration *etc.*
- Developing a democratic manner to enhance the extent of thinking.
- To broaden the tolerance and understanding capacity of various spiritual faiths.

Environmental training needs to be imparted through formal and informal methods to all sections of the society. Everyone needs to apprehend it because 'environment belongs to all' and 'every man or woman matters' when it comes to conservation and safety of surroundings.' Various stages and techniques that may be useful for elevating environmental awareness in specific sections of the society are as follows:

Among students through formal education: Environmental education have to be imparted to the students' right from the early life stage. It is a welcome step that now all around the country we are introducing environmental research as a topic in any respect ranges consisting of college and college degree, following the directives of the Supreme Court.

Among the loads via mass-media: Media can play an vital role to train the loads on environmental troubles through articles, environmental rallies, plantation campaigns, road plays, actual eco-disaster tales and success tales of conservation efforts.

Values-primarily based training for sustainability has been developing over the last several years to cope with this information-motion gap. In many locations around the arena, small-scale projects have been flourishing, supported by means of studies on methodologies and networking to proportion experience. The UN Decade of Education for Sustainable Development and its successor Global Plan of Action were one catalyst; the Partnership for Education and Research approximately Responsible Living (PERL) has been another. Efforts to increase education primarily based on concepts of world citizenship, sustainable lifestyles, and sustainable intake have also contributed. Faith-based totally and inter-faith agencies have also been active, along with corporations focusing on ethics, both spiritual and secular. The Pope's current encyclical, the Islamic Declaration on Climate Change, statements of the Bahá'í International Community, and others have proven a unanimity of idea when it comes to ethical responsibility for the environmental and social demanding situations facing the sector today. All of this provides material for values-based totally education to strengthen rapidly.

The project is to head from stating moral principles to enforcing them. We have no longer been very powerful in building ethical duty into our institutions. At the governmental degree, the considerable work load of the Human Rights Council and other human rights bodies demonstrates the ability with which governments can ignore fundamental human rights. Corporations easily place profits above standards of ethics and social and environmental responsibility. The flourishing of corruption and the unlawful economy in large part overreaches the efforts to control them. Globalization has made it easy to escape from national regulation and taxation and to make the most of the vacuum of global mechanisms for regulation and justice. While the UN machine need to logically cope with these issues, governments defending their country wide pastimes over the collective interest make progress difficult.

Values are first transmitted within the own family, with the aid of dad and mom and other close family members, consciously or unconsciously, by example and by means of word, including through traditional memories, tales and fables. Increasingly, however, values are also being transmitted via the media and advertising, regularly with pursuits which can be commercial instead than in the collective hobby. As children emerge as included into community lifestyles, values are modelled and communicated in a larger circle reflecting the nearby tradition, and perhaps a faith tradition. Then come schools, which may additionally encompass curriculum on citizenship and civic behaviour. Peer agencies start to have greater have an impact on on values than parents. Children and youth can be enrolled in associations and community corporations, one motive of which may be to carry values of service, accomplishment or care for nature. Sports also exemplify positive values. In this complex procedure of education for values, it isn't continually clear who is accountable, and whether they're conscious of that obligation. When there may be a failure to transmit values necessary for the proper insertion of the kid in society, who's responsible and how can the error be corrected? This is an increasing hassle in societies in which parents have rejected a conventional faith or tradition, or where those were lost via migration, and not anything is transmitted to children other than the superficial values of a materialistic society. Such children develop up with no experience of a better which means or purpose in life, and are left with self-hobby as the highest cost, if no longer a entire denial of any social norms. The costs for society may be very high.

7.9 HIV/AIDS

HIV (human immunodeficiency virus) is a plague that attacks cells that assist the body fight contamination, making someone greater at risk of different infections and diseases. It is spread by way of contact with positive physical fluids of someone with HIV, most usually at some stage in unprotected intercourse (sex without a condom or HIV medication to save you or deal with HIV), or through sharing injection drug equipment.

HIV is a deadly disease spread via certain body fluids that attacks the frame's immune machine, in particular the CD4 cells, often referred to as T cells. Over time, HIV can break so lots of these cells that the frame can't combat off infections and disorder. These special cells assist the immune machine fight off infections. Untreated, HIV reduces the variety of CD4 cells (T cells) within the body. This damage to the immune system makes it more difficult and harder for the body to fight off infections and some different diseases. Opportunistic infections or cancers take gain of a completely weak immune system and signal that the character has AIDS. Learn greater about the stages of HIV and how to realize whether you have HIV.

AIDS stands for Acquired Immune Deficiency Syndrome and is caused by HIV. The names HIV and AIDS may be confusing due to the fact both terms describe the same sickness. Think of AIDS as superior HIV disease. A man or woman with AIDS has an immune device so weakened via HIV that the individual normally turns into unwell from one of several opportunistic infections or cancers including PCP (a type of pneumonia) or KS (Kaposi sarcoma, a form of cancer that impacts the skin and inner organs in HIV), losing syndrome (involuntary weight loss), reminiscence impairment, or tuberculosis. If a person with HIV is identified with this sort of opportunistic infections (even though the CD4 depend is above 200), she or he is said to have AIDS. AIDS normally takes time to develop from the time a person acquires HIV–typically among 2 to 10 years or extra.

Transmission of the virus occurs:

- Through sexual touch – inclusive of oral, vaginal, and anal intercourse.
- Through blood – thru blood transfusions (now extremely rare inside the U.S) or needle sharing.
- From mother to child – a pregnant female can passively transmit the virus to her fetus, or a nursing mother can transmit it to her baby.
- Other transmission techniques are uncommon and consist of accidental needle injury, synthetic insemination with donated semen, and thru a donated organ.

HIV contamination isn't spread through casual contact (which include hugging and touching), by touching dishes, doorknobs, or toilet seats formerly touched by using a person infected with the virus, all through participation in sports, or by means of mosquitoes. It isn't transmitted to someone who DONATES blood or organs in the U.S. due to the fact hospitals do not re-use syringes and sterilize all gadgets worried in such procedures. However, HIV may be transmitted to the person RECEIVING blood or organs from an inflamed donor. This is why blood banks and organ donor programs display screen donors, blood, and tissues thoroughly.

Those at highest threat include homosexual or bisexual guys carrying out unprotected intercourse, intravenous drug customers who share needles, the sexual companions of folks who participate in excessive-threat activities, toddlers born to moms with HIV, and those who obtained blood transfusions or clotting merchandise among 1977 and 1985 (prior to standard screening for the virus inside the blood). AIDS begins with HIV contamination. People infected with HIV may also have no symptoms for ten years or longer, but they could nevertheless transmit the infection to others during this symptom-loose period. Meanwhile, their immune system gradually weakens until they develop AIDS.

Acute HIV infection progresses over the years to asymptomatic HIV contamination and then to early symptomatic HIV infection. Later, it progresses to AIDS (very superior HIV contamination with T-mobile rely beneath 200).Most people inflamed with HIV will development to AIDS, if now not treated. However, there is a tiny institution of patients who broaden AIDS very slowly or in no way in any respect. These patients are known as non-progressors and many seem to have a genetic difference which prevents the virus from attaching to certain immune receptors.

Signs and Symptoms

The symptoms of HIV vary relying on the stage of infection. Though humans dwelling with HIV tend to be maximum infectious within the first few months after being inflamed, many are blind to their status till the later degrees. In the first few weeks after initial contamination human beings can also revel in no signs and symptoms or an influenza-like contamination which include fever, headache, rash or sore throat.

As the contamination step by step weakens the immune device, they are able to expand other signs and symptoms and symptoms, along with swollen lymph nodes, weight loss, fever, diarrhoea and cough. Without treatment, they could additionally broaden excessive illnesses consisting of tuberculosis (TB), cryptococcal meningitis, severe bacterial infections, and cancers which include lymphomas and Kaposi's sarcoma.

Symptomatic HIV Infection

As the virus continues to multiply and destroy your immune cells – the cells in your body that help fight off germs – you may develop mild infections or chronic signs and symptoms such as:

- Fever
- Fatigue
- Swollen lymph nodes – often one of the first signs of HIV infection
- Diarrhea
- Weight loss
- Oral yeast infection (thrush)

What are the Stages of HIV?

When people get HIV and don't receive treatment, they will typically progress through three stages of disease. Medicine to treat HIV, known as antiretroviral therapy (ART), helps people at all stages of the disease if taken as prescribed. Treatment can slow or prevent progression from one stage to the next. Also, people with HIV who take HIV medicine as prescribed and get and keep an undetectable viral load have effectively no risk of transmitting HIV to an HIV-negative partner through sex.

Stage 1: Acute HIV infection

Within 2 to 4 weeks after infection with HIV, people may experience a flu-like illness, which may last for a few weeks. This is the body's natural response to infection. When people have acute HIV infection, they have a large amount of virus in their blood and are very contagious. But people with acute infection are often unaware that they're infected because they may not feel sick right away or at all. To know whether someone has acute infection, either an antigen/antibody test or a nucleic acid (NAT) test is necessary. If you think you have been exposed to HIV through sex or drug use and you have flu-like symptoms, seek medical care and ask for a test to diagnose acute infection.

Stage 2: Clinical latency (HIV inactivity or dormancy)

This period is sometimes called asymptomatic HIV infection or chronic HIV infection. During this phase, HIV is still active but reproduces at very low levels. People may not have any symptoms or get sick during this time. For people who aren't taking medicine to treat HIV, this period can last a decade or longer, but some may progress through this phase faster. People who are taking medicine to treat HIV (ART) as prescribed may be in this stage for several decades. At the end of this phase, a person's viral load starts to go up and the CD4 cell count begins to go down. As this happens, the person may begin to have symptoms as the virus levels increase in the body, and the person moves into Stage 3.

Stage 3: Acquired immunodeficiency syndrome (AIDS)

AIDS is the most severe phase of HIV infection. People with AIDS have such badly damaged immune systems that they get an increasing number of severe illnesses, called opportunistic illnesses.

Without treatment, people with AIDS typically survive about 3 years. Common symptoms of AIDS include chills, fever, sweats, swollen lymph glands, weakness, and weight loss. People are diagnosed with AIDS when their CD4 cell count drops below 200 cells/mm or if they develop certain opportunistic illnesses. People with AIDS can have a high viral load and be very infectious.

- Shingles (herpes zoster)
- Pneumonia
- Progression to AIDS

The signs and symptoms of some of these infections may include:

- Sweats
- Chills
- Recurring fever
- Chronic diarrhea
- Swollen lymph glands
- Persistent white spots or unusual lesions on your tongue or in your mouth
- Persistent, unexplained fatigue
- Weakness
- Weight loss
- Skin rashes or bumps

How Can AIDS be Prevented?

- Prevention of AIDS requires foresight and self-discipline. The requirements often seem personally restrictive, but they are effective and can save your life.
- Do not have sexual intercourse with:
- People known or suspected to be infected with AIDS
- Multiple partners
- A person who has multiple partners
- People who use IV drugs
- Do not use intravenous drugs. If IV drugs are used, do not share needles or syringes. Many communities now have needle exchange programs where used syringes can be disposed of and new, sterile needles obtained for free. These programs can also provide referrals to addiction treatment.
- Avoid exposure to blood from injuries or nosebleeds where the HIV status of the bleeding individual is unknown. Protective clothing, masks, and goggles may be appropriate when caring for people who are injured.
- Anyone who tests positive for HIV may pass the disease on to others and should not donate blood, plasma, body organs, or sperm.
- From a legal, ethical, and moral standpoint, they should warn any prospective sexual partner of their HIV positive status. They should not exchange body fluids during sexual activity and must use whatever preventative measures (such as a latex condom) will afford the partner the most protection.

- HIV positive women should be counselled before becoming pregnant about the risk to unborn children and medical advances which may help prevent the foetus from becoming infected. Use of certain medications can dramatically reduce the chances that the baby will become infected during pregnancy.
- Other information on sexual risk reduction: The riskiest sexual behaviour is unprotected receptive anal intercourse - the least risky sexual behaviour is receiving oral sex. Performing oral sex on a man is associated with some risk of HIV transmission, but this is less risky than unprotected vaginal intercourse.
- Female-to-male transmission of the virus is much less likely than male-to-female transmission. Performing oral sex on a woman who does not have her period carries low risk of transmission.
- HIV-positive patients who are taking anti-retroviral medications are less likely to transmit the virus. For example, pregnant women who are on treatment at the time of delivery transmit HIV to the infant about 5 per cent of the time, compared to approximately 20 per cent if medications are not used.
- If you believe you have been exposed to HIV, seek medical attention IMMEDIATELY. There is some evidence that an immediate course of anti-viral drugs can reduce the chances that you will be infected. This is called post-exposure prophylaxis (PEP), and it has been used to treat health care workers injured by needle sticks for years.

There is less information on the effectiveness of PEP for people exposed via sexual activity or intravenous drug use - however, if you believe you have been exposed, you should discuss the possibility with a knowledgeable specialist (check local AIDS organizations for the latest information) as soon as possible. All rape victims should be offered PEP and should consider its potential risks and benefits in their particular case.

7.10 Women and Child Welfare

Overall infant mortality declined significantly inside the 1990s, however environmental dangers nevertheless kill as a minimum three million children under age five each yr. (1) Such young kids make up more or less 10 per cent of the world's population, however comprise greater than forty per cent of the population suffering from health problems associated with the surroundings. (2) Children international require special protection from long-standing risks consisting of smoke from conventional fuels and from emerging risks together with exposure to more and more dangerous chemical substances.

There are many factors that may have an impact on girl fertility, from nutrition, weight and physical health to psychological pressure and use of medications.

One key influencer that girls increasingly need to take into account is the environment wherein they live. Environmental elements have been strongly related to girls's capability to conceive and feature a wholesome pregnancy

More than three million children below five die each yr from environment-related reasons and conditions. This makes the surroundings one of the most important members to the global toll of extra than ten million baby deaths annually – as well as a very essential factor within the health and well-being in their moms.

Polluted indoor and outdoor air, contaminated water, lack of adequate sanitation, toxic hazards, disorder vectors, ultraviolet radiation, and degraded ecosystems are all vital environmental threat factors for kids, and in maximum cases for their mothers as nicely. Particularly in developing countries, environmental dangers and pollutants are a major contributor to adolescence deaths, ailments and incapacity from acute respiration disease, diarrhoeal sicknesses, physical injuries, poisonings, insect-borne diseases and perinatal infections. Childhood loss of life and contamination from reasons inclusive of poverty and malnutrition also are related to unsustainable patterns of development and degraded city or rural environments

Air Pollution

Air pollutants is posing a developing danger to fitness all around the world. According to a World Health Organisation (WHO) report, more than nine out of ten of the world's populace (92 per cent) live in places in which air pollution exceeds secure limits.

There is a sturdy link among this fitness chance and fertility. Research has proven air pollutants is associated with a numerous set of outcomes, from altered production of sperm and eggs to epigenetic changes and beginning defects.

Exposure to Chemical Substances

Exposure to probably harmful chemical compounds is any other increasingly common fitness concern inside the cutting-edge world, in particular for girls who are looking to conceive or are presently pregnant.

Organic pollutants and endocrine-disrupting chemicals (EDCs) in the surroundings have an effect on each male and lady fertility.

EDCs are generally man-made and are observed in substances including pesticides, metals, meals components and personal care products. Human publicity usually occurs thru the ingestion of meals, dirt and water, the inhalation of gases and particles inside the air, and via the skin.

As nicely as changing the reproductive characteristic in each men and girls, EDCs can be transferred from mothers to kids throughout the placenta and through breast milk.

Children's Vulnerability

Due to their size, physiology, and behaviour, kids are more prone than adults to environmental hazards. Children are more heavily uncovered to pollution in share to their frame weight, and have more years of life beforehand of them wherein they will go through long-term effects from early exposure.3 Perinatal conditions, which may be influenced by using environmental conditions, purpose 20 percentage of deaths international in children beneath age 5. Furthermore, fatal exposure to chemicals together with lead will increase a infant's probabilities of having brain damage or developmental issues. Children at all ages, no longer just the very young, are at greater risk than adults. Children below age five breathe more air, drink greater water, and eat greater food in step with unit of frame weight than adults do, so they'll revel in higher costs of exposure to pathogens and pollutants. Typical early life behaviours, consisting of crawling and putting objects within the mouth, can also cause increased risks. Children between ages 5 and 18 can also face higher dangers of injuries, including publicity to unsafe chemicals, because of their growing participation in household chores and paintings outside of the home. Many faculty-age children attend schools without sanitation facilities, making them much more likely to contract various diseases and less possibly to go to faculty. According to UNICEF, approximately 10 per cent of faculty-age African women both do now not attend faculty in the course of menstruation or drop out at puberty because of the lack of sanitary facilities.

Other Foremost Environment-Associated Killers

Diarrhoea kills an anticipated 1.6 million children each 12 months, caused particularly by using hazardous water and terrible sanitation.

Indoor air pollutants associated with the still-great use of biomass fuels kills almost one million youngsters annually, ordinarily due to acute breathing infections. Mothers, in rate of cooking or resting close to the fireplace after having given start, are maximum vulnerable to growing persistent breathing sickness.

Malaria, which may be exacerbated due to terrible water management and storage, insufficient housing, deforestation and lack of biodiversity, kills an predicted one million kids underneath 5 annually, mostly in Africa.

Unintentional physical injuries, which can be associated with household or network environmental risks, kill almost 300 000 kids annually: 60 000 are attributed to drowning, forty 000 to fires, sixteen 000 to falls, sixteen 000 to poisonings, 50 000 to road visitors incidents and over one hundred 000 are because of different accidental injuries.

Women and Baby Welfare Programmes

The department of girls and child development is one of the main departments inside the district imposing variety of programmes for the development of ladies and kids, protecting youngsters from being neglected, exploited, save you destitution and delinquency, offer opportunity shelter based totally rehabilitation through institutional treatment, improve health nutritional popularity of kids and ladies, create focus among rural ladies to improve overall residing conditions.

The branch has evolved many programmes in fulfilment of the above objectives. ICDS scheme covers all the 7 talukas of the district. The branch is also answerable for prevention of youngster linguency, destitution and negligence of youngsters through parents and is jogging an statement home, juvenile home below authorities sector and seven destitute cottages run by way of NGO's are given financial assistance. All the offerings are supplied thru an anganwadi centre and there are 1303 anganwadi employees in the district providing offerings to the pregnant and lactating girls with dietary meals 300 days in a year, immunizing youngsters and pregnant ladies and other basic offerings.

Schemes of the Department

- ✰ Attendance scholarship for girls up to tenth standard.
- ✰ Assistance to women in activity oriented courses 7 encouragement to absorb self-employment through bank loans below schemes like Vikasini, Navajeevan, Mane Belaku, Udyogini *etc.*
- ✰ Special college for infant labour, bodily handicapped childrens.
- ✰ Providing scholarship to disabled college students and inspiring them for self-employment.
- ✰ Widow/Devadasi marriage

Assistance to Mahila Mandals and Mahila Federations

Integrated child development services scheme: Integrated infant improvement scheme launched on second Oct 1975 in 33 network improvement locks. Government of India proclaimed a National Policy on children in August, 1974 declaring kids as "supremely essential asset". The policy provided the specified frame paintings for assigning priority to one of a kind needs of the toddler.

It was launched in 1975 searching for to offer an integrated package deal of services in a convergent way for the holistic development of the child.

Objectives of ICDS

- ✰ Way the foundation for proper psychological improvement of the infant.
- ✰ Improve nutritional and health fame of kids 0-6 yrs.
- ✰ Reduce occurrence of mortality, morbidity, malnutrition and school dropouts.
- ✰ Enhance the functionality of the mother and family to look after the fitness, dietary and development wishes of the baby.
- ✰ Achieve powerful co-ordination of coverage and implementation amongst numerous departments to sell infant improvement.

Services

The scheme offers an integrated technique for converging simple offerings via public based people and helpers. The offerings are furnished at a centre referred to as anganwadi. A package of following six offerings is furnished under the ICDS scheme.'

Supplementary Nutrition

- ☆ Non-formal preschool education.
- ☆ Immunization
- ☆ Health check up
- ☆ Referral services
- ☆ Nutrition and health education.

7.11 Role of Information Technology in Environment and Human Health

Information technology (IT) is one of the firmest rising new technology. Numerous software have been industrialised for environment and health care which are operator approachable and help in healthier sympathetic for the topic. A lot of methods are used under IT for growth and application of computational tools to acquire, store, examine and imagine satellite data which is used for observation, and defence of environment Due to the growth of the internet and information through the satellites a wide database is produced which is the collection of numerous unified articles.

When you talk about information technology, it has a noteworthy hand in refining the status in the arenas of environmental education and human health as associated to that of the other individual areas such as business, economics, and culture or politics. The emerging growth of the internet services and facilities, geographic information system or GIS, and the data that gets conveyed through satellites, *etc.* have produced a higher wealth of the well-organized information on numerous features of the environment as well as health. When you look for the diversity of software in the market, you will come across a number of them that shaped for the health and environment studies in a healthier way.

The information of database can easily excerpt in a computer. When it comes to comprehensive databases, things that comprise in it are wildlife database, forest cover database, and conservation database, *etc.* the databases are also obtainable for some illnesses which comprise malaria, fluorosis, HIV/AIDS, *etc.*

- ☆ National management information system (NMIS).
- ☆ Environmental information system (ENVIS).
- ☆ Remote sensing and geographical information system (GIS).
- ☆ Geographical information system (GIS).

☆ The World Wide Web (WWW).

They are quite user-friendly and certainly help a learner to understand the respective subject with ease.

The important roles of information technology in environment are as follows:

1. Remote Sensing

Remote sensing technique through satellite can be used to assess ongoing changes in the environment and to predict natural calamities like droughts, floods and volcanic eruptions *etc.* The technique is of great use in exploring the possible availability of crude oils, mineral deposits and location of geothermal power sources.

2. Database

Database is the collection of inter-related data on various subjects. It is usually in computerized form and can be retrieved whenever required. There are several Distribution Information Centres (DICs) in our country that are linked with each other and with the central information network having access to international database.

The Ministry of Environment and Forests, Government of India has taken up the task of compiling a database on issues like wildlife, forest cover, wastelands *etc.*

3. Environmental Information System (ENVIS)

ENVIS established in 1982 aims on providing environmental information to decision makers, policy planners, engineers and scientists all over the country. ENVIS centres work for generating a network of database in areas like pollution control, clean technologies, biodiversity, renewable energy, wildlife, environmental management and remote sensing.

4. National Management Information System (NMIS)

NMIS of the Department of Science and Technology has compiled a database on Research and Development projects along with information about research scientists and personnel involved.

5. Geographical Information System (GIS)

GIS has proved to be a very effective tool in environmental management. GIS is a technique of superimposing various thematic maps using digital data on a large number of inter-related or inter-dependent aspects. Different thematic maps containing digital information on a number of aspects like water resources, forest land, soil type, crop land, industrial growth, human settlement *etc.* are superimposed in a layered form in computer using soft-wares.

Disaster Management

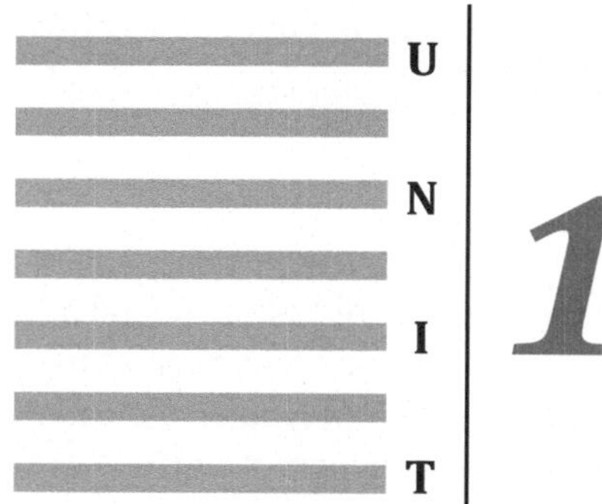

Introduction to Disasters

A disaster is a sudden, disastrous event that extremely disturbs the operative of a community or civilisation and reasons human, material, and economic or environmental fatalities that exceed the community's or society's ability to cope using its own capitals. Though often caused by nature, tragedies can have human origins.

(VULNERABILITY+ HAZARD)/CAPACITY = DISASTER

Strictly speaking, there aren't any such things as natural disasters, however there are herbal hazards. A disaster is the result of a hazard's impact at the society. So the effects of a disaster are decided by means of the volume of a community's vulnerability to the risk (or conversely, its capability, or potential to deal with it). This vulnerability isn't herbal, however the end result of a whole range of constantly changing physical, social, financial, cultural, political, and even psychological factors that shape human beings's lives and create the environments in which they live. 'Natural' screw ups are nature's judgement on what people have wrought. A natural phenomenon like a hellfire, or a person using a primus stove aren't dangers till they pose a hazard to human life, property, infrastructure or the environment. If the hellfire receives out of hand and rages right into a populated area, or the primus falls over and causes a hearth inside the home, then each will become a chance. The hellfire or primus fire may want to then spread uncontrollably and burn down many homes and homes and the environment could be significantly damaged. Many lives will be lost and those should turn out to be in clinic having sustained extreme burns. The financial system and livelihoods of human beings might be adversely affected and if assets are not within the means of the humans to manage the situation, then a catastrophe has occurred. Hazards, do no longer necessarily end up failures. It is dependent on the extent to which damage is finished and how all of the humans worried can deal with the situation.

Definitions

Hazard

Hazard may be defined as a probably destructive physical occasion, social and economic disruption or environmental degradation. Typical examples of risks may be absence of rain (leading to drought) or the abundance thereof (main to floods). Chemical manufacturing flowers close to settlements and incorrect agricultural techniques, can also been visible as risks which can lead to viable screw ups. Hazards may be the introduction of guy or the environment.

Natural Hazards

Natural dangers include phenomena which includes: earthquakes; volcanic hobby; landslides; tsunamis, tropical cyclones and other intense storms; tornadoes and excessive winds; river floods and coastal flooding; wildfires and associated haze; drought; sand/dust storms; insect infestations ns organic ones. Other styles of dangers consist of human-caused stemming from technological hazards and environmental degradation.

Risk

Risk is usually related to the human lack of ability to cope with a particular situation. In phrases of disaster management it can be described as the probability of harmful effects, or predicted losses demise, injury, damage to belongings and the environment, jobs, disruption of monetary activity or social systems. Hazards will have an effect on groups in a different way in terms of potential and assets with which to cope. Poorer groups will be greater at risk than others.

Vulnerability

Vulnerability may be visible as, the potential someone or community has, to predict, deal with, or avoid and get better from, the results of a chance or disaster. Marginalised, poorer and over-populated communities are extra prone and less capable to address failures.

Disaster

A disaster is the critical disruption of the functioning of a society, inflicting or threatening to motive, full-size human, fabric or environmental losses which render the affected network unable to cope the usage of its own sources. Disasters are most effective screw ups because they effect and have an effect on the manner wherein we live. They can be as a result of the effect of a herbal occurrence on humans or by way of people themselves. From the above definitions, it becomes apparent that it is incorrect to refer to “herbal screw ups”, since herbal phenomena in themselves aren’t disasters unless they effect negatively on populated areas.

Disaster Management

Disaster mitigation refers to structural and non-structural measures which are undertaken to restriction the adverse impact of natural dangers, environmental degradation and technological risks on prone areas, communities and households. These efforts can target the risk or danger itself (for instance, the location of fire-breaks at the urban/wild land interface). This is frequently cited as 'structural mitigation', since it requires infrastructure or engineering measures to keep the chance away from the ones at danger.

Disaster mitigation efforts can also goal individuals who are at danger, by way of reducing their vulnerability to a specific danger (for instance, promoting network responsibility for controlling fire hazard in an casual settlement). This is frequently known as 'non-structural mitigation', as it promotes hazard-avoidance behaviours and attitudes.

1.1 Meaning and Nature of Natural Disasters

A natural danger is a threat of a evidently occurring occasion could have a poor effect on people. This negative effect is what we name a herbal disaster. In other phrases when the hazardous hazard definitely takes place and harms people, we name the event a herbal disaster. Earthquakes, landslides, flood, drought, fires, and hailstorms are a number of the natural calamities that occur each year, at any point, and anywhere, causing threats to the livelihoods of smallholder farmers and their food security. Disasters can purpose lack of human and animal existence, field crops, saved seeds, agricultural equipment/materials, and their deliver systems (*e.g.* Infrastructure) as well as related indigenous knowledge, as a consequence disrupting not simplest the immediate developing season however also future seasons.

Natural screw ups encompass all styles of intense weather, which have the capability to pose a widespread threat to human fitness and safety, property, important infrastructure, and place of birth security. Natural failures arise both seasonally and without warning, subjecting the kingdom to common durations of insecurity, disruption, and financial loss. These assets serve to put together IHEs for a number of herbal screw ups, along with wintry weather storms, floods, tornados, hurricanes, wildfires, earthquakes, or any aggregate thereof. Natural disasters are excessive, sudden events because of environmental factors that injure human beings and harm assets. Earthquakes, windstorms, floods, and disease all strike anywhere on earth, frequently with out warning.

Risk is characteristic of the connection between people and geologic techniques. We all take dangers each day. The danger from herbal dangers, while it cannot be eliminated, can, in a few instances be understood in such a manner that we will limit the hazard to humans, and hence minimize the risk. To do this, we want to apprehend something approximately the tactics that operate, and apprehend the energy required for the technique. Then, we can develop an motion to take to limit the risk. Such minimization of threat is referred to as risk mitigation.

Although human beings can once in a while influence natural failures (as an instance while terrible levee design results in a flood), different failures which are without delay generated by way of humans, which includes oil and poisonous material spills, pollution, large vehicle or train wrecks, airplane crashes, and human triggered explosions, are taken into consideration technological failures, and will now not be taken into consideration on this course, except when they arise as a secondary result of a natural disaster.

Natural disasters are produced by way of techniques that have been working for the reason that Earth formed. Such tactics are beneficial to us as human beings due to the fact they may be chargeable for things that make the Earth a liveable planet for existence. For instance:

Throughout Earth history, volcanism has been chargeable for generating tons of the water present on the Earth's surface, and for producing the atmosphere.

Earthquakes are one of the tactics chargeable for the formation of mountain stages which direct water to flow downhill to form rivers and lakes.

Erosional strategies, inclusive of flooding, landslides, and windstorms replenishes soil and facilitates maintain existence.

Such approaches are simplest taken into consideration hazardous when they adversely have an effect on humans and their activities.

1.2 Types and Effects of Natural Disasters

Natural disasters can be labelled as "acute" or "slow" in their onset. They are predictable due to the fact they cluster in geographic areas. Natural dangers are unpreventable and, for the most part, uncontrollable. Natural failures with acute onsets encompass events such as earthquake, flood, hurricane or hurricane, tornado, fire, tsunami or typhoon surge, avalanche, volcanic eruption, intense cold or blizzard, and heat wave. Natural dangers with a slow or gradual onset include drought, famine, desertification, deforestation, and pest infestation

Natural failures are large-scale geological or meteorological occasions which have the capability to cause lack of life or assets. These sorts of screw ups include:

- ☆ Tornadoes and Severe Storms
- ☆ Hurricanes and Tropical Storms
- ☆ Floods
- ☆ Wildfires
- ☆ Earthquakes
- ☆ Drought

Severe storms and floods are the maximum common sorts of natural disasters reported inside the United States. These meteorological activities are sometimes preceded by means of presidential "emergency declarations" requiring state and local making plans previous to the event, including evacuations and protection of public assets. Disaster Distress Helpline personnel are available to speak to individuals who name or text before, for the duration of, and after a natural disaster. The natural catastrophe is the outcome while a herbal danger impacts humans in an detrimental manner. Human vulnerability, because of the dearth of appropriate preparedness frequently results in financial, environmental, or human effect in a bad manner. A natural disaster continually brings along loss of existence and property harm, and commonly leaves economic harm in its wake, the severity of which depends on the intensity of the disaster. At times there may be a few minor losses most effective while human existence has also witnessed few failures that have induced a major lack of life and property as well.

Different varieties of natural failures may be:

- Tsunamis
- Earthquakes
- Avalanches
- Volcanoes
- Landslides
- Floods
- Droughts
- Forest fires
- Hurricanes
- Thunderstorms
- Tornadoes
- Winter storms
- Heat Waves

Effects of Natural Disasters

Every year, thousands and thousands of humans are affected by each human-brought on and herbal screw ups. Disasters can be explosions, earthquakes, floods, hurricanes, tornados, or fires. There are quite a number challenges, together with climate change, unplanned-urbanization, under-development/poverty as well as the hazard of pandemics, that will form humanitarian assistance in the future. These aggravating factors will result in elevated frequency, complexity and severity of disasters.

Displaced Populations

One of the maximum immediate effects of natural disasters is population displacement. When nations are ravaged by way of earthquakes or different effective forces of nature, many human beings have to abandon their homes and searching for shelter in different regions. A big influx of refugees can disrupt accessibility of health care and education, in addition to meals elements and easy water.

Health Risks

Aside from the obvious immediate chance that herbal disasters gift, the secondary consequences may be just as negative. Severe flooding can bring about stagnant water that allows breeding of waterborne bacteria and malaria-wearing mosquitos. Without emergency relief from worldwide aid companies and others, demise tolls can upward thrust even after the immediate threat has passed.

Food Scarcity

After herbal screw ups, meals often turns into scarce. Thousands of people around the world cross hungry due to destroyed crops and loss of agricultural resources, whether or not it takes place in a typhoon or progressively in a drought. As a end result, meals prices upward thrust, decreasing families' purchasing strength and increasing the threat of excessive malnutrition or worse. The impacts of hunger following an earthquake, typhoon or typhoon can be tremendous, causing lifelong harm to youngsters' development.

Environmental Problems

In March 2011, a tsunami following the 9.0-magnitude Tôhoku earthquake in Japan prompted what got here to be known as the Fukushima Daiichi nuclear catastrophe, where radioactive material was launched in Japan and into the Pacific Ocean. This was the biggest nuclear disaster seeing that Chernobyl, and it brought about a cascade of troubles within the atmosphere and surrounding waters, spreading radioactive material through far-ranging ocean currents.

Emotional Aftershocks

Natural disasters can be in particular annoying for young youngsters. Confronted with scenes of destruction and the deaths of pals and cherished ones, many children develop post-worrying pressure disorder (PTSD), a severe psychological situation resulting from intense trauma. Left untreated, youngsters stricken by PTSD can be prone to lasting psychological damage and emotional distress.

Severity of Publicity

The quantity of publicity to the disaster is highly related to risk of future mental problems. At highest danger are those that go through the disaster themselves. Next are those in near contact with victims. At lower chance of lasting impact are folks who most effective had indirect publicity, which include information of the intense harm. Injury and lifestyles threat are the elements that lead most usually to mental fitness issues. Studies have checked out excessive natural disasters, including the Armenian earthquake, mudslides in Mexico, and Hurricane Andrew in the US. The findings show that at least half of these survivors suffer from distress or mental health issues that need scientific care.

Natural Disasters of every type can have number one, secondary, and tertiary outcomes.

- Primary Effects arise due to the technique itself. For example water damage all through a flood or collapse of buildings in the course of an earthquake, landslide, or typhoon.
- Secondary Effects occur simplest due to the fact a number one impact has brought on them. For example, fires ignited due to earthquakes, disruption of electrical power and water service due to an earthquake, flood, or storm, or flooding resulting from a landslide right into a lake or river.
- Tertiary Effects are long-term results which can be set off because of a number one occasion. These include such things as loss of habitat because of a flood, permanent changes inside the role of river channel as a result of flood, crop failure caused by a volcanic eruption *etc.*

Types of Natural Disasters

Floods

It is a herbal occasion or occurrence where a piece of land (or location) that is generally dry land, unexpectedly receives submerged below water. Some floods can arise abruptly and recede speedy. Others take days or maybe months to build and discharge. Floods are unpredictable and detrimental, and they can happen in regions that have never seen rain.

They can purpose death and injuries, isolate groups, damage main infrastructure, cut essential services, ruin property and livelihoods.

Causes of Flooding

Many conditions bring about flooding. Hurricanes, clogged drainages, and rainfall are a number of the conditions which have led to flooding in numerous areas across the globe. Here are the leading reasons of flooding.

1. Rain

Rain is the leading contributor to maximum of the flooding instances witnessed throughout the world. Too a great deal rain reasons water to drift overland contributing to flooding. In particular, it's miles because of excessive rainfall depth over a extended duration.

Depending on the rainfall distribution, the quantity of rain, and soil moisture content, brief rainfall length also can bring about flooding. Light rains for longer periods – numerous days or weeks, can also bring about floods. The rain water erosive pressure can weaken the rules of buildings, causing tumbles and cracks.

2. River Overflows

Rivers or streams can overflow their banks. This occurs while the river or move holds greater water upstream than usual, and it flows downstream to the neighbouring low-lying areas, typically referred to as the floodplains. As a consequence, this creates a surprising discharge of water into the adjacent lands main to flooding.

Dams in rivers may additionally at instances overwhelm rivers when the carriage capacity is exceeded, inflicting the water to burst and get into the floodplains. Flood as a result of river overflow has the capability of sweeping the whole lot in its course downstream.

3. Lakes and Coastal Flooding

Lake and Coastal flooding takes place whilst massive storms or tsunamis reasons the water frame to surge inland. These overflows have detrimental electricity since they could smash ill-prepared structures to resist water's power which includes bridges, houses, and cars.

In the coastal areas, robust and massive winds and hurricanes force water onto the dry coastal lands and give rise to flooding. The scenario is even worsened when the winds blowing from the sea convey rains in them. Sea waters from the tsunami or typhoon can reason substantial harm.

4. Dam Breakage

Dams are man-made structures used to preserve water from flowing down from a raised ground. The potential electricity stored within the dam water is used to generate electricity. At instances, the partitions can grow to be vulnerable and break due to overwhelming carriage ability. Due to this reason, breakage of the dam can cause massive flooding in the adjoining regions.

Flooding happens while the embankments built alongside the perimeters of the river to forestall excessive water from flowing onto the land breaks. Sometimes, the excess water from the dam is deliberately released from the dam to save you it from breaking thereby causing floods.

5. Melting of the Glaciers and Mountain Tops

In the bloodless regions, ice and snows build up all through the winters. When the temperature rises in summer, the accumulated snows and ice are subjected to melting resulting in considerable actions of water into lands that are commonly dry. Regions with mountains that have ice on top of them also experience the identical outcome while the atmospheric temperature rises. This sort of flooding is usually termed as snowmelt flood.

6. Clogged Drainages

Flooding also takes region while snowmelt or rainfall runoff cannot be channeled appropriately into the drainage structures forcing the water to waft overland. Clogged or lack of proper drainage machine is commonly the reason of this kind of flooding.

The regions continue to be flooded till the stormwater systems or waterways are rectified. Instances in which the structures or water ways are not rectified, the regions continue to be flooded till the extra water evaporates or is transpired into the surroundings with the aid of vegetation.

Droughts

Drought implies a lack of moisture for an extended period of time which in turns causes a deficit of moisture in the soil. This can mean different things for different regions. For areas which get hold of high quantities of precipitation (rain, snow), a circumstance of drought can develop more unexpectedly than in a place which doesn't. A drought is a duration of abnormally dry climate that persists long enough to supply a severe hydrologic imbalance, inflicting, for example, crop damage and shortages inside the water supply. The severity of a drought relies upon at the degree of moisture deficiency, the duration, and the dimensions of the affected region. Drought may be defined 4 ways:

1. Meteorological Drought–when a place gets less precipitation than normal. Due to climatic differences, what is taken into consideration a drought in one region won't be a drought in another area.
2. Agricultural Drought–while the quantity of moisture inside the soil now not meets the needs of a specific crop.
3. Hydrological Drought–whilst the surface and subsurface water components are below normal.
4. Socioeconomic drought –while water supply is unable to satisfy human and environmental wishes can upset the stability among deliver and demand.

What Reasons a Drought?

Drought has many causes. It can be caused by not receiving rain or snow over a time frame. We learned in the discussions about the water cycle and weather that changes within the wind styles that pass clouds and moisture thru the surroundings can reason an area to not obtain its everyday amount of rain or snow over an extended period of time.

If you stay in a place wherein maximum of the water you operate comes from a river, a drought in your place can be caused by locations upstream from you now not receiving sufficient moisture. There might be less water in the river for you and other humans who live alongside the river to use.

People can also play a large function in drought. If we use too much water at some point of instances of normal rainfall, we might not have sufficient water whilst a drought happens.

A drought is due to drier than ordinary situations which can eventually cause water deliver issues. Really hot temperatures can make a drought worse with the aid of inflicting moisture to evaporate from the soil. Just due to the fact a place is warm and dry would not always mean it's far going via a drought. Droughts best occur when a place is abnormally dry. Here's why:

Rain and snow don't fall evenly throughout Earth. Some regions are mechanically wet and others are mechanically dry. From season to season – and from year to year – the amount of rain or snow in a area can vary.

However, over a period of many years, the average quantity of precipitation in a area within reason consistent. For example, in the deserts of the American Southwest, the common precipitation is much less than three inches according to 12 months. But, the average every year precipitation in Atlanta is set 50 inches.

When a particular vicinity gets much less rain than usual, the soil receives an awful lot much less moisture, too. The soil starts drying out and flowers die. When this sample keeps for numerous weeks, months or years, the go with the flow of streams and rivers decreases and water levels in lakes, reservoirs and wells fall. Eventually, the uncommon dry weather reasons water deliver issues, and the dry length will become a drought.

Effects of Drought

Drought will have many devastating consequences on communities and the encircling environment. The amount of devastation relies upon on the energy of the drought and the duration of time an area is taken into consideration to be in drought situations. Drought has greater influences on poorer groups than on more prosperous groups who have better possibilities to deliver in resources from other regions. Drought will have extreme health, social, monetary, and political influences with far-reaching effects.

Water is one in every of the maximum essential commodities for human survival, second handiest to breathable air. When there is a drought–which with the aid of definition means having too little water to fulfil cutting-edge demands–conditions can become hard or risky very quick.

Hunger and Famine

Drought conditions frequently offer too little water to aid food crops, through either herbal precipitation or irrigation using reserve water substances. The identical trouble affects grass and grain used to feed cattle and poultry. When drought undermines or destroys food sources, human beings move hungry. When the drought is severe and maintains over an extended duration, famine may occur. Many of us bear in mind the 1984 famine in Ethiopia, which became the result of a deadly aggregate of a extreme drought and a dangerously useless government. Hundreds of hundreds died as a result.

Not Enough Drinking Water

All dwelling things must have water to survive. People can live for weeks without food, however only some days without water. In places like California, drought is experienced in particular as an inconvenience, perhaps with a few financial losses, however in very negative international locations the results are plenty more direct. When desperate for water to drink, human beings will flip to untreated sources that can cause them to sick.

Disease

Drought regularly creates a loss of smooth water for ingesting, public sanitation and private hygiene, which can result in a wide range of life-threatening diseases. The trouble of water access is critical: every year, millions are sickened or die due to lack of easy water access and sanitation, and droughts only make the trouble worse.

Wildfires

The low moisture and precipitation that frequently represent droughts can fast create hazardous situations in forests and across rangelands, putting the level for wildfires that may also motive injuries or deaths as properly as huge damage to property and already shrinking food elements. In addition, even flowers usually adapted to dry conditions will drop needles and leaves throughout a drought, contributing to a layer of dead flora on the ground. This dry duff then will become a dangerous gasoline for detrimental wildfires.

Wildlife

Wild flowers and animals suffer from droughts, even if they have a few diversifications to dry conditions. In grasslands, sustained lack of rain decreases forage production, affecting herbivores, grain-eating birds, and indirectly, predators and scavengers. Droughts will cause expanded mortality and decreased reproduction, which is especially complex for populations of at-danger species whose numbers are already very low. Wildlife needing wetlands for breeding (for example, ducks and geese) revel in drought as a decline in to be had nesting sites.

Social Conflict and War

When a precious commodity like water is in brief deliver because of drought, and the shortage of water creates a corresponding loss of meals, humans will compete–and subsequently fight and kill–to secure sufficient water to survive. Some accept as true with that the present day Syrian civil struggle in the long run commenced after 1.five million rural Syrians fled the drought-troubled rural areas for the towns, triggering unrest.

Electricity Generation

Many regions within the world rely on hydroelectric initiatives for electricity. Drought will reduce the quantity of water saved in reservoirs behind dams, reducing the quantity of strength produced. This trouble may be very tough for the numerous small communities counting on small-scale hydro, wherein a small electric powered turbine is established on a local creek.

Migration or Relocation

Faced with the other influences of drought, many people will flee a drought-stricken area in search of a new home with a higher supply of water, enough food, and without the ailment and struggle that were present in the region they are leaving.

Cyclones

Tropical cyclones are climate phenomena which shape over the sea via the release of strength generated by using evaporation and saturation of water on the ocean's surface. This system results in heavy rain and robust winds and when those storms technique land they could reason damage and flooding to inhabited areas. 20 per cent of Charter activations are a result of tropical cyclones. A cyclone describes a climate system characterized by way of swirling winds around a low-pressure centre; wind route around the low is counter clockwise within the Northern Hemisphere, clockwise in the Southern Hemisphere. Coming in a wide sort of sizes and settings, cyclones reason a number of the most dramatic and outright violent climate on the planet, along with the tropical cyclones referred to as hurricanes and typhoons. The science in the back of cyclones will assist you recognize why, where and how this weather phenomena exists.

Tropical cyclones are revolving storms that begin inside the tropics. Storms of this type are called hurricanes within the North Atlantic and eastern Pacific and typhoons in South East Asia and China. They are known as tropical cyclones inside the southwest Pacific and Indian Ocean vicinity. In the southern hemisphere storms rotate clockwise (see photographs above). Tropical cyclones shape close to the equator and advantage their energy from the warmth that is launched whilst water vapour condenses into rain. They are about 500 km throughout and can have a central place with tremendously little cloud and mild winds known as the eye.

Categories of Cyclones

Cyclones are categorized consistent with wind speeds and the harm they reason.

1. **Category 1 Cyclone:** Wind speeds between ninety and a hundred twenty five kilometres per hour, a few noticeable damage to houses and trees.
2. **Category 2:** Wind speeds between a hundred twenty five and 164 kilometres according to hour, harm to homes and considerable damage to crops and trees.
3. **Category 3**: Wind speeds between 165-224 kilometres consistent with hour, structural damage to houses, full-size damage to plants and uprooted trees, upturned automobiles and destruction of buildings.
4. **Category 4:** Wind speeds between 225 and 279 kilometres in keeping with hour, energy failure and a good deal damage to towns and villages.
5. **Category 5:** Wind speeds over 280 kilometres per hour, huge harm.

Causes of Cyclones

A tropical cyclone desires two most important ingredients: a cluster of thunderstorms and a warm frame of water – at the least 27°C – from which the storm gathers its electricity. The warm, tropical ocean below a growing hurricane evaporates then condenses to shape clouds, liberating heat during the system.

The heat power combined with the rotation of the Earth, receives the cyclone spinning and propels it forward. While the cyclone looks savage from the outside, its low-strain centre, commonly called the eye, is deceptively calm. This belies the hazard of the dense wall of cloud that surrounds it, that is the deadliest part of a cyclone.

Here the strongest winds and finest rainfall are found. The eye is normally 40km in diameter, but can variety in length from less than 10km to over 100km – as is the case with Yasi – relying on the dimensions of the cyclone itself.

Tropical cyclones acquire strength through the temperatures of the sea-floor which means that they disappear when they stumble upon land or less warm water. Tropical Cyclones can persist for a median of a week.

Tropical Cyclones generate floods, storms and damaging winds. Its winds can damage the belongings and houses of citizens and airborne particles can grow to be ability chance. The heavy rainfall comprised of the eye-wall is capable of causing extreme floods. Floods can cause extra destruction and deaths. The winds can produce generate storm surges that is a sudden upward thrust of seal-level near the shore as a result of cyclones. The sea water inundates low-lying coastal areas and this could cause erosion of embankments, destruction of plant life and infertility of soil. Strong winds harm communication systems for misery mitigation which makes it greater difficult to alert human beings inside the place of chance. Storm surges may contaminate ingesting water which can reason an epidemic of disease. After the cyclone passes, roads can nonetheless be flooded which can cause troubles for citizens to travel to dry regions.

Earthquakes

An earthquake is what occurs when blocks of the earth slip past one another. The surface in which they slip is known as the fault or fault plane. The region beneath the earth's floor where the earthquake begins is referred to as the hypocentre, and the region at once above it on the surface of the earth is called the epicentre.

The length of the faults varies greatly – from large scars that reduce throughout our entire planet, marking the factors of contact between the numerous lithospheric plates, to small surfaces of only few rectangular metres. The power of an earthquake, however, not only depends on the extension of the floor but also on the amount of movement and the quantity of energy that had accumulated before the breakage. Generally, however, seism of a more depth are placed close to the bigger faults. Together with these, typically smaller superficial actions arise, which may initiate seism of a minor intensity, at instances on account of a more important seism, as within the case of the ground settling aftershocks that follow the principle event.

Sometimes an earthquake has foreshocks. These are smaller earthquakes that happen in the same region as the bigger earthquake that follows. Scientists can't inform that an earthquake is a foreshock until the bigger earthquake happens. The largest, foremost earthquake is known as the main shock. Main shocks constantly have aftershocks that follow. These are smaller earthquakes that arise afterwards within the same place as the main shock.

What reasons earthquakes and wherein do they show up?

Earthquakes are induced while rocks underground all at once break, releasing strength wherein reasons the ground to shake. The two plates that are formed do not slide smoothly, alternatively the rocks capture on to every different. The rocks sooner or later breaks because of the constructed up stress, which reasons the earthquake.

The earth has four important layers: the internal core, outer core, mantle and crust. The crust and the top of the mantle make up a thin skin at the surface of our planet.

But this skin isn't always multi-functional piece – it's far made of many portions like a puzzle covering the floor of the earth. Not handiest that, but these puzzle portions keep slowly transferring around, sliding past each other and bumping into each other. We call those puzzle portions tectonic plates, and the rims of the plates are referred to as the plate boundaries. The plate limitations are made up of many faults, and maximum of the earthquakes around the world arise on these faults. Since the rims of the plates are rough, they get stuck even as the rest of the plate keeps moving. Finally, whilst the plate has moved a ways enough, the edges unstick on one of the faults and there is an earthquake.

Effects of Earthquake

Ground Shaking

Ground shaking is a time period used to describe the vibration of the ground during an earthquake. Ground shaking is caused by body waves and floor waves. As a generalization, the severity of floor shaking increases as importance increases and decreases as distance from the causative fault increases. Although the physics of seismic waves is complex, floor shaking can be defined in terms of frame waves, compressional, or P, and shear, or S, and floor waves, Rayleigh and Love.

Surface Faulting

Surface faulting is the differential motion of the 2 facets of a fracture at the Earth's surface and may be strike-slip, every day, and reverse (or thrust). Combinations of the strike-slip kind and the opposite two kinds of faulting can be observed. Although displacements of those types can end result from landslides and different shallow processes, surface faulting, because the term is used here, applies to differential movements caused by deep-seated forces in the Earth, the sluggish movement of sedimentary deposits towards the Gulf of Mexico, and faulting associated with salt domes.

Landslides

Past experience has proven that several kinds of landslides take location along with earthquakes. The most abundant styles of earthquake induced landslides are rock falls and slides of rock fragments that form on steep slopes. Shallow particles slides forming on steep slopes and soil and rock slumps and block slides forming on mild to steep slopes additionally take region, but they may be much less abundant. Reactivation of dormant slumps or block slides by earthquakes is rare.

Damage to Homes

High importance earthquakes can lead to finish collapse of buildings. Debris from collapsing homes is the primary threat in the route of an earthquake due to the fact the falling outcomes of huge, heavy objects can be lethal to human beings. High importance earthquakes result in shattering of mirrors and windows, which also present danger to human beings.

Damage to Infrastructure

Earthquakes can reason electricity lines to fall. This is risky because the exposed stay wires can electrocute humans or start fires. Major earthquakes can reason rupturing of roads, gasoline lines, and water pipelines. Broken gas traces can purpose gas to get away. Escaping gasoline can bring about explosion and fires, which may be tough to contain.

Leads to Liquefaction

Liquefaction is a phenomenon in which the soil turns into saturated and loses it electricity. When sediments consisting of excessive water content material are subjected to steady trembling, water pressure held within the sediment pores slowly boom. Ultimately, the sediments lose nearly all cohesive electricity and begin acting like liquids. Buildings and different systems constructed on pinnacle of this liquefied soil overturn or sink into the floor. Earthquakes are chargeable for most of the liquefaction taking place across the world. A typical instance of liquefaction phenomenon is the earthquake of 1692 in Jamaica that resulted inside the devastation of the city of Port Royal.

Landslides

A landslide is a mass motion of material, along with rock, earth or debris, down the slope of a hill or cliff. They can manifest all at once or flow slowly over long periods of time. A landslide is described because the motion of a mass of rock, particles, or earth down a slope. Landslides are a type of “mass wasting,” which denotes any down-slope motion of soil and rock below the direct impact of gravity. The term “landslide” encompasses 5 modes of slope motion: falls, topples, slides, spreads, and flows. These are further subdivided via the kind of geologic material (bedrock, debris, or earth). Debris flows (commonly called mud-flows or mudslides) and rock falls are examples of not unusual landslide types.

Landslides are categorised by means of their type of movement. The 4 primary forms of motion are:

- ☆ Falls
- ☆ Topples
- ☆ Slides
- ☆ Rotational slides
- ☆ Translational slides
- ☆ Flows

Landslides may be categorized as just this type of moves or, more commonly, can be a mixture of several. The phrases landslide or mudslide seek advice from the downward motion of massive hundreds of rocks, soil, mud and organic particles. Areas with steep slopes, for instance mountainous regions, are particularly liable to landslide dangers. Most landslides are because of multiple elements that act collectively to destabilize the slope.

Effects of Landslides

Lead to Monetary Decline

Landslides were proven to result in destruction of property. If the landslide is giant, it may drain the economic system of the area or country. After a landslide, the region affected commonly undergoes rehabilitation. This rehabilitation involves huge capital outlay. For instance, the 1983 landslide at Utah within the United States led to rehabilitation value of about $500 million. The annual loss because of landslides in U.S. Stands at an estimated $1.5 billion.

Decimation of Infrastructure

The pressure flow of mud, particles, and rocks because of a landslide can purpose extreme damage to property. Infrastructure together with roads, railways, enjoyment destinations, homes and verbal exchange systems may be decimated via a unmarried landslide.

Loss of Life

Communities dwelling at the foot of hills and mountains are at a more hazard of dying with the aid of landslides. A considerable landslide contains along massive rocks, heavy particles and heavy soil with it. This sort of landslide has the potential to kills plenty of humans on impact. For instance, Landslides within the UK that took place some years ago prompted rotation of debris that destroyed a faculty and killed over 144 human beings consisting of 116 college children aged between

7 and 10 years. In a separate event, NBC News stated a dying toll of 21 humans within the March 22, 2014, landslide in Oso, Washington.

Affects Splendour of Landscapes

The erosion left in the back of through landslides leaves behind rugged landscapes which might be unsightly. The pile of soil, rock and particles downhill can cover land used by the community for agricultural or social purposes.

Impacts River Ecosystems

The soil, particles, and rock sliding downhill can find way into rivers and block their herbal waft. Many river habitats like fish can die because of interference of natural go with the flow of water. Communities relying at the river water for household sports and irrigation will suffer if drift of water is blocked.

Avalanches

An avalanche is a mass of snow that slides swiftly down a willing slope, together with a mountainside or the roof of a building. Avalanches are prompted by way of both herbal forces (*e.g.* Precipitation, wind drifting snow, speedy temperature changes) and human activity. In mountainous terrain, they may be among the maximum serious hazards to human life and property. Avalanches are sometimes called snow slides. Avalanche, a French word actually meaning 'snow slide', is a mass of snow set in motion because of disturbances in the chemical equilibrium. Simply speaking, whilst heavy chunk of snow slides down the mountains, they cause avalanches.

Causes and Effects of Avalanches

An avalanche is a unexpected downhill movement of snow. It is a good sized chance to humans dwelling in, or visiting, glacial areas. A slab avalanche is the most dangerous shape of motion. It may be resulting from:

- Heavy snowfall
- Deforestation (as an instance because of new ski runs) making the slope much less stable
- Steep slopes, as this allows to growth the rate of movement
- Vibrations (as an instance from an earthquake, noise or off-piste skiers)
- Layering of snow - for example wherein snow is already at the mountain and has was ice, and then sparkling snow falls on pinnacle that can effortlessly slide down
- The wind path piling snow that could overhang a mountain

Effects of Avalanches

Damage to Life and Property

A huge quantity of casualties takes vicinity after avalanches hit closely populated areas. Infrastructure is broken and the blockage caused, influences the livelihood of many. People who enjoy skiing, snowboarding and snowmobiling are at a greater risk of losing their lives. A effective avalanche may even spoil homes and strength supplies can be cut off.

Flash Floods

When an avalanche happens, it brings down all the particles with it and can purpose havoc in low mendacity areas. Flash floods are visible to manifest after avalanches, that is a protracted time period problem many villagers and townspeople must deal with. They can also change weather styles and cause crop failure in farms gift at the decrease fields.

Economic Impact

An avalanche can block anything in its course and even restrict the normal movement of traffic. Various ski accommodations depend upon travellers to run their enterprise successfully. Ski hotels and different organizations are compelled to close till the avalanche decreases and weather conditions grow to be suitable.

Volcanic Eruptions

A volcano is an opening within the earth's crust through which lava, volcanic ash, and gases escape. Volcanic eruptions are partially driven via pressure from dissolved gas, tons as escaping gases pressure the cork out of a bottle of champagne. Beneath a volcano, liquid magma containing dissolved gases rises through cracks inside the Earth's crust. As the magma rises, pressure decreases, allowing the gases to form bubbles. How the magma (lava) behaves when it reaches the surface relies upon on each its gasoline content and chemical composition. Lavas with low silica contents have low viscosities and go with the flow freely, allowing any gas bubbles to escape readily, at the same time as lavas with excessive silica contents are greater viscous (immune to flow), so that any trapped gases can't break out gradually.

The maximum not unusual type of volcanic eruption happens when magma (the time period for lava whilst it's miles under the Earth's floor) is launched from a volcanic vent. Eruptions may be effusive, wherein lava flows like a thick, sticky liquid, or explosive, in which fragmented lava explodes out of a vent. In explosive eruptions, the fragmented rock can be accompanied through ash and gases; in effusive eruptions, degassing is common but ash is generally not.

How are Volcanoes Formed?

Volcanoes are formed by way of eruptions of lava and ash whilst magma rises thru cracks or weak-spots inside the Earth's crust. The fundamental concept of an eruption is that an boom in strain on the chamber lid causes the magma to be launched from under it. However, there are variances inside the cause of this magma movement and the form of eruption generated.

Volcanoes are commonly determined near the limits of Earth's tectonic plates. These can either spread apart or depart a gap inside the surface, or they could push under one another – a process referred to as subduction.

When the plates separate, magma rises slowly that allows you to fill the space through a mild explosion of skinny basaltic lava, which is at temperatures from 800 to 1200 stages Celsius.

What Determines the Dimensions of an Eruption?

Ultimately, the dimensions of an eruption will depend upon the thickness of the magma, the density of gases it contains and the quantity of latest magma being driven into the magma chamber.

Basaltic lava allows gas to escape without difficulty, ensuing in smaller eruptions, while andesitic and rhyolite lava makes it harder for the gas to break out, main to larger eruptions.

Typical Volcanic Risks

Lava is regularly idea to be the main chance of a volcanic eruption, but this is not the case. Numerous hazards result from eruptions and they could have a range of consequences. The most risky are the pyroclastic clouds, which damage something in their route. Other hazards consist of ash clouds, ash rain, mudslides, earthquakes, tsunamis, odd weather styles and glacial flooding.

Heat and Cold Waves

Heat waves are periods of abnormally high temperatures and warmth index. Definitions of a warmth wave vary due to the version of temperatures in one-of-a-kind geographic locations. Excessive heat is regularly observed via high stages of humidity, however also can be catastrophically dry. Because warmth waves aren't visible as different kinds of excessive climate are, like hurricanes, tornadoes, and thunderstorms, they're one of the much less known forms of extreme weather. The evaporation of our bodies of water can be devastating to marine populations, decreasing the scale of the habitats available as well as the quantity of vitamins present in the waters. Livestock and other animal populations may also decline as well.

A bloodless wave is a climate phenomenon that is prominent by means of a cooling of the air. A bloodless wave is a speedy fall in temperature within a 24-hour length requiring substantially expanded safety to agriculture, industry, commerce, and social activities. The particular criterion for a cold wave is decided by using the price at which the temperature falls, and the minimum to which it falls. This minimum temperature is depending on the geographical location and time of year. Cold waves normally are able to going on any geological location and are formed by means of huge cool air loads that collect over certain regions, due to actions of air streams. A cold wave can purpose demise and damage to cattle and wildlife. Extreme winter cold frequently causes poorly insulated water pipes to freeze. Even some poorly included indoor plumbing may additionally rupture as frozen water expands inside them, causing belongings damage. Fires, paradoxically, come to be greater hazardous throughout extreme cold. Water mains may break and water supplies may additionally come to be unreliable, making fire-fighting greater difficult.

Global Climate Change

Climate exchange is a change in the statistical houses of the climate device that persists for several a long time or longer–usually at the least 30 years. These statistical residences include averages, variability and extremes. Climate trade may also be because of herbal processes, which include changes inside the Sun's radiation, volcanoes or inner variability within the weather gadget, or due to human influences which include adjustments in the composition of the atmosphere or land use.

Climate alternate describes a change within the common situations – together with temperature and rainfall – in a area over an extended duration of time. For example, 20,000 years ago, a great deal of America was protected in glaciers. In America today, we've got a warmer weather and less glaciers.

Global weather alternate refers to the common long-term adjustments over the complete Earth. These encompass warming temperatures and modifications in precipitation, in addition to the consequences of Earth's warming, including:

- Rising sea stages
- Shrinking mountain glaciers
- Ice melting at a faster rate than standard in Greenland, Antarctica and the Arctic
- Changes in flower and plant blooming times.

Earth's climate has constantly been changing – even long before people came into the picture. However, scientists have observed unusual adjustments recently. For example, Earth's average temperature has been growing plenty extra speedy than they might expect over the past 150 years.

Causes of Climate Change

Global warming is resulting from the greenhouse impact, a herbal method via which the surroundings retains some of the Sun's warmness, allowing the Earth to preserve the necessary situations to host lifestyles. Without the greenhouse effect, the average temperature of the planet could be -180C.

The trouble is that daily human activities maximize the greenhouse effect, causing the planet's temperature to increase even extra.

Consequences of Climate Change

The worldwide temperature increase brings disastrous results, endangering the survival of the Earth's flora and fauna, consisting of human beings. The worst climate exchange affects include the melting of the ice mass at the poles, which in turn causes rising sea level, generating flooding and perilous coastal environments thru which small island states danger disappearing entirely.

Climate trade also increases the appearance of extra violent climate phenomena, drought, fires, the dying of animal and plant species, flooding from rivers and lakes, the creation of weather refugees and destruction of the food chain and monetary resources, especially in developing countries.

1. Global Warming
2. Sea Level Rise
3. Ozone Depletion

Global Warming

Global warming affords the gravest threat to lifestyles on Earth in all of human history. The planet is warming to a point past what many species can handle, changing or disposing of habitat, decreasing meals sources, inflicting drought and other species-harming intense climate activities, and even at once killing species that truely can't stand the warmth. In fact, scientists are expecting that if we keep going alongside our present day greenhouse gasoline emissions trajectory, weather alternate will purpose extra than a 3rd of the Earth's animal and plant species to face extinction through 2050 – and up to 70 percentage by the stop of the century. Such a catastrophic loss could irreversibly lessen biodiversity, seriously disrupt ecosystems, and motive immense hassle for human societies worldwide.

Global warming refers back to the anthropogenic warming of the planet. This warming comes from intensifying the greenhouse effect, an impact that exists on many different planets and has existed on Earth for almost as long as the planet has. When utilized in conversation, weather alternate and global warming are sometimes used interchangeably, as if there is no difference. However, international warming refers back to the anthropogenic warming of the Earth's climate, over a long length of time, whereas climate change includes each natural and man-made causes. It is more accurate to mention that worldwide warming is one of the many signs of climate alternate.

Global warming is most effective one in all many troubles as a result of weather change. At the moment, among the outcomes of global warming have been particularly benign, and sometimes even beneficial, however as worldwide warming intensifies the outcomes will preserve getting worse.

Global warming happens when carbon dioxide (CO_2) and other air pollution and greenhouse gases gather inside the environment and soak up sunlight and sun radiation that have bounced off the earth's surface. Normally, this radiation could break out into space–but these pollutants, which can closing for years to centuries within the atmosphere, trap the heat and reason the planet to get hotter. That's what's known as the greenhouse effect.

Impact of Global Warming

Global warming, due to greenhouse gasoline pollution, is inflicting instant and direct adjustments to the planet.

The Earth's temperature had already warmed via 1°C in comparison to pre-industrial degrees. This temperature rise may appear small, however small rises in temperature translate into large changes for the world's climate. This is because the quantity of greater electricity wished to boom the world's temperature, even by means of a little, is vast. This extra electricity is like force-feeding the global weather machine.

Hotter Days

2015 become the hottest 12 months on report, the previous report turned into broken in 2014, and 2016 is anticipated to set a new record for the third yr. in a row. In the past few years records have being broken for longest heat waves and the Bureau of Meteorology has added purple and magenta to the forecast map for temperatures up to 54°C.

Rising Sea Stages

Increased ocean temperatures are melting glaciers and ice caps all over the world. Melted ice will increase the quantity of water in our oceans. Warmer temperatures also bring about the growth of the water's mass, which reasons sea stages to rise, threatening low-mendacity islands and coastal towns.

More Common and Extreme Climate Activities

Extreme weather occasions like bushfires, cyclones, droughts and floods are getting greater common and greater extreme due to global warming.

Oceans are Warming and Acidifying

The oceans have absorbed maximum of more warmness and carbon dioxide (CO_2) so far – extra than the air – making the seas both warmer and greater acidic. Warming waters are bleaching coral reefs and driving stronger storms. Rising ocean acidity threatens shellfish, consisting of the tiny crustaceans without which marine food chains could collapse.

Other Effects

1. Melting glaciers, early snowmelt, and excessive droughts will reason more dramatic water shortages and increase the chance of wildfires in the American West.
2. Rising sea stages will cause coastal flooding on the Eastern Seaboard, mainly in Florida, and in different regions along with the Gulf of Mexico.
3. Forests, farms, and cities will face troublesome new pests, warmness waves, heavy downpours, and expanded flooding. All those elements will harm or destroy agriculture and fisheries.

4. Disruption of habitats such as coral reefs and Alpine meadows could power many plant and animal species to extinction.
5. Allergies, asthma, and infectious sickness outbreaks becomes more common due to increased growth of pollen-generating ragweed, better degrees of air pollution, and the unfold of conditions beneficial to pathogens and mosquitoes.

Sea Level Rise

Rising global sea level is a crucial result of climate alternate. As ocean waters warm, they expand. Also, as air temperatures warm, water from melting ice sheets, polar ice caps, and glaciers enters into our ocean basins. Sea stage rise (SLR) is considered one of the most severe influences of weather change, with growing waters threatening to inundate small-island countries and coastal regions by the stop of the century.

Sea level rise is not uniform across the globe. Coastal groups are suffering from their nearby sea degree rise, which displays worldwide sea stage upward thrust, changes in nearby land elevation, tides and winds. In Louisiana, for example, local sea level is rising approximately four inches in step with decade due to the fact the land is sinking and sea level is rising.

Global average sea degree has risen by approximately eight inches (approximately 21 cm) due to the fact that 1900, with about 3 of those inches (about 7.five cm) taking place considering that 1993. Human-caused climate change has made a huge contribution to sea degree upward thrust on account that 1900, contributing to a fee of upward push greater than throughout any preceding century in at the least 2,800 years. In addition to the worldwide average sea stage upward push, nearby sea stage upward push – from time to time called "relative sea level upward thrust" – takes place at extraordinary charges in exceptional places. Local sea degree rise is stricken by the global sea level rise, however also by means of nearby land motions, and the consequences of tides, currents, and winds. Many places alongside America coast have seen their local sea levels upward push faster than the average international rate. As sea ranges have risen, the wide variety of tidal floods each year that purpose minor affects, often called "nuisance floods," have multiplied five- to 10-fold since the Sixties in numerous U.S. Coastal cities (very excessive confidence). Rates of growth are accelerating in over 25 Atlantic and Gulf Coast cities (very excessive confidence).

Sea Degree Rise: Predominant Reasons

The first is the thermal expansion: water, while heated by temperature upward push, tends to expand, *i.e.*, oceans take up more space.

In addition, the melting of the iced territories of Greenland and West Antarctica accelerates international warming. This technique is negatively influenced by way of freshwater seepage from the floor, which acts as a lubricant for the ice streams and facilitates them to slide quicker. That is, sparkling filtered water to the base of the ice sheets melt, weaken and slide them into the sea.

Sea Degree Rise: Important Consequences

The outcomes of sea level upward thrust are already being felt, and the forecasts are not very hopeful. First, water is increasingly invading coastal areas, causing soil erosion and dangerous farmland, housing or undertaking areas. The flooding of wetlands and pollution of aquifers additionally occur, affecting the plants and fauna of each place, causing the loss of habitat for fish, birds, plants and plenty of other species.

On the social aspect, the regular danger of sea level rise menaces hundreds of hundreds of thousands of people living in coastal communities. If water keeps to rise, they may be compelled to desert their houses and move to some other area, with the corresponding demographic problem. This is known as compelled migration ensuing from weather change.

Ozone Depletion

The atmosphere of the Earth is divided into 5 layers. In order of closest and thickest to farthest and thinnest the layers are listed as follows: troposphere, stratosphere, mesosphere, thermosphere and exosphere. The majority of the ozone inside the environment resides in the stratosphere, which extends from six miles above the Earth's floor to 31 miles. Humans rely heavily at the absorption of ultraviolet B rays by the ozone layer due to the fact UV-B radiation causes pores and skin most cancers and can result in genetic damage. The ozone layer has historically covered the Earth from the harmful UV rays, despite the fact that in latest a long time this protection has diminished due to stratospheric ozone depletion.

Atmospheric ozone has two outcomes at the temperature balance of the Earth. It absorbs solar ultraviolet radiation, which heats the stratosphere. It additionally absorbs infrared radiation emitted by the Earth's floor, successfully trapping heat within the troposphere. Therefore, the climate effect of changes in ozone concentrations varies with the altitude at which these ozone changes occur. The essential ozone losses that have been observed inside the decrease stratosphere because of the human-produced chlorine- and bromine-containing gases have a cooling effect on the Earth's floor. On the opposite hand, the ozone increases which can be anticipated to have befell in the troposphere due to floor-pollution gases have a warming impact at the Earth's surface, thereby contributing to the "greenhouse" impact.

Effects of Ozone Depletion

Effects on Human Health

Ozone layer depletion will increase the amount of UVB that reaches the Earth's floor. Laboratory and epidemiological research display that UVB causes non-melanoma pores and skin cancer and plays a primary position in malignant melanoma development. In addition, UVB has been related to the development of cataracts, a clouding of the eye's lens.

Effects on Plants

UVB radiation influences the physiological and developmental methods of plants. Despite mechanisms to reduce or restore these effects and a capacity to adapt to extended stages of UVB, plant boom can be directly laid low with UVB radiation.

Indirect adjustments because of UVB (such as adjustments in plant shape, how vitamins are distributed in the plant, timing of developmental phases and secondary metabolism) may be similarly or sometimes greater essential than damaging consequences of UVB. These changes can have vital implications for plant competitive balance, herbivore, plant diseases, and biogeochemical cycles.

Effects on Marine Ecosystems

Phytoplankton form the inspiration of aquatic meals webs. Phytoplankton productiveness is confined to the euphotic zone, the higher layer of the water column wherein there is sufficient daylight to support internet productiveness. Exposure to solar UVB radiation has been shown to affect both orientation and motility in phytoplankton, ensuing in decreased survival charges for these organisms. Scientists have demonstrated an immediate reduction in phytoplankton production due to ozone depletion-associated will increase in UVB.

Effects on Biogeochemical Cycles

Increases in UVB radiation could affect terrestrial and aquatic biogeochemical cycles, therefore altering each resources and sinks of greenhouse and chemically vital trace gases (*e.g.*, carbon dioxide, carbon monoxide, carbonyl sulphide, ozone, and in all likelihood other gases). These capability modifications would contribute to biosphere-atmosphere feedbacks that mitigate or make bigger the atmospheric concentrations of these gases.

Effects on Materials

Synthetic polymers, naturally going on biopolymers, in addition to some different substances of commercial interest are adversely stricken by UVB radiation. Today's substances are fairly covered from UVB with the aid of special additives. Yet, will increase in UVB ranges will accelerate their breakdown, proscribing the length of time for which they are useful outdoors

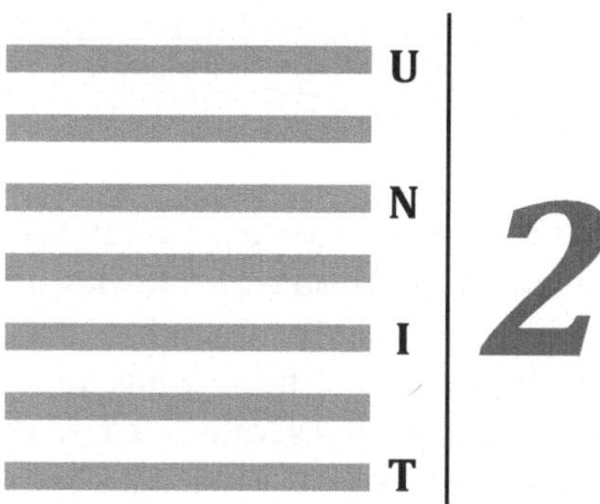

Man-Made Disasters

Introduction

Disasters are unlucky events that almost continually end within the loss of lifestyles and destruction of property. They show up randomly and suddenly, and have been happening since earlier than recorded time – well as a minimum natural screw ups like earthquakes, floods, landslides, hurricanes and tornadoes. Natural screw ups are dictated by way of natural forces that humans have little to no affect over. These are the screw ups humans analyze to prepare for and survive because very little may be executed to save you them. Man-made failures are difficult to predict, however they're preventable. With a bit vigilance, they shouldn't occur inside the first place.

Man-made failures can include hazardous fabric spills, fires, groundwater infection, transportation injuries, structure failures, mining injuries, explosions and acts of terrorism. There are movements that we will take to put together to react accurately to those activities. The key to acting accurately to those man-made threats is to discover what you need to do and what you need to have earlier of a man-made emergency and to prepare to respond as it should be.

Man-made failures can encompass hazardous cloth spills, fires, groundwater infection, transportation injuries, shape failures, mining accidents, explosions and acts of terrorism. There are movements that we are able to take to put together to react accurately to those occasions. The key to acting appropriately to those man-made threats is to find out what you want to do and what you need to have in advance of a synthetic emergency and to put together to respond as it should be. As new era develops, unseen dangers or flawed precautions can result in tragedy. Even when every safety degree is taken, extreme climate can cause disastrous consequences. Upon the discovery of trains, we had teach derailments. Upon splitting the atom, we had nuclear disasters.

2.1 Nuclear Disasters

Of all of the environmental catastrophe events that human beings are capable of causing, nuclear disasters have the greatest harm potential. The radiation release associated with a nuclear disaster poses considerable acute and chronic dangers in the on the spot environs and chronic chance over a huge geographic area. Radioactive infection, which typically turns into airborne, is long-lived, with half-lives guaranteeing contamination for hundreds of years.

Without doubt, the main drawback of nuclear energy is the possibility of nuclear accidents. Despite the safety of nuclear energy vegetation and that nuclear screw ups are not common, whilst they show up, the outcomes may be extraordinarily serious. Nuclear failures can be resulting from technical or human failure and are characterized through releasing radioactive products into the surroundings, in the form of radioactive count number or radiation. Nuclear accidents are events that emit a positive stage of radiation that can harm public health.

One way to outline the different sorts of nuclear accidents is in keeping with their severity. To decide the severity of an accident, an International Nuclear Event Scale has been defined (it's also known as the INES scale).

In reality, nuclear accidents are a form of nuclear occasion. Nuclear occasions are classified between nuclear injuries and nuclear incidents consistent with their severity. This classification includes both nuclear injuries and radioactive accidents. To understand each other, a nuclear accident could be a fault in a nuclear power plant reactor and a radiation twist of fate can be the discharge of a source of radiation to a river.

The worst nuclear failures within the world to this point are the following:

- ✰ Explosion of the Chernobylnuclear power plant, Ukraine.
- ✰ Earthquake and tsunami in primary Fukushima, Japan.
- ✰ Mayak nuclear catastrophe, Russia.
- ✰ Radiological coincidence in Goiânia, Brazil.
- ✰ Radioactive particle emission at Three Mile Island Nuclear Power Plant, USA.
- ✰ Accident inside the laboratories of Chalk River, Canada.
- ✰ Nuclear twist of fate in Windscale Pile, United Kingdom.
- ✰ Nuclear catastrophe at the uranium fuel treatment plant in Tokaimura, Japan.

A nuclear and radiation twist of fate is defined with the aid of the International Atomic Energy Agency as "an event that has led to big consequences to people, the environment or the facility." Examples consist of lethal effects to individuals,

NUCLEAR ACCIDENT OF THE CHERNOBYL NUCLEAR POWER PLANT, UKRAINE - 1986

Classified with level 7 on the INES scale, serious nuclear accident.

The Chernobyl nuclear accident is considered the worst nuclear accident in history.

In April 1986, during a series of tests in the Chernobyl nuclear reactor, a series of explosions occurred in the reactor core. The accident was due to a succession of human errors in the course of previously planned tests.

A cloud of radioactive material spilled from the reactor and fell over large areas around the plant, contaminating them heavily and making it necessary to evacuate and resettle some 336,000 people in other areas. Radioactive clouds also reached Eastern Europe, Finland and Scandinavia with gradually lower levels of pollution, which also affected Italy, France, Germany, Switzerland, Austria and the Balkans, to parts of the eastern coast of North America.

large radioactivity release to the environment, or reactor centre melt." The prime example of an "important nuclear coincidence" is one in which a reactor middle is broken and enormous quantities of radioactivity are released, such as inside the Chernobyl Disaster in 1986.

Impact of Nuclear Disasters

Apart from the damage caused by fires and explosions, accidents also release radioactive materials which can cause radiation sickness. Radiation exposure above a certain threshold, usually only received by workers and emergency teams in a stricken plant, causes acute radiation syndrome within hours of exposure. Depending on the dose of radiation this ranges from skin rashes, vomiting and diarrhoea, to coma and death.

Radiation damages DNA, especially as it assembles in dividing cells. That means tissues which contain many dividing cells, such as the gut lining, skin and bone marrow, are most at risk of damage. High enough doses also damage brain cells and such doses are invariably fatal.

Less severe damage can be treated, however. Gut damage disturbs fluid balance and can lead to blood infection; marrow damage means no blood cells are produced for clotting and fighting infection. If those problems can be managed, people can be kept alive long enough for gut and marrow to regenerate. A cloned human hormone that boosts white blood cell production sometimes helps; there is little else.

Fears of nuclear terrorism have recently inspired more funding for research into new treatments, most aiming to limit cell death in damaged tissues.

2.2 Chemical Disasters

Chemical, being at the core of contemporary manufacturing schemes, has reached a very thoughtful concern for disaster management within government, private sector and community at large. Chemical disasters may be traumatic in their impacts on human beings and have occasioned in the casualties and also damages nature and property. Chemical disasters are occurrence of emission, fire or bang involving one or more dangerous chemicals in the course of industrial activity *i.e.* handling storing or transport or due to natural events.

Chemical disasters can be demarcated as the unintentional issue of large quantities of poisons into the environment. The effects suffered by people in the area are determined by the toxicity of the chemical, its speed in dispersal, its composition (liquid, solid, or gaseous), and the spill site, esp. its nearness to a water supply or buildings. The major result may be due to the biochemical itself or to a resulting fire or explosion. The disastrous release of substances may overpower, at least provisionally, local or regional health care resources.

The Indian chemical industry donates 2.11 per cent to country's GDP. Between 2000 and 2015, the Indian chemical industry witnessed foreign direct asset worth 10.49 billion US dollars. This industry is expected to grow at 11 per cent per annum to US$ 154bn by 2020. However, major chemical disaster disturbs its development and these tragedies are easily avoidable.

Types of Major Chemical/Industrial Hazards

Major trade hazards are generally associated with the possible for fire, explosion or spreading of toxic chemicals and usually include the unintentionally announcement of substances from containment. Accidents involving major dangers could include:

- ☆ Leakage of flammable material, mixing of material with air, formation of flammable vapour cloud to a source of ignition, leading to a fire or an explosion affecting the site and possibly a populated area.

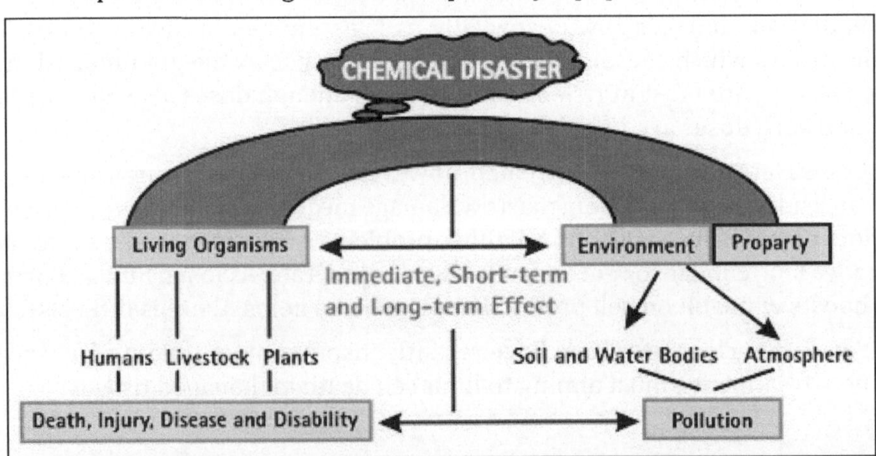

Major Chemical Accidents in India

Following the Bhopal Gas Disaster in 1984, major incidences of chemical disasters in India include

- A fire in an oil well in Andhra Pradesh (2003);
- A vapour cloud explosion in the Hindustan Petroleum Corporation Limited Refinery (HPCL), Vishakhapatnam (1997); and
- An explosion in the Indian Petrochemicals Corporation Limited (IPCL) Gas Cracker Complex, Nagothane, Maharashtra (1990).

Over 20 major chemical accidents have been reported in MAH units during 2002–06.

- Leakage of toxic material, formation of a toxic vapour cloud and drifting the cloud, affecting directly the site and possibly populated area.
- Depending upon the state of released chemical, cause and on its consequences, the major hazards in chemical process industry are classified as:
- Fire

Causes of Chemical Disasters

Ageing of the method plant life and insufficient steps to tempo with contemporary technologies in Indian chemical industry has accelerated vulnerability to chemical failures.

- A majority of the industrial injuries occurs due to human errors, because of non-compliance with standard operating procedure (SOP's) that have been put into area by the agency. Ex: Piper alpha accident is a classic instance of human error precipitated disaster.
- Chemical disaster have additionally occurred because of defects in design; absence of SOP's to mitigate an early warning in the procedure, terrible co-ordination between distinctive departments within the chemical employer.
- Natural failures including flood and earthquakes have also caused predominant screw ups in chemical industry. The release of acrylonitrile at Kandla Port in Gujarat, during an earthquake in 2001, is one of the examples.

2.3 Biological Disasters

Biological catastrophe is defined as, "the devastating results caused by an significant unfold of a certain type of dwelling organism that can unfold a disorder, viruses or infestation of plant, animal or insect lifestyles on an endemic or pandemic degree".

According to National Disaster Management Agency, Biological failures are natural scenarios involving disorder, incapacity or demise on a huge scale among people, animals and flowers due to micro-organisms like bacteria, or virus or pollutants. Biological failures may be within the shape of:

- Epidemic affecting a disproportionately big variety of individuals inside a population, network, or area at the same time, examples being Cholera, Plague; or,
- Pandemic is a virulent disease that spreads across a big vicinity, that is, a continent, or maybe global of existing, emerging or re-emerging illnesses and pestilences, instance being Influenza H1N1 (Swine Flu).

The biological element of this hazard can in many cases occur naturally inside the environment however they most usually impact human populations when they're brought or favourable situations exist.

Biological substances are labelled as dangerous due to toxicity. Toxicity is one of 4 factors used to classify hazardous wastes and refers to the degree of harm the substance can do to a dwelling organism *i.e.* The impact on the complete organism. Biological dangers, additionally known as biohazards, refers to biological substance or organic subjects produced by using parasites, viruses, bacteria, fungi and protein that pose a risk to the fitness of residing organisms, generally that of humans.

Occupations/Workplace Where People May Come Into Contacts with Biological Hazards

- Medical group of workers, cleansing team of workers and laboratory technicians inside the medical profession.
- Employees in environmental hygiene services together with liquid waste and rubbish collection and disposal.
- Agriculture, fishery, veterinary services and manufacturing industries that use plant or animal based raw materials.
- Indoor places of work in enclosed areas with central air con like inns and restaurants.

The United States Center for Disease Control (CDC) categories organic dangers that might potentially reason a organic disaster, into four biosafety tiers or BSL 1-4.

1 *BSL-1*: Bacteria and Viruses including Bacillus subtilis, dog hepatitis, some cell cultures and non-infections bacteria. Protection is minimum at this stage concerning gloves and facial protection.

2. *BSL-2*: Bacteria and viruses that purpose handiest slight disease to human beings, or are tough to contract thru aerosol in lab placing such as hepatitis

A, B, C, mumps, measles, HIV *etc.* Precaution is little excessive with use of autoclaves for sterilizing and organic safety cabinets.

3. *BSL-three*: Bacteria and viruses that can motive intense to deadly disorder in humans including anthrax, west Nile Virus, MERS coronavirus. Precaution is tons extra stringent safety protocols together with the usage of respirators to prevent airborne infection.
4. *BSL-four*: Viruses which might be doubtlessly deadly to human beings for which there may be no acknowledged remedy or vaccine inclusive of Marburg virus, Ebola virus, Lassa fever virus *etc.* Protective measures include using a positive pressure employee's suit, with a segregated air deliver.

Biological screw ups are causative of technique or phenomenon of natural beginning or conveyed by organic vectors, along with exposure to pathogenic micro-organisms, pollution and bioactive materials which could cause loss of existence, injury, illness or other health effects, belongings harm, loss of livelihoods and services, social and economic disruption, or environmental harm.

2.4 Building Fire

Building Fires can be natural or man-made, depending on what originally prompted the catastrophe. Lightning can also purpose a natural fireplace, but leaked gasoline or faulty mechanical device is taken into consideration a man-made cause.

For example, in 1944 leaked gas from a Cleveland fuel corporation found its manner into a sewer, where it ignited. The explosion blew manholes into the air and ignited funnels of fire, which set homes on fire. The disaster ended with a hundred thirty dead and plenty of extra homeless.

In 1991, a food chicken processing plant in North Carolina saw 25 deaths because of a hearth because of a faulty hydraulic piece. This fire was specifically eye-opening because the employees could have survived were it no longer for fireplace doorways that have been locked. The plant had operated for eleven years without submitting to a protection inspection.

It is hard in charge a natural catastrophe on everybody because it's miles an 'act of God,' however shoddy safety and prevention measures aren't any excuse for the loss of life in man-made failures. Safety is maintained via system management, operational planning, diligent instruction and precise prevention. Unfortunately, man-made screw ups have many more causes now than ever earlier than. Fortunately, there are individuals who dedicate their lives and careers to help others while failures arise, man-made and herbal alike.

Buildings want to be designed to provide an acceptable degree of fireplace protection and minimise the dangers from warmness and smoke. The number one objective is to reduce to inside proper limits the potential for dying or injury to the occupants of a constructing and others who may additionally come to be involved, such as the fireplace and rescue service, as well as to defend contents and make certain that as plenty as possible of a constructing can maintain to feature after a fire and that it may be repaired. The hazard to adjoining houses additionally needs to be taken into consideration, as well as feasible environmental pollutants.

Fire occurs because of a chain of very rapid chemical reactions among a gas and oxygen that releases heat and light. For combustion to occur, oxygen, warmth and a gas supply should all be present; this is the 'fire triangle'. Flames are the seen manifestation of combustion.

Fire and Rescue Provider

The hearth and rescue provider:

- ✰ Promote fire protection and put in force hearth protection regulation.
- ✰ Undertake firefighting.
- ✰ Ensure country wide resilience.
- ✰ Offer other unique services, along with rescuing humans from road visitors' injuries.

In certain buildings, it may be tough for the fire and rescue provider to safely reach and work near fires. Under such circumstances additional centers are required to make certain that there may be no delay and to provide a secure running base. This might encompass:

- ✰ Firefighting routes.
- ✰ Firefighting lifts.
- ✰ Firefighting stairs.
- ✰ Firefighting shafts

2.5 Coal Fires

Coal mine fires are insidious, persistent, and as tremendous as the incidence of coal itself, but their capability adverse human health affects were poorly characterised. These fires are known as coal seam fires. They occur underground while a layer of coal in the earth's crust is ignited. Due to the out-of-sight nature of the fires, they are regularly hard to discover at first, and even tougher to extinguish. At any given second there are hundreds of coal seam fires burning anywhere big coal deposits exist, from Pennsylvania to China to Germany. Due to their in reality infinite deliver of fuel and oxygen, many coal seam fires are predicted to hold burning nicely into the next century.

In all around the world, mining is taken into consideration as a high-danger activity that is pregnant with critical screw ups no longer handiest for miners, engineers, and other humans into it, but additionally for folks that stay near the mines.

Coal fires, both man-made or spontaneous combustion, no longer simplest reason losses of natural resources, however also motive environmental problems.

The surface feature and by way of-merchandise of coal fires consist of pyro-metamorphic rocks, fumarolic minerals, burnt pits and trench, subsidence and cracks, and surface thermal anomalies. These capabilities can be detected from seen, close to infrared, short-wave infrared, radar and thermal infrared far flung sensing images. The ability to locate those features is limited by using the spectral, spatial and temporal decision of the far off sensing data. The advances of new far flung sensing structures will beautify the capability to locate coal hearth related capabilities.

Coal mine fires purpose physical risks and negative air great. Proximity to the source of pollutants and smouldering combustion traditional of coal mine fires increase the danger of network publicity to high concentrations of recognized pollution together with aerosolised particles, and merchandise of incomplete combustion. Coal mine fireplace smoke is probable to have short-time period detrimental respiratory impacts. Adverse cardiovascular results and improved mortality are also plausible depending upon the magnitude of publicity and the quantity of people affected. There is inadequate proof to determine the chance of other fitness consequences. There are foremost gaps in the available proof for health consequences associated with exposure to bad air first-rate for time durations of weeks to months. The incomplete proof base hampers moves to mitigate harms in a timely, scientifically-knowledgeable manner. The want to in addition recognize the fitness influences of coal mine fires is pressing, especially as they disproportionately affect susceptible and disadvantaged groups and are possibly to come to be greater frequent and extreme therefore of weather exchange.

Unlike woodland fires, coal seam fires also release large portions of toxic fumes inclusive of mercury, methane, carbon monoxide and carbon dioxide which are present inside the coal and can be deadly. Carbon monoxide, referred to as the silent killer, can motive flu-like signs and symptoms and regularly loss of life with out warning. Metal toxins like mercury also can settle out of the air and into the ecosystem, contaminating food deliver chains. At the moment, the total quantity of greenhouse gases released from coal seam fires isn't properly studied, however scientists agree that the quantities may be substantial. Coal seam fires can burn for years, and lots of them span hectares underground, so if only a part of one breaches the surface, it will possibly be detected at the GFW platform.

2.6 Forest Fires

Wildfire, additionally called wooded area, bush or plants hearth, may be described as any out of control and non-prescribed combustion or burning of plant life in a herbal setting including a woodland, grassland, brush land or tundra, which consumes the herbal fuels and spreads primarily based on environmental conditions (*e.g.*, wind, topography). Wildfire can be incited with the aid of human movements, along with land clearing, excessive drought or in rare instances with the aid of lightning (IRDR).

Forest hearth, out of control hearth taking place in plant life greater than 6 feet (1.8 m) in height. These fires frequently attain the proportions of a first-rate conflagration and are now and again begun by combustion and heat from floor and floor fires. A huge forest fire may additionally crown–this is, unfold rapidly thru the topmost branches of the trees before concerning undergrowth or the woodland floor. As a result, violent blow-ups are commonplace in woodland fires, and they will anticipate the traits of a firestorm.

Causes of Forest Fire

Forest fires are resulting from Natural reasons in addition to Man-made reasons:

1. Natural reasons - Many wooded area fires begin from natural causes inclusive of lightning which set timber on fireplace. However, rain extinguishes such fires without causing an awful lot harm. High atmospheric temperatures and dryness (low humidity) provide favourable circumstance for a fire to start.
2. Man-made causes - Fire is precipitated when a source of hearth like naked flame, cigarette or bidi, electric powered spark or any supply of ignition comes into contact with inflammable material.

Impact of Forest Fires

Wildfires can have instantaneous and long-time results at the pleasant of rivers, lakes, and streams. The maximum noticeable effect of wildfires is typhoon water runoff. After the loss of vegetation, the ground's soil turns into hydrophobic and forestalls the absorption of water. This inability to soak up water promotes the transportation of debris and sediment into larger our bodies of water, similarly polluting valuable and crucial resources. Post-hearth flash floods grow to be a danger and permit the creation of heavy metals from ash and soil to infiltrate waterways. Filtering these water sources can be costly as well as time consuming.

Wildfires have both immediate and long-term effects on air quality. As a wooded area burns, huge amounts of smoke are launched into the atmosphere. These smoke particles are typically small and made up of gases and water vapour. Air pollutants from fires have the capability to travel first-rate distances and in many instances may additionally pose a risk to human fitness. These small particles can turn out to be lodged deep within our lungs, making it tough to breathe in addition to placing additional pressure on our hearts. Additionally, wildfires produce an multiplied amount of carbon monoxide, which too can cause lots of health implications.

The Ecological Benefits of Wildfires

Although wildfires go away an immense quantity of destruction of their path, they do depart in the back of some beneficial qualities too. Many flora require everyday burns in an effort to unfold seeds and survive. Fires can also kill sicknesses and insects that may be affecting the livelihood of flora, remove immoderate debris from the forest floor, and permit extra access to the nutrients supplied by way of exposed sunlight. Low depth fires clear underbrush and save you the harm of destiny fires from spreading.

After a wildfire, new grasslands are created and permit grazing animals to enjoy the change. This growth in the natural order of species allows a change in ecology that promotes increase and the continual cycle of life. Vegetation just like the fireweed, requires the disturbance of fire to bloom and allow the regrowth of vegetation which have perished due to the fireplace. As flora and flora die away, new existence starts to heal and spring forth.

2.7 Field Fires–Burning of Straw, Stables and Residues Oil Fire

Open burning or Field fires consult with a not unusual agricultural practice found today at some stage in the world, including essential cryosphere regions including northern Eurasia, the Andes and the Himalayas. It is the ordinary and periodic burning of cultivated fields for the cause of cost effectively and quickly removing excess vegetation, which includes crop residue which includes straw, weeds and waste, previous to sowing a brand new crop. In assessment to the traditional understanding of the Andes and Himalayas mentioned above, today's burning is notion by using a few farmers and maximum city dwellers to enrich the soil. In reality, however, burning damages soil and reduces its productive ability by way of destroying the humus (natural matter) and soil structure important to correct production, at the same time as at the identical time destroying a capability resource – the straw. With every successive burn, soils emerge as less fertile and water retentive and prone to erosion, growing the want for fertilizer, local irrigation and run-off systems. Many farmers are well aware about those impacts however lack the gear and knowledge to adopt alternatives. Nevertheless, alternatives abound– mainly Conservation Agriculture (CA) practices (broadly defined as practices that preserve the soiled covered), but also the use of straw for gas, or

Although intentional use of fires to transform land has decreased globally (1, 2), specifically among surprisingly capitalized countries through regulatory and market-oriented processes and moral suasion, regulatory strategies were less powerful in southern and jap Asia (see desk S21). Some densely populated agricultural regions in China and India buck the worldwide trend, showing will increase in agricultural fires (2). This is in particular actual in north western India, in which rice residue burning makes a substantial contribution to air pollution and brief-lived climate pollutants (3, 4). Regulations are in area to lessen agricultural fires, but burning continues because of uncertainty concerning coverage implementation and regarding get entry to and returns to alternative technologies. With the sector burning season soon upon us, we synthesize emerging evidence on options to burning, clarify the commercial enterprise case for opportunity practices, identify closing uncertainties, and speak procedures to growth their large adoption. Often, there are difficult trade-offs between environmental improvement and profitable monetary opportunities. The case of crop residue management in north western India does not appear to suit this sample and provides training that may be useful elsewhere.

In India, Wheat stubble burning is a extraordinarily new trouble which started with mechanised harvesting the use of integrate harvesters. In the last four to 5 years, farmers from UP's Ghazipur district, mainly Zamania and Chandauli regions, have been burning wheat stubble at a massive scale. The government officials are not paying attention towards this. Ramnagina Kushwaha, a farmer from Zamania, Ghazipur, currently died in a field after being set ablaze even as burning wheat stubble.

The state authorities has now not carried out the National Policy for Management of Crop Residues to defend the parali (crop residue). On December 10, 2015, the National Green Tribunal (NGT) had banned crop residue burning in the states of Rajasthan, Uttar Pradesh, Haryana and Punjab. Burning crop residue is against the law beneath Section 188 of the IPC and beneath the Air and Pollution Control Act of 1981. However, authorities's implementation lacks strength. The Delhi excessive courtroom had additionally ordered towards burning residues, whilst Punjab government imposed a penalty of Rs 73.2 lakh farmers in 2016 for burning of crop residue.

Stubble (parali) burning is the act of setting fireplace to crop residue to get rid of them from the sector to sow the next crop.

In order to plant next wintry weather crop (Rabi crop), farmers in Haryana and Punjab need to move in a completely short interval and if they may be late, because of short winters these days, they may face full-size losses. Therefore, burning is the cheapest and fastest way to do away with the stubble.

If parali is left in the discipline, pests like termites may attack the upcoming crop.

The precarious monetary situation of farmers doesn't allow them to use luxurious mechanised techniques to eliminate stubble.

Causes of the Stubble Burning

Technology

The problem arises because of the use of mechanised harvesting which leaves numerous inches of stubble within the fields.

1. Earlier, this excess crop was used by farmers for cooking, as hay to hold their animals heat or at the same time as greater insulation for homes.
2. But, now the stubble use for such functions has become outdated.

Adverse Impact of Laws

Implementation of the Punjab Preservation of Subsoil Water Act (2009) made the term of stubble burning coincident with the onset of winter in Northern India.

- Late transplanting of paddy throughout Kharif season to prevent water loss as directed with the aid of PPSW Act (2009) had left farmers with little time among harvesting and getting ready the field for the next crop and for this reason farmers are resorting to the burning of stubble.

High Silica Content

Rice straw is considered vain as fodder in the case of non-basmati rice, because of its high silica content.

Effects of Stubble Burning

Pollution

Open stubble burning emits massive quantities of toxic pollution within the surroundings which contain harmful gases like methane (CH_4), Carbon Monoxide (CO), Volatile natural compound (VOC) and carcinogenic polycyclic aromatic hydrocarbons.

After the release within the surroundings, these pollution disperse in the surroundings, may undergo a bodily and chemical transformation and in the end adversely affect human health through inflicting a thick blanket of smog.

1. Soil Fertility: Burning husk on floor destroys the nutrients inside the soil, making it less fertile.
2. Heat Penetration: Heat generated by using stubble burning penetrates into the soil, main to the lack of moisture and useful microbes.

Advantages of Burning

- Cheap
- Quick and easy
- Can help weed, insect and ailment control
- Reduced nitrogen tie-up

Disadvantages of Burning

- Loss of vitamins
- Loss of carbon
- Impact on soil microbes and fauna
- Reduction in soil structure (soil aggregate stability)
- Increase in erosion (wind and water)
- Can growth acidity over time

2.8 Air Pollution

Air pollution may be defined because the presence of poisonous chemical substances or compounds (along with those of biological origin) within the air, at tiers that pose a fitness chance. In a fair broader sense, air pollutants manner the presence of chemicals or compounds in the air which are usually now not present and which decrease the high-quality of the air or motive unfavourable modifications to the fine of lifestyles (inclusive of the dangerous of the ozone layer or causing international warming).

Air pollution is a mixture of solid debris and gases within the air. Car emissions, chemicals from factories, dust, and pollen and mould spores can be suspended as debris. Ozone, a gas, is a essential a part of air pollution in towns. When ozone paperwork air pollutants, it's also known as smog. Air pollutants is now considered to be the world's largest environmental fitness threat, accounting for 7 million deaths around the world every year. Air pollutants reasons and exacerbates a number of diseases, starting from allergies to cancer, pulmonary ailments and heart disorder. Outdoor air pollutants and particulate matter, certainly one of its main components, were classified as carcinogenic to people by way of the International Agency for Research on Cancer.

Causes of Air Pollution

- ☆ The burning of fossil fuels
- ☆ Agricultural activities
- ☆ Exhaust from factories and industries
- ☆ Mining operations
- ☆ Indoor air pollutants
- ☆ Natural Disasters

Effects of Air Pollution

It has been proven that air best influences human fitness. Air pollution causes around seven million deaths a 12 months worldwide. Growth and awareness of the population in cities, as well as the manner wherein we consume strength in urban areas thru delivery or heating and aircon systems, among others, result in the emission of big quantities of gases which can be harmful to our fitness.

Our physical and psychological well-being is affected otherwise via the form of air pollution we are exposed to. There are many organs and bodily features that can be harmed, the effects which include:

- ☆ Respiratory diseases
- ☆ Cardiovascular damage
- ☆ Fatigue, headaches and anxiety
- ☆ Irritation of the eyes, nose and throat
- ☆ Damage to reproductive organs
- ☆ Harm to the liver, spleen and blood
- ☆ Nervous system damage.

Urban populations are more exposed to go through the results of air pollution and, in this context, folks who are already unwell are especially vulnerable, as are children and the elderly.

2.9 Water Pollution

Water pollution is any contamination of water with chemicals or other foreign substances which might be unfavourable to human, plant, or animal fitness. These pollutants encompass fertilizers and insecticides from agricultural runoff; sewage and food processing waste; lead, mercury, and different heavy metals; chemical wastes from business discharges; and chemical contamination from risky waste sites. Worldwide, almost 2 billion human beings drink contaminated water that would be dangerous to their health.

When poisonous substances enter lakes, streams, rivers, oceans, and other water bodies, they get dissolved or lie suspended in water or get deposited at the bed. This results within the pollution of water whereby the great of the water deteriorates, affecting aquatic ecosystems. Pollutants also can seep down and affect the groundwater deposits.

Causes of Water Pollution

Water pollutants can come from a number of sources. Pollution can input water directly, through each legal and illegal discharges from factories, for example, or imperfect water remedy plants. Spills and leaks from oil pipelines or hydraulic fracturing (fracking) operations can degrade water substances. Wind, storms, and littering–specifically of plastic waste–also can ship particles into waterways.

Most water pollution doesn't begin inside the water itself. Take the oceans: around 80 percentage of ocean pollutants enters our seas from the land. Virtually any human interest can have an effect at the quality of our water environment. When farmers fertilize the fields, the chemical compounds they use are gradually washed through rain into the groundwater or floor waters nearby. Sometimes the causes of water pollution arc quitc surprising. Chemicals released by way of smokestacks (chimneys) can enter the surroundings after which fall back to earth as rain, getting into seas, rivers, and lakes and inflicting water pollutants. That's called atmospheric deposition. Water pollutants has many one-of-a-kind reasons and that is one of the motives why it is such a tough problem to solve.

Effects of Water Pollution

Pollution of water impacts each humans and aquatic existence. Most water assets near to cities and concrete centres are polluted with the aid of rubbish and dumping of chemical compounds, legally or illegally. Below are a number of the commonplace in addition to adverse results of polluting water bodies. Water pollution can bring about human health problems, poisoned wildlife, and long-time period environment damage. When agricultural and commercial runoff floods waterways with extra vitamins which includes nitrogen and phosphorus, those nutrients often fuel algae blooms that then create dead zones, or low-oxygen areas wherein fish and other aquatic existence can no longer thrive.

Sewage is another good instance of how pollutants can affect us all. Sewage discharged into coastal waters can wash up on seashores and cause a health chance. People who bathe or surf in the water can fall sick in the event that they swallow polluted water–but sewage can have other harmful results too: it could poison shellfish (which include cockles and mussels) that grow near the shore. People who devour poisoned shellfish danger laid low with an acute–and every so often fatal–illness referred to as paralytic shellfish poisoning. Shellfish is now not caught along many shores because it's miles sincerely too polluted with sewage or toxic chemical wastes which have discharged from the land nearby.

The results of water pollutants are not most effective devastating to people but also to animals, fish, and birds. Polluted water is flawed for drinking, recreation, agriculture, and industry. It diminishes the aesthetic first-rate of lakes and rivers. More seriously, infected water destroys aquatic life and reduces its reproductive ability. Eventually, it is a risk to human health. Nobody can break out the effects of water pollution.

Globally, unsanitary water resources additionally exact a health toll within the form of disorder. At least 2 billion people drink water from assets contaminated by means of faeces, in line with the World Health Organization, and that water can also transmit risky diseases such as cholera and typhoid.

2.10 Deforestation

Deforestation has substantially altered landscapes around the arena. About 2,000 years ago, 80 percentage of Western Europe become forested; nowadays the parent is 34 percentage. In North America, about half of the forests inside the eastern part of the continent had been reduce down from the 1600s to the 1870s for wooden and agriculture. China has lost super expanses of its forests over the last 4,000 years and now just over 20 per cent of its far forested. Much of Earth's farmland turned into as soon as forests.

Deforestation additionally threatens the world's biodiversity. Tropical forests are home to tremendous numbers of animal and plant species. When forests are logged or burned, it is able to drive many of the ones species into extinction. Some scientists say we are already in the midst of a mass-extinction episode.

More immediately, the lack of trees from a forest can go away soil more susceptible to erosion. This causes the ultimate flowers to emerge as extra at risk of fire because the woodland shifts from being a closed, moist surroundings to an open, dry one.

Deforestation is one of the most pressing environmental troubles that the world is going through currently. It is the conversion of forested land to non-forested land via humans. Deforestation takes place while a land dominated by means of naturally taking place bushes is transformed to offer positive services in reaction to the human demand. The indiscriminate felling of timber has resulted in a discount of three.16 per cent within the international forest cowl from 1990 to 2015. Although India has visible an increment in the total wooded area cowl of ca. 1 per cent, still there are positive regions inside the USA which have sought a decrease in the wooded area cowl. The main reasons attributed to the reduction in forest cowl are moving cultivation, rotational felling, different biotic pressures, diversion of wooded area lands for developmental activities, *etc.*

The Impact of Deforestation

Deforestation takes place for some of reasons, together with: farming, usually livestock due to its brief turnaround; and logging, for materials and development. It has been taking place for lots of years, arguably given that man began changing from hunter/gatherer to agricultural based totally societies, and required larger, unobstructed tracks of land to accommodate livestock, crops, and housing. It became handiest after the onset of the current generation that it became an epidemic. Following are a number of the adverse effects of deforestation at the surroundings and its life paperwork:

- **Loss of Habitat:** One of the most dangerous and unsettling effects of deforestation is the loss of animal and plant species due to their loss of habitat; no longer most effective will we lose the ones recognized to us, but additionally those unknown, potentially an even more loss.
- **Increased Greenhouse Gases:** In addition to the lack of habitat, the lack of timber also allows a more quantity of greenhouse gases to be released into the atmosphere.
- **Water in the Atmosphere:** The bushes additionally assist control the extent of water inside the surroundings with the aid of supporting to modify the water cycle. With fewer timber left, because of deforestation, there is much less water in the air to be returned to the soil. In turn, this reasons dryer soil and the incapacity to grow crops, an ironic twist when considered towards the reality that 80 per cent of deforestation comes from small-scale agriculture and cattle ranching.
- **Soil Erosion and Flooding:** Further results of deforestation include soil erosion and coastal flooding. In addition to their previously cited roles, bushes also feature to keep water and topsoil, which offers the rich nutrients to maintain additional forest existence.
- **Destruction of Homelands**: The loss of forests has an immediate and direct impact on their lifestyle that we inside the modern world, regardless of our very own dependency on what the rainforest provides, will never know. The degree of immediacy is exponentially greater.

2.11 Industrial Wastewater Pollution

Water pollutants is any infection of water with chemicals or different foreign substances which are unfavourable to human, plant, or animal health. These pollutants include fertilizers and insecticides from agricultural runoff; sewage and meals processing waste; lead, mercury, and other heavy metals; chemical wastes from commercial discharges; and chemical contamination from dangerous waste sites. Worldwide, nearly 2 billion people drink infected water that could be harmful to their fitness.

When toxic substances enter lakes, streams, rivers, oceans, and other water bodies, they get dissolved or lie suspended in water or get deposited at the bed. This results in the pollution of water whereby the exceptional of the water deteriorates, affecting aquatic ecosystems. Pollutants can also seep down and affect the groundwater deposits.

Industry is a massive source of water pollution, it produces pollutants that are extremely harmful to people and the environment.

Many industrial facilities use freshwater to hold away waste from the plant and into rivers, lakes and oceans.

Pollutants from Industrial Sources include

- ✰ **Asbestos** – This pollutant is a serious fitness risk and carcinogenic. Asbestos fibres may be inhaled and reason illnesses which includes asbestosis, mesothelioma, lung most cancers, intestinal cancer and liver cancer.
- ✰ **Lead** – This is a metallic element and can purpose health and environmental problems. It is a non-biodegradable substance so is tough to clean up as soon as the environment is contaminated. Lead is dangerous to the health of many animals, consisting of humans, as it may inhibit the motion of bodily enzymes.
- ✰ **Mercury** – This is a steel detail and can reason fitness and environmental problems. It is a non-biodegradable substance so is hard to easy up as soon as the surroundings is infected. Mercury is also dangerous to animal health as it can cause contamination thru mercury poisoning.
- ✰ **Nitrates** – The elevated use of fertilisers' method that nitrates are greater often being washed from the soil and into rivers and lakes. This can cause eutrophication, which may be very problematic to marine environments.
- ✰ **Phosphates** – The improved use of fertilisers' method that phosphates are more often being washed from the soil and into rivers and lakes. This can reason eutrophication, which may be very complex to marine environments.
- ✰ **Sulphur** – This is a non-steel substance this is dangerous for marine lifestyles.

- ☆ **Oils** – Oil does not dissolve in water, as an alternative it bureaucracy a thick layer on the water surface. This can prevent marine vegetation receiving enough mild for photosynthesis. It is also harmful for fish and marine birds.
- ☆ **Petrochemicals** – This is fashioned from fuel or petrol and can be toxic to marine life.

Effects of Industrial Wastewater Pollution

Some humans agree with pollutants is an inescapable end result of human activity: they argue that if we need to have factories, cities, ships, vehicles, oil, and coastal resorts, a few diploma of pollutants is sort of certain to end result. In different words, pollution is a important evil that human beings need to placed up with if they want to make progress. Fortunately, no longer each person is of the same opinion with this view. One purpose humans have woken as much as the problem of pollution is that it brings expenses of its own that undermine any monetary blessings that come about by means of polluting.

Take oil spills, for instance. They can take place if tankers are too poorly built to survive injuries at sea. But the financial benefit of compromising on tanker satisfactory brings a monetary cost while an oil spill occurs. The oil can wash up on nearby seashores, devastate the ecosystem, and severely affect tourism. The main trouble is that the people who undergo the price of the spill (normally a small coastal community) aren't the folks that brought about the problem within the first place (the people who function the tanker). Yet, arguably, every person who places gasoline (petrol) into their automobile–or makes use of nearly any sort of petroleum-fuelled shipping–contributes to the problem in some way. So oil spills are a hassle for all people, no longer simply people who stay by the coast and tanker operates.

Sewage is another proper instance of how pollutants can affect us all. Sewage discharged into coastal waters can wash up on seashores and reason a health danger. People who shower or surf in the water can fall unwell in the event that they swallow polluted water–yet sewage could have other dangerous effects too: it may poison shellfish (such as cockles and mussels) that develop close to the shore. People who devour poisoned shellfish threat laid low with an acute–and every now and then fatal–infection called paralytic shellfish poisoning. Shellfish is not caught alongside many shores due to the fact it's far truly too polluted with sewage or toxic chemical wastes that have discharged from the land nearby.

The results of water pollution are not handiest devastating to human beings however also to animals, fish, and birds. Polluted water is mistaken for drinking, recreation, agriculture, and industry. It diminishes the aesthetic satisfactory of lakes and rivers. More seriously, contaminated water destroys aquatic existence and decreases its reproductive capability. Eventually, it's far a threat to human fitness. Nobody can break out the consequences of water pollutants.

2.12 Road Accidents

Road accidents are absolutely the most frequent and, overall, the reason of the most harm. The motives for this are the extraordinarily dense avenue site visitors and the relatively exquisite freedom of motion given to drivers. Accidents related to heavy goods automobiles (specifically coaches and lorries with trailers) arise all too frequently notwithstanding calls for responsible behaviour, for respect of the loading rules and the motorway code, in addition to the obligation for drivers to conform their pace, which affects preventing distances, to the visitors and weather conditions (rain, ice, fog, *etc.*). The prevention of street injuries is also extraordinarily vital and will be ensured by strict laws, by way of technical and police controls, ongoing schooling for drivers (particularly those involved within the transport of dangerous materials) and, if need be, through criminal and administrative penalties for those responsible.

A road accident refers to any accident related to at the least one road vehicle, happening on a street open to public circulation, and in which as a minimum one individual is injured or killed. Intentional acts (murder, suicide) and natural failures are excluded. Accidents are the most not unusual cause of loss of life. Accidents occur, even to folks that are cautious, however many accidents can be avoidable if easy precautions are taken.

The number of cars at the roads is usually increasing. Roads are usually more secure than they have been, excluding occasional terrible design, poor renovation and velocity bumps, which may damage tyres/steering/suspension and can make contributions to accidents, especially for motor cyclists. Cars are tons safer for occupants with seat belts, air bags, superior breaking systems, electronic balance manipulate and posted crash ratings. However, protection for pedestrians still lags behind.

Causes of Road Accidents

- ☆ **Over Speeding** – Increase in velocity multiplies the chance of coincidence and severity of injury for the duration of twist of fate. Faster automobiles are extra at risk of coincidence than the slower one and the severity of coincidence may also be extra in case of quicker the severity of coincidence can even be greater in case of faster vehicles. Higher the velocity, greater the risk. At high velocity the vehicle needs greater distance to prevent *i.e.* Braking distance.
- ☆ **Drunk Driving** – When you drink, you lose the potential to attention and function nicely and it's very dangerous while working a automobile. Driving beneath the have an effect on of alcohol causes car injuries every day, even whilst they may be one the pinnacle reasons that can be avoided. Always use a designated driver if you go out and drink.
- ☆ **Distraction to Driver** – Distractions can be out of doors or within the vehicle. The predominant distraction now days are speaking on mobile

telephone while driving. Act of talking on phone occupies essential part of brain and the smaller part handles the using skills. This division of brain hampers response time and ability of judgment. This will become one of the motives of crashes. Some of the distractions on road are Adjusting mirrors whilst using, Stereo/Radio in car, Animals at the street, Banners and billboards.

- **Running Red Lights** – When you're riding your vehicle, red light approach stop and not doing so commonly leads to automobile accidents. Drivers that run crimson lights, run the threat of causing wrongful death due to the fact they often reason side-effect collisions at high speeds. To keep away from a automobile accident, appearance both ways for oncoming cars as you technique a green light.
- **Teenage Drivers** – Youth is wasted on the young, however careful riding is by no means wasted on young drivers. Unfortunately, teens aren't frequently acknowledged for his or her carefulness. When youngster drivers hit the roads they don't continually recognise what to do and that lack of enjoy ends up inflicting automobile accidents.
- **Over loading** – many people are doing over loading and doing many accidents.

Effects of Road Accidents

- Poor avenue condition quantity for high percentage of street site visitors, injuries every yr. consisting of loss of life.
- Innocent human beings are dead in avenue coincidence.
- Many humans are handicap in street coincidence.
- Thousands human beings losing their and households losing their love ones.
- This not leaves effect on humans but also affects the US's gross domestic product.
- Some human beings can't preserve get back to their everyday lives.

2.13 Rail Accidents

Train accidents are not usually considered a primary threat, but once they do occur, they can bring about any range of significant accidents or even death. Trains are nonetheless typically used by means of human beings all over the USA, mainly commuter trains and subways, as a cheaper, more efficient means of getting to and from work. Train accidents aren't as commonplace as different transportation accidents, which perhaps is why they aren't considered as a chief threat. Although railroads aren't used as frequently as they have been in centuries past, they nonetheless remain pretty active. Unfortunately, whilst train accidents manifest, they regularly bring about serious injuries and fatalities.

A train accident may be a devastating incident ensuing in serious injury or even demise to the sufferers involved. Whether it's miles a derailment, a educate-automobile collision, a railroad crossing twist of fate, an twist of fate concerning risky substances or any different sort of railroad coincidence, many human beings may be harmed. Where negligence on the part of railroad employees, the teach company or a governmental agency, or equipment producers has been the purpose of such an accident, they'll be held responsible for damages linked with ensuing injuries or deaths.

Causes of Train/Rail Accidents

Train accidents occur as a result of derailment, collisions with other trains and automobiles, and the interference of bystanders or cars at teach crossings. Here are a number of the causes of these injuries:

- ✰ **Negligence –** Negligence may be assessed on the part of the railway itself, the railroad employees, a government agency, or equipment producers. Examples of railway negligence include the defective operation of a crossing arm or the failure of a signal mild to offer adequate warning. Unfortunately, there are many instances of previous railway protection technology; even though the science is ready and available, the railway won't have gotten round to enforcing it.
- ✰ **Conductor Error, Inexperience, or Fatigue –** Human error is always an ability element in any twist of fate. Poor judgement and impaired reactions have contributed too many avoidable educate failures.
- ✰ **Mechanical Failure –** Despite the railway's pleasant efforts to keep the infrastructure of the tracks, trains, and operating systems, mechanical failure, but slim, can result in a teach derailment or collision.
- ✰ **High Speeds –** As shown by using the scary Amtrak teach twist of fate in 2015 that killed 8 people, the faster the teach is moving, the more severe the effects are if it crashes. Derailments are much more likely if the train is travelling at high speeds and the severity of accidents are exponentially increased.
- ✰ **Irresponsible or Negligent Pedestrian or Motor Vehicle Operator –** In uncommon cases, a pedestrian or vehicle may be on or close to the tracks at the wrong time, inflicting the conductor to react all of sudden or the teach to derail.
- ✰ **Obstruction on the Tracks –** In a few cases, foreign objects, along with automobiles or equipment, may be left inadvertently in the route of an oncoming teach; if the conductor fails to look them or goes to fast, he or she may not be able to keep away from hanging them, causing an coincidence. This is what triggered the Chester Amtrak teach disaster on April 3 of this year, whilst a passenger educate collided with a backhoe, killing the 2 railroad construction workers operating it.

2.14 Air Accidents

As in keeping with Wikipedia, An aviation twist of fate is defined by the Convention on International Civil Aviation Annex 13 as an occurrence associated with the operation of an plane, which takes area from the time any person boards the plane with the intention of flight till all such men and women have disembarked, and wherein a) someone is fatally or critically injured, b) the aircraft sustains sizeable harm or structural failure, or c) the aircraft goes missing or turns into absolutely inaccessible. Annex 13 defines an aviation incident as an occurrence, other than an accident, associated with the operation of an aircraft that influences or could have an effect on the safety of operation.

Aviation safety experts have found out for some time that plane incidents and injuries almost usually end result from a series of events, every of which is associated with one or greater motive elements. Thus, the cause of a coincidence or incident has many aspects. Some the world over universal definitions in the context of the research of an aircraft accident or incident are listed below (ICAO, 1994):

- Accidents are occurrences related to the operation of plane, from the time any character boards an plane with the aim of flight till the time all humans have disembarked, that effects in one or greater of the following:
  - A person is fatally or significantly injured.
  - The aircraft sustains damage or structural failure that adversely affects the structural strength, performance, or flight characteristics of the plane and would generally require predominant repair or replacement of the affected component.
  - The plane is missing or completely inaccessible.
- Incidents are occurrences, aside from accidents, associated with the operation of aircraft that affect or could affect the safety of operation.

Causes of Air Accidents

Some not unusual factors that may make contributions to accidents include:

Equipment Failure/Malfunction

Defects in aircraft may be because of design flaw, producer flaw, or put on and tear from use. Although inspections are accomplished at the airplane before and after flights, injuries still result from faulty device and malfunctioning components.

Human Error

From mechanics and ground group to flight attendants and air visitors' controllers, these professionals all play a critical role within the protection and secure operation of the aircraft. The Federal Aviation Administration (FAA) seems on the training anyone has received, the attention to detail that was supplied before the flight took off, the communication that exists before air visitors manipulate, the ground group, and the pilot to determine if human errors contributed to the reason of the coincidence.

Mid-air Collisions

Mid-air collisions are less common than different forms of aviation injuries, due to the fact the FAA uses the most up-to-date generation for its air traffic manage system, which monitors airliners inside the United States. In addition, airliners are required to have TCAS II collision-avoidance systems, which detect potential collisions with other transponder-prepared plane and endorse pilots to climb or dive in response. When these gadgets malfunction or are ignored, injuries can occur.

Pilot Error

Pilots are chargeable for the secure transportation in their passengers. At times, rarely, pilots fail to conform to right procedures in the operation of an aircraft.

Weather

Unpredictable weather, together with shear gusts of winds, freezing ice storms, thunderstorms and lightning, and greater, can have an effect on both small and huge airplanes. Before airplanes can fly in freezing weather, they're de-iced, and all airplanes are required to have forward-searching radar wind-shear detectors on-board.

Measures to Be Taken

[Priority Measures and newly-introduced measures in the Eighth Program]

- ☆ Improving the New Air Navigation System
- ☆ Enhancing air traffic services
- ☆ Expanding airspace capability and selling the powerful use of airspace
- ☆ Strengthening catastrophe control measures for airports and aviation protection facilities
- ☆ Strengthening supervision over air carriers Shifting the emphasis to prevention in visitors protection administration

Because accidents are rare, analyses of accident facts can offer steerage on broad areas of concern but are inherently incapable of stopping different sorts of accidents. Incidents are more frequent and are a rich supply of protection statistics, but the amount of the records is so huge that it is hard to discover meaningful risks and keep away from unfruitful diversions. The manner is complex because some accidents are in reality particular and might not be indicative of future hazards, whereas a few seemingly inconsequential incidents are disasters waiting to happen.

2.15 Sea Accidents

- ☆ Accidents can arise anywhere that there are humans operating machinery. The world's oceans aren't any exception. Sea-going accidents have been a

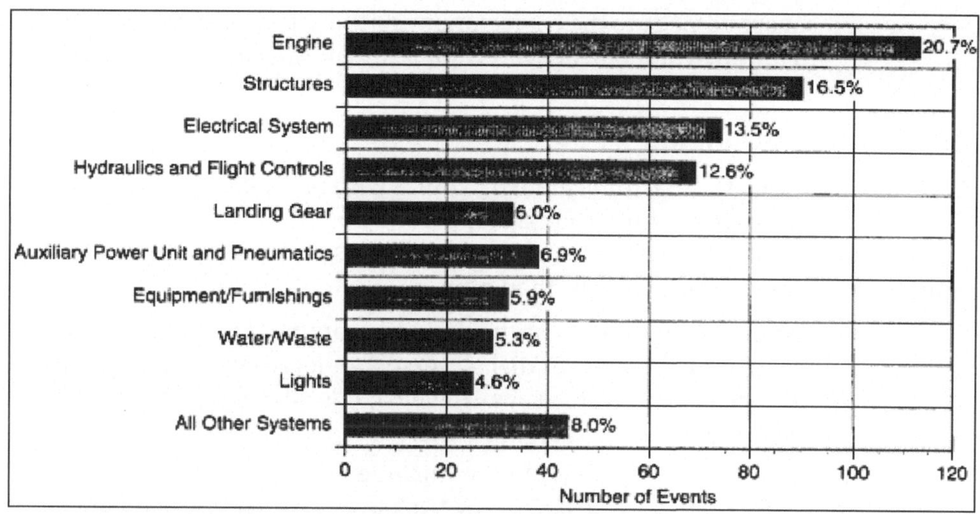

Related cause Factors in Worldwide Incidents involving Large Commercial Jet Aircraft Produced by a Particular Manufacturer (About 25 per cent of the worldwide commercial jet fleet), 1987 through 1996.

consequence of maritime way of life ever considering the fact that boats positioned out to sea thousands of years ago.

- Even inside the early days of seafaring, there was hobby in figuring out how a coincidence passed off and in looking for ways of preventing comparable occurrences. These investigations sooner or later became legitimate responsibilities of the government, for example, in Britain formal marine accident inquiries were mounted about a hundred and fifty years ago. Today, the investigations of sea accidents can reply at the abilities of forensic scientists and the numerous technologies they could bring to bear.
- Marine casualty means an event, or a series of events, that has led to any of the subsequent which has occurred directly in reference to the operations of a ship:
  - the death of, or extreme injury to, a person
  - the loss of someone from a deliver
  - the loss, presumed loss or abandonment of a deliver
  - material harm to a deliver
  - the stranding or disabling of a ship, or the involvement of a ship in a collision;
  - material harm to marine infrastructure outside to a ship, that could significantly endanger the protection of the ship, another ship or an individual

- severe damage to the environment, or the capacity for severe damage to the environment, delivered approximately by means of the damage of a deliver or ships.

Causes of Sea Accidents

Working within the maritime industry comes with many dangers and dangers. Many times, injuries and injuries happen apparently out of nowhere, leaving in the back of devastating results. There are several elements that contribute to maritime injuries, a few more commonplace than others.

- **Equipment Failure** – Maritime device is commonly well-designed and built to tolerate the stresses that all vessels undergo even as navigating the world's oceans and waterways. However, it is also increasingly complex and carries a myriad of moving parts.
- Common reasons of device failure include:
  - Lack of proper and consistent upkeep to save you gadget failure
  - Failure to monitor a deliver's systems to be able to ferret out ability problems earlier than they arise
  - Lack of the vital gadget and/or protecting parts or gear
  - Breakdowns in working procedures
  - Breakdowns in communications procedures
- **Lack of Proper Training and Human Error** – According to the Mariners Alerting and Reporting Scheme (MARS), human errors accounts for almost all of maritime accidents. Human blunders usually occurs because of improper schooling and/or lack of training, inexperience, fatigue from being overworked, and overexertion. In some cases, including the Costa Concordia grounding which brought on 32 deaths in January of 2012, human blunders may be resulting from the incompetence of a deliver's captain.
- **Collisions at sea** – Collisions at sea are more likely to occur in busy ports and crowded shipping lanes, especially near the coast or in narrow straits and inlets. These accidents are regularly the result of inexperienced watch standers on the bridge, navigational errors, malfunctioning or misused radar systems, and terrible decision-making via overworked personnel.
- **Poor Maintenance** – Maritime vessels are basically a set of interlinked systems that need to work together which will perform and navigate safely on the water. Every part of a ship is essential, regardless of if it's the deliver's engines or a watertight hatch. Even the most solid pieces of a deliver are laid low with wear and tear, so sticking to a right protection plan facilitates to save you maritime accidents.

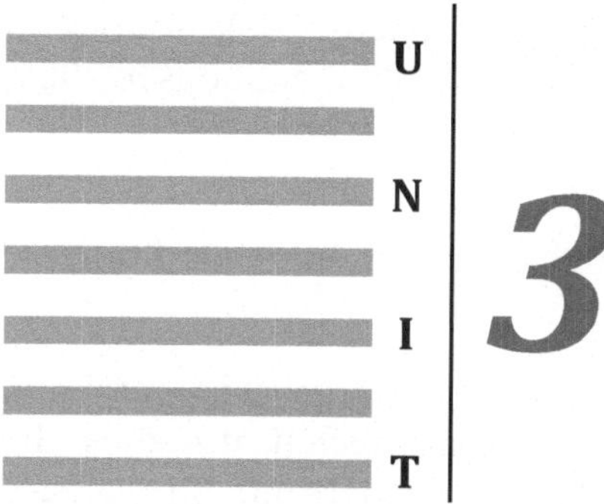

Disaster Management

Introduction

A tragedy is an unintended event in which the wants of the pretentious public overshadow the available resources. A tragedy occurs anywhere in the world nearly every day, but these events differ significantly in scope, size, and context. Large-scale tragedies with many fatalities are comparatively rare events. Convinced extensively exposed tragedies, counting events such as the terrorist attacks on September 11, 2001, Hurricanes Katrina and Sandy, and the Boston Marathon bombing, have absorbed people's attention on tragedy preparation and readiness. Disasters are becoming more recurrent, and the amount of persons pretentious is also cumulative. This better illness is attributable not only to the better amount of events, but also to populace dynamics, location, and vulnerabilities.

3.1 Efforts to Migrate Natural Disaster at National and Global Levels

3.2 International Strategy for Disaster Reduction

The International Strategy for Disaster Risk Reduction (UN/ISDR) is the focal point in the UN System to indorse links and synergies between, and the coordination of, disaster discount doings in the socio-economic, humanitarian and development fields, as well as to provision policy integration. It helps as a global information clearinghouse on disaster decrease, developing awareness campaigns and creating articles, journals, and other publications and promotional materials related to disaster reduction.

The United Nations International Strategy for Disaster Risk Reduction (UNISDR) is a planned outline, adopted by United Nations Member States in 2000, aiming to guide and organise the efforts of a wide range of partners to:

- Achieve substantive reduction in disaster losses
- Build resilient nations and communities as an essential condition for sustainable development.

UNISDR works with frequent organizations, States, intergovernmental and non-governmental organizations, financial organisations, technical bodies, the reserved sector and civil society, which work composed and share info. To decrease disaster risk. The UN delivers the secretariat of the strategy.

What is ISDR?

- The United Nations International Strategy for Disaster Reduction (ISDR) is a world-wide outline recognised within the United Nations for indorsing action to reduction social susceptibility and natural hazards risks and related technological and environmental disasters.
- It shapes upon the knowledge gained during 1990 – 1999 which had been chosen as the International Decade for Natural Disaster Reduction.
- The ISDR integrates the values which were accepted in the aforementioned Decade including Yokohama Strategy for a Safer World: Guidelines for Natural Disaster Prevention, Preparedness and Mitigation and its Plan of Action, and the Geneva Mandate on Disaster Reduction.

Vision

To allow all communities to end up resilient to the consequences of natural, technological and environmental risks, reducing the compound risks they pose to social and monetary vulnerabilities within cutting-edge societies.

To continue from safety in opposition to hazards to the control of hazard thru the mixing of danger prevention into sustainable development.

Goals

1. Increase public awareness of the risks that herbal, technological and environmental risks pose to trendy societies.
2. Obtain commitment by public government to lessen dangers to people, their livelihoods, social and financial infrastructure, and environmental assets.
3. Engage public participation at all levels of implementation to create catastrophe-resistant communities thru extended partnership and expanded danger discount networks at all tiers.
4. Reduce the monetary and social losses of disasters as measured, as an example, by using Gross Domestic Product.

Objectives

1. Stimulate research and application, offer knowledge, bring experience, build skills and allocate important sources for reducing or stopping extreme and recurrent influences of hazards, for those humans most vulnerable.
2. Increase opportunities for corporations and multi-disciplinary relationships to foster more clinical and technical contributions to the general public decision-making technique in topics of hazard, risk and catastrophe prevention.
3. Develop a extra proactive interface between management of herbal resources and risk discount practices.
4. Form a worldwide community devoted to making threat and disaster prevention a public value.
6. Link risk prevention and economic competitiveness issues to enhance possibilities for greater monetary partnerships.
7. Complete comprehensives hazard tests and integrate them within improvement plans.
8. Develop and apply chance discount techniques and mitigation measures with assisting arrangements and assets for catastrophe prevention in any respect levels of activity.
9. Identify and engage designated authorities, experts drawn from the widest possible range of expertise, and community leaders to broaden improved partnership sports.
10. Establish chance monitoring abilities, and early caution structures as integrated processes, with unique attention being given to emerging hazards with worldwide implications together with those related to weather variant and change, at all stages of responsibility.
11. Develop sustained packages of public facts and institutionalized educational components pertaining to risks and their effects, chance management practices and catastrophe prevention activities, for all ages.
12. Establish the world over and professionally agreed standards/ methodologies for the analysis and expression of the souci-economic influences of screw ups on societies.
13. Seek innovative funding mechanisms devoted to sustained chance and catastrophe prevention activities.

3.3 Concept of Disaster Management

Disaster Management may be defined as the company and management of sources and duties for coping with all humanitarian elements of emergencies, especially preparedness, reaction and recovery to be able to reduce the impact of disasters.

Disaster Management is a strategic making plans and procedure this is administered and employed to defend crucial infrastructures (also regarded as "vital assets") from excessive damages when herbal or human made calamities and catastrophic even occur. In the United States, Executive Order 13407 is set up as policy for America to have an powerful, reliable, integrated, flexible, and comprehensive system to alert and warn the overall public, which is referred to as "Integrated Public Alert and Warning System (IPAWS) (FEMA, 2011). In the later year of 2010, Europe began to broaden a strategic National Disaster Management after so many natural catastrophes passed off inside the year of 2010. According to European Academy (2010), there are 725 extremely climate phenomena brought about billions of Euro damage and hundreds of humans's lifestyles.

Disaster Prevention

These are sports designed to provide permanent protection from failures. Not all screw ups, particularly herbal screw ups, may be prevented, but the danger of loss of life and injury may be mitigated with accurate evacuation plans, environmental planning and design standards. In January 2005, 168 Governments adopted a 10-12 months global plan for natural disaster risk discount referred to as the Hyogo Framework. It gives guiding principles, priorities for action, and realistic method for accomplishing disaster resilience for vulnerable groups.

Disaster Preparedness

These sports are designed to minimise loss of life and harm – for example via removing humans and belongings from a threatened vicinity and by using facilitating well timed and powerful rescue, alleviation and rehabilitation. Preparedness is the main way of lowering the impact of disasters. Community-based totally preparedness and control need to be an excessive priority in physical therapy practice control.

Disaster Comfort

This is a coordinated multi-company reaction to lessen the effect of a disaster and its long-time period results. Relief sports consist of rescue, relocation, presenting food and water, stopping ailment and disability, repairing essential offerings which include telecommunications and transport, providing transient refuge and emergency health care.

Disaster Recovery

Once emergency needs had been met and the initial crisis is over, the people affected and the groups that assist them are nevertheless vulnerable. Recovery activities include rebuilding infrastructure, health care and rehabilitation. These should combination with improvement sports, including building human assets for health and developing rules and practices to avoid similar situations in future.

Disaster management is linked with sustainable improvement, especially when it comes to vulnerable humans consisting of those with disabilities, elderly human beings, children and other marginalised groups. Health Volunteers overseas publications cope with some of the commonplace misunderstandings about catastrophe control.

3.4 National Disaster Management Framework

Disaster Management is the coordination and integration of all activities necessary to construct, preserve and improve the capability to prepare for, guard in opposition to, respond to and get over threatening or actual natural or human-brought on disasters. It is a multijurisdictional, multi-sectorial, multi-disciplinary and multi-resource initiative. Therefore, it's far critical that the Federal, State and Local Governments, Civil Society Organisations (CSOs) and the non-public sector discharge their respective roles and obligations and complement every other in accomplishing shared desires of catastrophe control.

The National Disaster Management Framework targets to make India disaster resilient and significantly reduce the loss of lives and belongings. The plan is primarily based on the four priority topics of the "Sendai Framework," namely: understanding catastrophe risk, improving catastrophe danger governance, investing in catastrophe discount (thru structural and non-structural measures) and catastrophe preparedness, early warning and constructing back better in the aftermath of a disaster.

Organisations and Policies related to Disaster Management Framework at National Level

National Disaster Management Authority of India (NDMA)

- It was set up in 2005, underneath the Disaster Management Act 2005.
- The objective of NDMA is, to construct a safer and catastrophe resilient India with the aid of a holistic, proactive, era driven and sustainable improvement strategy.
- The NDMA is chaired through the Prime Minister of India and has a vice chairman with the reputation of Cabinet Minister and 8 members with the status of Ministers of State.
- The NDMA Secretariat is headed via a Secretary and deals with mitigation, preparedness, plans, reconstruction, network attention and monetary and administrative factors.

National Disaster Management Plan (NDMP)

- It was released in 2016, it is the first ever countrywide plan prepared within the USA for catastrophe control.
- With National Disaster Management Plan (2016) India has aligned its National plan with the Sendai Framework for Disaster Risk Reduction 2015-2030, to which India is a signatory.

- The objective of the plan is to make India disaster resilient, achieve significant disaster hazard reduction. It goals to noticeably decrease the losses of life, livelihoods, and belongings in phrases of financial, physical, social, cultural, and environmental. To maximize the capability to deal with failures at all degrees of administration as well as among communities.

State Disaster Management Authority (SDMA)

- At State level, State Disaster Management Authorities are set up underneath Disaster Management Act 2005.
- SDMA is chaired via the Chief Minister of the State and has now not extra than 8 participants who're appointed by the Chief Minister.
- The SDMA prepares the kingdom catastrophe management plan and implements the National Disaster Management Plan.

District Disaster Management Authority (DDMA)

Under Disaster Management Act 2005, each State authorities shall set up a DDMA for each district inside the State.

- There are not extra than seven different individuals in DDMA.
- The Disaster Management Committee governed beneath District Magistrate will formulate village level catastrophe control plans for problem villages.
- The DDMA makes District Disaster Management Plan and implements the kingdom Disaster Management Plan.

Powers and Functions of NDMF

The key obligations of NDMF consist of laying down the policies, plans and tips for catastrophe control for making sure well timed and powerful response to catastrophe. It lays down the policies on disaster management, approve the countrywide plan, and approve plans developed by means of various ministries of the union, lay down the suggestions to be observed with the aid of the kingdom government in drawing up country plans. NDMF is also mandated to advocate pointers for the minimal requirements of remedy to be provided to persons affected by catastrophe. The minimal standards are as follows:

- Minimum requirements to be supplied in the comfort camps when it comes to safe haven, meals, ingesting water, medical cowl and sanitation;
- Special provisions to be made for widows and orphans;
- Ex gratia help resulting from lack of lifestyles as additionally assistance because of harm to homes and for recuperation of manner of livelihood;
- Such different alleviation as may be vital.

In case of a disaster of excessive magnitude, the NDMF is empowered to advise comfort in repayment of loans or for grant of fresh loans to the persons stricken by disaster.

Responsibilities of the Central Government

The DM Act places on central government the responsibility to take all measures important and expedient for the cause of catastrophe control which includes coordination among ministries and department, country governments, diverse home and international organizations *etc.* It is likewise obliged to make right allocation of funds.

National Executive Committee

Apart from NDMF, the central government also constitutes a National Executive Committee which is liable for supporting NDMF in execution of various capabilities for catastrophe management. The secretary of the ministry which is answerable for disaster management is chairperson of NEC. The secretaries in the ministries of agriculture, atomic energy, defence, ingesting water supply, surroundings and forests, finance (expenditure), health, power, rural development, technological know-how and technology, space, telecommunication, urban improvement, water assets and the Chief of the Integrated Defence Staff of the Chiefs of Staff Committee can be members of the NEC.

Powers and Functions of NEC

- ☆ Assist NDMA in its features
- ☆ Implementing the plans and policies of NDMF
- ☆ Ensuring compliance with the directives of Central Government.
- ☆ To act as a coordinating and monitoring body for disaster management
- ☆ Prepare the National Plan to be approved by the NDMF
- ☆ Prepare guidelines for exceptional ministries with recognize to disaster management.
- ☆ Provide technical assistance to nation governments and authorities.
- ☆ Monitor the implementation of the National Plans and plans of numerous ministries
- ☆ Monitor implementation of the suggestions laid down by means of the NDMA
- ☆ Coordinate reaction in case of a disaster.
- ☆ Advise and assist diverse ministries and departments.

3.5 Financial Arrangements for Disaster Management

Financial help to fulfill the rescue and alleviation expenditure at some stage in any notified disaster occasion is governed by using notified recommendations on State Disaster Response Fund (SDRF) and National Disaster Response Fund (NDRF). These funds have been created below the legal frame work of 48(1) (a) and phase forty six of Disaster Management Act, 2005 respectively. Each kingdom has a corpus of price range, known as State Disaster Response Fund, administered by means of a kingdom stage committee headed via the Chief Secretary of the State Government. The size of the corpus is determined as regards to the expenditure generally incurred by means of the nation on remedy and rehabilitation over the last ten years.

In case the funds beneath State Disaster Response Fund aren't sufficient to meet the specific requirements, State Governments can are seeking assistance from the National Disaster Response Fund–a fund created at Central Government degree. Both these budget, because the names suggest, are intended for alleviation and rehabilitation and do now not cowl either mitigation or reconstruction works, which must be funded one at a time with the aid of the State or Union Government.

Natural calamities of cyclone, drought, earthquake, fire, flood, tsunami, hailstorm, landslide, avalanche, cloud burst, pest assault and bloodless wave and frost considered to be of extreme nature by using Government of India and requiring expenditure with the aid of a State Government in excess of the balances to be had in its own State Disaster Response Fund (SDRF), will qualify for fast remedy assistance from NDRF. Ministry of Agriculture is chargeable for dealing with natural calamities of Drought, hailstorm, pest Attack, Cold Wave and Frost. All other notified herbal calamities are managed by using Ministry of Home Affairs. Ministry of Home Affairs is nodal ministry for disaster control inside the country.

There is not any categorisation of calamities noted as National Calamity in Disaster Management Act, 2005. As in line with the DM Act 2005 and 14th Finance Commission (FC) suggestions, any notified calamity of a excessive nature will qualify for assistance from NDRF.

The State Governments are primarily accountable for execution of alleviation operations in the wake of natural calamities. The Government of India supplements the efforts of the State Governments by way of extending extra economic help from NDRF. Funds from SDRF and NDRF are released to assist States to provide on the spot relief. Assistance for long term reconstruction of assets is supplied via overall development plans of the Centre and the States, and is not covered beneath the norms and recommendations of SDRF and NDRF.

Assistance furnished from NDRF is on 100ntral furnish basis while in case of SDRF, states also want to make a contribution their proportion as per successive Finance Commission's guidelines.

NDRF is maintained through Government of India in the Public Account. A restrained length cess (NCCD-National Calamity Contingency Duty) subsidized the fund earlier. After implementation of GST, most of the cesses are subsumed in GST, for this reason the size of cess is narrowed down and no longer sufficient to fulfil the developing wishes of catastrophe funding underneath NDRF. Therefore, now as in line with requirement, essential budgetary assist is being supplied for NDRF.

Thirteenth Finance Commission (FC-XIII) had encouraged differential State shares, with fashionable category States contributing 25 according to cent and unique category States contributing 10 per cent, and the stability being contributed via the Union Government as grants-in-useful resource. Fourteenth Finance Commission (FFC) has recommended an quantity of Rs. 61,219 crore as aggregated corpus of State Disaster Response Fund(SDRF) for all States for award duration 2015-20 with kingdom contribution of 10 per cent (6122 crore) to SDRF, the closing 90 per cent (Rs. 55097 crore) coming from Central Government. The Government has universal the above hints of FFC with modification that the share share of the States will stay as before (*i.e.* throughout FC-XIII award length) and as soon as GST is in place, the advice of FFC on catastrophe alleviation might be completely implemented. Thereafter, from the yr 2018-19, the percentage of states in SDRF has been decided at 10 per cent and relaxation 90 per cent is being contributed with the aid of the centre.

The furnish under SDRF is to be released via Ministry of Finance in two instalments in June and December in each economic yr at the pointers of Home Ministry. In case of severe calamity, there may be a provision of boost release of these instalments. State-smart and Year-wise allocation of SDRF has been decided with the aid of the 14th FC for its award length 2015-16 to 2019-20. Further, inside the occasion of a calamity of a severe nature while the SDRF is insufficient to meet the relief requirements, extra central assistance is supplied from NDRF. The quantum of help from NDRF is subjected to an adjustment of 50 per cent of the stability in SDRF as on 1st April the financial 12 months.

3.6 Role of NGOs in Disaster Management

The collaborative catastrophe threat governance framework promises better collaboration between governments, the private sector, civil society, academia, and groups at risks. In the context of contemporary disaster risk reduction systems, the key triadic institutions, namely government (state), the non-public sector (business/ market), and NGOs (civil society), were gradually reworking their ecosystem to utilize extra proactive disaster reaction strategies, equipped with expert body of workers and technical experts and armed with social and humanitarian imperatives to lessen the dangers of failures. While the roles of governments and public actions have received more interest in disaster and emergency control studies, current shifts in interest to promote bolder engagements of both non-governmental groups and business groups in danger discount can be seen as a necessary condition for the future resilience of society.

Be it an organisation serving within a rustic or an worldwide association, all NGOs observe a single goal of offering specialized and expert useful resource to needy people on every occasion required. It is vital to word that the nature of an NGO can vary relying on how they are functioning. This is due to the fact there are various styles of not-for-earnings organisations, together with worldwide, regional, local and national. While the worldwide ones act like advocators or donors, the others receive the offered assistance that allows you to deliver help to folks who are in urgent need of it.

In the current past, the position of NGOs in disaster management (DM) has began converting from presenting post-disaster remedy to strengthening pre-catastrophe preparedness and mitigation thru potential building, public focus campaigns, mock exercises, workshops and conferences, *etc.* NGOs have additionally started to collaborate with company entities in Public-Private Partnership (PPP) initiatives and Corporate Social Responsibility (CSR) projects in the field of DM at State, District and Sub-district levels. Till recently, the paintings of NGOs within the area of DM has been usually sporadic, reactive, responsive and driven by using neighborhood level compulsions in the geographic regions where they're imposing improvement initiatives and very frequently they faced widespread challenges in coordinating with the authorities machinery and even amongst NGOs themselves.

Global catastrophe threat discount initiatives have developed as a device that has been continuously expanding and developing in phrases of proliferation of actors which includes governments, Red Cross societies, the United Nations, the private sector, coverage industries, academia, non-governmental organizations (NGOs), and civil society. This machine has turn out to be a collective humanitarian imperative whereby each actor has a stake in catastrophe response and risk discount. Year by means of yr, humanitarian actors were expanding their mandates from single reactive emergency reaction closer to greater various units of mandates. One of the urgent mandates is to ensure that proactive disaster threat reduction materializes in all prone localities.

Besides generating cognizance, NGOs also provide advocacy for disaster control. In fact, those organisations regularly acquire accreditation for serving as partners of advocacy at one of a kind levels. Mostly, they interact governments for reviewing exclusive policies with a view to cope with concerns pertaining to catastrophe management. They also can broaden in setting up an advocacy roadmap which predicts possibilities that are available for a set or country so that it will spend money on preparation. They invite participation of different companions as well to gather finances and use it for developing interventions for the catastrophe. Such a shape of advocacy is very critical because it improves approach improvement, policy formulation, and drafting of disaster control programmes at distinctive levels.

One of the NGOs' comparative benefits is operability at grassroots degree, wherein they paintings with the most at-threat and prone groups. NGOs regularly take inclusive and consensual approaches to nearby catastrophe making plans and resilience building (Maskrey, 1989; Lassa, 2018). Such operability frequently lets in NGOs to understand and respond to the priorities and agendas of vulnerable communities. Their quite small organizational size allows them to be spark off and efficient, as they operate in a greater bendy and less bureaucratic environment (UNISDR, 2006). This is the purpose NGOs are without problems formed when wanted and removed while no longer needed. This argument is in favour of the added fee of NGOs and their complementarity to acquire shared objectives and dreams of DRR through public, community, and personal moves.

3.7 Role of Community-Based Totally Organizations and Media in Disaster Management

Role of Community-Based Totally Agencies

A community-based agency is exceptional positioned to efficaciously perform activities and plans affecting the lives of the communities: socio-financial improvement, herbal resources management, environmental conservation or catastrophe control. CBOs are made up of network representatives and one in all their pursuits is to make certain that decision makers take word of community worries.

Community-primarily based disaster control (CBDM) is an technique to constructing the capability of groups to assess their vulnerability to both human brought about and natural risks and broaden techniques and sources vital to prevent and/or mitigate the impact of diagnosed risks as well as respond, rehabilitate, and reconstruct following its onset. CBDM strategies have grown to be increasingly important within the face of global climate change, elevated populations expanding into greater vulnerable regions, and the heightened popularity of a want for more linkages between top-down governmental and network stage responses. CBDM empowers communities to be pro-lively in catastrophe control and creates a space for them to broaden techniques on their personal terms in place of looking forward to already overstretch governments and NGO's.

All governments are chargeable for defensive their residents and endorsing the 2005 Hyogo Declaration which states that: "strengthening network degree capacities to reduce catastrophe threat at the neighbourhood stage is specially wished, thinking about that appropriate disaster reduction measures at that degree permit the groups and people to lessen drastically their vulnerability to hazards."

Members of a community are the instant victims of adverse outcomes of a catastrophe. They have the fine understanding approximately their neighborhood surrounding in terms of the maximum disaster-prone areas, the demography of their community and their social and traditional company. It is vital that they've the capability to cope with the impacts of a disaster and are worried inside the improvement of disaster management sports proper from the initial planning stages. Community participation can also make them greater confident in their skills to act inside the event of a catastrophe main to a self-reliant community.

Moreover, the community based approach to catastrophe control will become crucial because of the following shortcomings of the present device:

- ☆ The equal plan, regardless of the regional characteristics, is applied or imposed everywhere.

- ✰ Local knowledge, experiences, skills, sources and techniques aren't given due importance. Rather external sources and techniques are proposed to be utilized.
- ✰ Negligence about nearby cultural instincts and heritage.
- ✰ Prioritisation is determined with the aid of an outsider and not the stakeholders or the community itself.
- ✰ Local community does not have any information about the catastrophe control plans for their vicinity and the role of different sectors in helping the network in the course of disasters.

The main tasks of community-based totally organizations in disaster Management are to:

- ✰ Provide a platform for the community to come together and talk improvement troubles;
- ✰ Analyse the causes of problems and solutions;
- ✰ Communicate with choice makers to share deliberate action and gain their endorsement;
- ✰ Galvanize network individuals to put in force deliberate solutions;
- ✰ Mobilize inner or locally to be had resources and external sources (out of doors the community) to put in force recognized solutions;
- ✰ Raise community recognition and sell community movement on environmental and disaster troubles and livelihood options;
- ✰ Build ability of community contributors;
- ✰ Carry out monitoring and evaluation.

Role of Media

The media performs an essential role in catastrophe management through the insurance of herbal and man-made screw ups, drawing close the events to the broader society, growing public discourse and mobilizing donations. In doing so it's miles crucial to pay attention to the ways wherein media portrays representations of warfare and human suffering as well as the size of the scope inside the coverage of such occasions. Recent examples of media coverage in humanitarian screw ups consisting of the Asian tsunami in 2014 have highlighted the manner in which the media favours positive activities towards others depending on traits which includes the monetary and political effect or the belief of cultural proximity. As a result of that, there's an imbalance inside the stage of media attention which complicates disaster control in those cases where media insurance is a whole lot lower.

The function of the media in disaster control has improved considerably in recent years with the aid of the usage of new era and introducing the understanding of technical professionals who're in a position to better describe the reasons and results of failures. The truth that technology has made feasible things such remote television transmission or satellite communique have brought about a proliferation of data and the record of humanitarian activities whenever and wherever they occur (Rattien, 1990). Not handiest has the evolution of technology allowed the media to cowl a wider variety of screw ups but additionally to reach a greater quantity of people. As actors involved at once in shaping public opinion, the way in which the media portrays pictures and describes actual scenarios is crucial to decide the extent of response.

The information media performs several roles in present day societies, ranging from primary capabilities together with informing and teaching to emergent capabilities that end result from interactions between the media and other social systems. It is these emergent features that dominate the media's dating to healing from herbal disasters. In the direction of reporting essential events as they spread the media presents metaphors that ineluctably promote particular readings of these activities. These Roles of the media vis-à-vis natural disasters are considered in turn.

- ☆ **Social software:** A basic know-how of the mass media's purpose is that it spreads records to the populace. In the case of disasters, the media can provide vital information to the ones in and round affected regions. In this role, the media follow the unfolding of events as roughly objective observers (for greater at the media and objectivity, see Anderson, 1997), reporting data to those at once stricken by the disaster (victims, relief workers, *etc.*) and more substantially – from a social point of view – to society at-large.
- ☆ **Frames and metaphors**: Linking occasions into narratives Information disseminated with the aid of the media is shaped with the aid of many factors, most extensively the traits of particular media forms in addition to via journalistic and editorial practices. Satellite-driven TV news and full-colour newspapers thrive on the fast transmission of dramatic pictures from occasion scenes. As Anderson (1997) notes, the intersection of complex social, political, economic, and environmental issues "require journalists to transform technical jargon into laypersons' language; this itself may additionally involve strategies of interpretation and selection."
- ☆ **Positive comments loops: media hype**: Positive comments loops operating in the media are enabled with the aid of the self-referential nature of mass communications. Media hypes are induced through uncommon or stunning activities...which might be framed in such a manner that the media shift into a higher gear, searching for 'newer' information at the topic. A new wave is created via those intensive news making sports of the media and are then reinforced again and again by considerable insurance of the social actors' reactions, responding to the large media attention to a topic. Once a topic profits a sure degree of interest within the media, it attracts greater interest, and, as it attracts greater interest, it will become more noteworthy.

☆ **Risk amplification and attenuation:** Modern mass media is a central pressure at the back of the social production of chance. This is due to the social software of the media, its narrative-forming tendency, in addition to the focusing strength of media hypes. The media's function as social glue and tv's unique psychological electricity (discussed below) serve to legitimize the records obtained via the media. Thus legitimized, the statistics is more likely to be understood in phrases of the narratives in which it's far presented.

3.7 Role of Central, State, District and Local Administration in Disaster Management

While the number one duty of catastrophe management rests with the States, the Central Government supports the efforts of State Governments by means of supplying logistical and economic guide.

On behalf of the Central Government, DM Division inside the Ministry of Home Affairs co-ordinates with disaster affected State Government(s), concerned line ministries/departments, National Disaster Management Authority (NDMA), National Disaster Response Force (NDRF), National Institute of Disaster Management (NIDM) and the Directorate General of Fire Services, Home Guards and Civil Defence, and Armed Forces for effective disaster threat reduction. The Division is answerable for legislation, policy, capability building, prevention, mitigation, response and long term rehabilitation.

Role of the Union Government

India with its federal system of government has unique roles for the principal and State governments. However, disaster management does not particularly discover point out in any of the 3 lists it1 the Seventh Schedule of the Indian Constitution, wherein subjects underneath the valuable, nation governments and underneath each are specified. There was till now, no specific valuable or country Act, which definitely defines the methods of managing screw ups. The critical authorities delivered the Disaster Management Bill within the winter consultation of Parliament.In 2005 which has been passed. The Governments of Gujarat and Madhya Pradesh have already enacted disaster management Acts. The important authorities has asked different kingdom government to legislate Disaster Management Acts.

Although the State Government involved has the primary obligation for disaster control, the Union Government performs a key supportive role in terms of bodily and monetary assets and providing complementary measures, including early warning and co-ordination of efforts of all union ministries, departments and establishments. At the apex stage, a Cabinet Committee on Natural Calamities opinions the disaster state of affairs.

An excessive degree committee of ministers beneath the chairmanship of Minister of Agriculture deals with the issue of financial assist to be furnished to the State Governments from the National Disaster Response Fund, if the funds to be had with the State Governments underneath State Disaster Response Fund are not adequate. Matters referring to nuclear, biological and chemical emergencies are regarded after by the Cabinet Committee on Security.

Role of State Government

In India, the basic responsibility to adopt rescue, remedy and rehabilitation measures in the occasion of natural disasters rests with the country authorities. Since the very beginning, the entire structure of crisis administration in the state governments were oriented toward post disaster comfort and rehabilitation.

At the nation degree, disaster response, remedy and rehabilitation are dealt with with the aid of Department of Relief and Rehabilitation. The attention of this department has normally been provision of post-calamity comfort, The Chief Secretary is answerable for remedy operations within the state and the Relief Commissioner and Additional Relief Commissioners characteristic beneath the Chief Secretary's course and manipulate. Most of the states have Relief Commissioners however in some states, the Secretary, Department of Revenue is in rate of relief operations. The Government of India is running with country governments to transform the Department of Relief and Rehabilitation into Department of Disaster Management with an increased vicinity of responsibility which include catastrophe preparedness. Uttaranchal is the first kingdom to have a separate Department of Disaster Management. Seven different country governments/union territories have additionally introduced this change. The authorities of Rajasthan has re-exact the department as Disaster and Relief. Other states are also bringing approximately these changes.

Most of the states have Relief Commissioners who're in fee of the relaxation and rehabilitation measures. Most of the states have converted to a Disaster Management Department with the required linkages with the numerous improvement and regulatory departments involved with prevention, mitigation and preparedness.

Every nation has a Crisis Management Committee beneath the chairpersonship of the Chief Secretary, together with secretaries in rate of involved departments, which opinions crisis situations on a every day basis on the time of disaster, coordinates the activities of all departments and provides selection guide system to the district management. At the ministers' degree, a Cabinet Committee on Natural Calamities below the chairpersonship of the Chief Minister takes stock of situations and is liable for all crucial coverage decisions.

Role of District Administration

The District Magistrate/Collector has the responsibility for standard management of disasters inside the district. He has the authority to mobilise the reaction equipment and has been given monetary powers to draw money under the provisions of the General Financial Rules/Treasury Codes. The district stage is the point of interest for disaster control activities at the sphere level. The Collector and District Magistrate is the chief administrator inside the district. The Collector/Deputy Commissioner is the focus for all disaster control activities. A District Level Coordination and Relief Committee is constituted and is headed by using tile Collector as Chairman with participation from all associated government and non-authorities establishments and departments similarly to the elected representatives. These committees are being reconstituted or re-distinct as Disaster Management Committees with officials from. Relevant departments being added as participants. The district heads and departments engaged in improvement shall additionally be part of this Committee so that catastrophe mitigation and preparedness are main-streamed into the district plan.

All departments of the State Government, along with the police, hearth services, public works, irrigation *etc.*, paintings in a coordinated way under the management of the Collector during a catastrophe, besides in metropolitan regions in which the municipal frame performs a chief position. The District Collector also enjoys the authority to request for assistance from the Armed Forces if circumstances so demand. NGOs have additionally been powerful in offering alleviation, rescue and rehabilitation in current times.

Role of Local Self-Governments

Local self-governments, each rural and urban, have emerged as vital stages of governance, after the 73rd and 74th Amendments to the Constitution. For the humans, they may be additionally the nearest units of administration and are most of the first responders to any crisis besides being intently knit with the communities. These gadgets can as a consequence play an crucial role in crisis management below the general management of the District Administration.

Their role is to:

- develop, overview and assess effective catastrophe control practices
- assist local government to prepare a local disaster control plan
- ensure the community knows the way to respond in a disaster
- identify and coordinate disaster resources
- manage local disaster operations
- ensure local catastrophe control and catastrophe operations integrate with country disaster management.

3.8 Role of Armed-Forces in Disaster Responses

From earthquakes to floods, when natural disasters strike, the army is often known as on to reinforce civilian responses. Disasters are of diverse sorts caused with the aid of herbal phenomena in addition to compounded by using human error. Natural catastrophe can be of various type consisting of earthquakes, cyclone, tornado, flood *etc.* And other occasions together with drought, vegetation failure, *etc.* whereas man-made disasters may additionally include radiation injuries, fuel leakage accidents bomb explosion, war time rioting, firing *etc.* Natural Disasters are in contrast to something else in human experience. These are in large part beyond control of individual resulting short changes in the lives of the sufferers. On the opposite hand, the cost of man-made catastrophe like battle is borne through the complete population.

The contribution which the defence force could make in coping with a catastrophe need not be over emphasized. This contribution has generally been made beneath some shape of reputable arrangement for useful resource to civil power, as laid down in the standing running procedures. Experience has proven that navy forces are fully powerful in counter-disaster function. The bendy organizational shape, functionality of sustained operation in all weathers via day and night time and properly trained control device of the defence force make them particularly nicely-suited for effective disaster remedy operations. In addition, lots of their everyday sports parallel to those in public emergency services. Thus armed forces can provide assistance to the sufferers with invaluable help in engineering, communications, transport, rescue, emergency clinical services, subject sanitation, water deliver and so on.

While the number one duty for disaster reaction lies with civilian companies at local, nation and federal levels, best the army has the manpower, equipment, education and corporation essential to accumulate the comfort attempt required in the course of catastrophic incident recuperation.

However, in recognition of the truth that humanitarian comfort have to stay a predominantly civilian function, worldwide norms location barriers on the use of foreign navy assets. Most current armies these days have made disaster comfort an crucial part of their programmes, supplying fast response and performing scientific assistance, logistical assist, air visitors response, resource distribution, safety and restoration in herbal disasters on a global scale.

In Asia, the position of the Indian Army particularly has changed in recent years, where it has end up one of the first responders in times of crises, with special interest to tsunami and earthquake remedy. But possibly the most vital role is played by way of the USA Army. The army's number one mission is to guard the nation's borders. In matters domestic, the navy is meant to be a second respondent, except in the case of chemical, organic, radiological and nuclear incidences. Theoretically, the principle is "final to enter and first to leave". However, when principle is matched with practice, this doesn't seem to be the case. According to the Administrative Reforms Commissions, the army needs to be taken off from the loop of disaster management gradually. While it can sound realistic on paper, it isn't always really possible in practice. The civil administration is usually now not properly ready for an effective reaction. It needs to be stated that subject and efficiency is the primary call for in disaster response and comfort tasks, which might be often risky missions and quite clearly the navy brings in order in post-disaster operations. Wherever there may be danger, the military has a constitutional obligation to undertake obligations and missions. The required wherewithal consisting of the command, manipulate and communication, are available with the sphere formations. Preparing for navy operations different than war (MOOTW), of which catastrophe is the main thing may be a vital part of navy schooling.

3.9 Role of Police and Other Organizations in Disaster Response

Man-made screw ups including bomb explosions and arson cause a sense of vulnerability. Natural screw ups inclusive of hurricanes, floods, tornadoes and fires all cause precise damage and chaos. A sturdy police presence in both conditions might also help ease fears and assist prevent others from taking gain of the situation. The role of law enforcement in responding to a disaster is very similar to the daily role of public protection and helping the community. In getting ready for a disaster, police officers consider in their training and capitalize on their knowledge of a community. Exercises portraying the situations (large- and small-scale occasions) help better put together officers and allow them to fully understand the resources wished for every event and practice that data to each network's needs. Law enforcement officials understand their communities first-class and have interaction with residents on an everyday foundation. This expertise offers them the potential to provide valuable situational cognizance to response and recuperation agencies coming in to assist. For example, where will there be language barriers? Does the network have unique challenges? Law enforcement can assist communicate this statistics to the emergency management crew and can provide aid to different individuals of the team with the aid of clearly being a presence in the neighbourhoods.

During a disaster, police officers play a key position in many operations together with: search and rescue, evacuations, door-to-door checks, and preserving basic public protection within the network. These are vital moves that assist now not only their own communities however neighbouring towns as properly.

Police chiefly performs the subsequent tasks at some stage in a catastrophe:

- ✰ **Crowd Control**: Disaster conditions draw crowds. There are humans hoping for a glimpse of the unfolding drama. Family participants and pals of the sufferers show up, hoping to glean data approximately their cherished ones. The human beings involved in the scenario also mill around, every now and then whilst injured or in shock. The police cordon off regions and paintings to keep crowds below manipulate.
- ✰ **Public Safety**: Officers might move from house to residence looking for the injured or people who want assist. They can offer transportation, emergency medical care and be a bridge between citizens and rescue personnel. They can check credentials and identity to peer if someone without a doubt needs to be inside the location. They may additionally

warn humans when electric wires are on the ground, when dangerous pests are lurking or when volatile people are nearby.

- ✰ **Directing Traffic**: Many disasters wipe out infrastructures along with electricity, making an risky state of affairs for traffic and pedestrians. The police can close dangerous roads in which downed wires, timber or flood situations would possibly make tour hazardous. They direct residents through nonworking site visitors lights. They reroute random visitors to more secure roads and strive to keep out all unnecessary visitors. They can put roadblocks within the areas wherein the crisis is underway.
- ✰ **Organizing Search Teams**: Disaster situations often lead to family participants being separated. Elderly and unmarried humans are at hazard if they haven't any one to check on them. If these residents wander around, problems may additionally result. The police can set up search groups and put together structures wherein loved ones can check on one another. They can set up command centres in which the victims and circle of relatives can be reunited. They work with emergency personnel to listing the numerous clinical facilities used for treating unwell or injured human beings.
- ✰ **Looters**: There are folks that take gain of a chaotic state of affairs. When human beings are evacuated from houses or neighbourhoods, houses are unsupervised and opportunities to dedicate crime may be rampant. Police presence may assist to deter looting. If looting does occur, the police are ready to understand suspects.

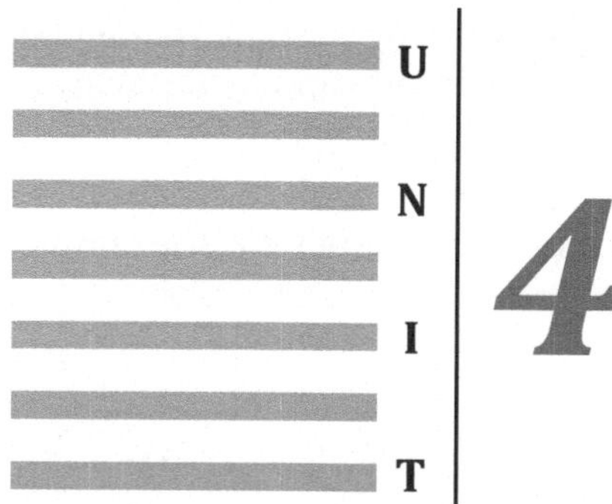

COVID-19 and its Impact on Environment

Worldwide spread of COVID-19 in a quite short time has brought a dramatic decrease in industrial activities, road traffic and tourism. Restricted human interaction with nature during this crisis time has appeared as a blessing for nature and environment. Reports from all over the world are indicating that after the outbreak of COVID-19, environmental conditions including air quality and water quality in rivers are improving and wildlife is blooming. India has always been a hub of pollution with huge population, heavy traffics and polluting industries leading to high air quality index (AQI) values in all major cities. But after declaration of lockdown due to COVID-19, quality of air has started to improve and all other environmental parameters such as water quality in rivers have started giving a positive sign towards restoring.

Introduction

- ☆ COVID-19 is an infectious disease caused by a newly discovered corona virus. Most people infected with the COVID-19 virus will experience mild to moderate respiratory illness and recover without requiring special treatment. Older people and those with underlying medical problems like cardiovascular disease, diabetes, chronic respiratory disease, and cancer are more likely to develop serious illness.
- ☆ The best way to prevent and slow down transmission is to be well informed about the COVID-19 virus, the disease it causes and how it

spreads. Protect yourself and others from infection by washing your hands or using an alcohol based rub frequently and not touching your face.

- The COVID-19 virus spreads primarily through droplets of saliva or discharge from the nose when an infected person coughs or sneezes, so it's important that you also practice respiratory etiquette (for example, by coughing into a flexed elbow).

Meaning of COVID-19

COVID-19 is a disease caused by a new strain of corona virus. 'CO' stands for corona, 'VI' for virus, and 'D' for disease. Formerly, this disease was referred to as '2019 novel corona virus' or '2019-nCoV.'

How do Viruses get their Name?

Viruses are named based on their genetic structure to facilitate the development of diagnostic tests, vaccines and medicines. Virologists and the wider scientific community do this work, so viruses are named by the International Committee on Taxonomy of Viruses (ICTV).

Why Corona Virus Infection is Called COVID-19?

Illness caused by SARS-CoV-2 was termed COVID-19 by the WHO, the acronym derived from "corona virus disease 2019." The name was chosen to avoid stigmatizing the virus's origins in terms of populations, geography, or animal associations.

Wildlife and COVID-19: The Good

- One major and predominately positive benefit of the pandemic for wildlife is less human travel.
- Due to the significant reduction in journeys, fewer people are hitting and injuring or killing wildlife on roadways.
- A study from March 2021 found that hedgehog roadkill rates in Poland were more than 50% lower compared with pre-pandemic years, saving tens of thousands of hedgehogs in Poland alone. This may help reverse the long-term decline of European hedgehog populations.
- Another study analyzing roadkill data from 11 countries found that roadkill rates fell by more than 40% during the first few weeks of the pandemic restrictions in Spain, Israel, Estonia, and the Czech Republic.
- In addition, fewer ships are traveling through the world's waterways and oceans for shipping, fishing, aquaculture, and tourism purposes.
- In November 2020, experts predicted that global maritime trade would have plunged by 4.1% by the end of that year. Other reports estimated a 10% decline in the container trade for 2020.

- A reduction in water travel and activity could reduce the risk of ships striking and injuring or killing marine animals. It may also reduce the marine disruption that occurs due to noise pollution from ships, fishing sonar, and recreational boats.
- Birds might also be benefitting from the sharp decline in air travel, which may have vastly reduced the risk of bird strikes.
- According to the Federal Aviation Administration, between 1990 and 2019, there were about 227,005 wildlife strikes with civil aircraft in the U.S. In addition, U.S. airplanes reported some 4,275 more wildlife strikes at foreign airports. These strikes resulted in injury to 327 people.
- The pandemic has also led to a in industry supply chains, reducing demand for commercial activities that exploit natural resources in many parts of the world. For example, lower fishing demand and activity may reduce the removal of animals from the wild.
- And in India, anecdotal reports that reduced fishing and vehicle traffic at nesting beaches may be boosting populations of the critically endangered olive ridley sea turtle.
- The pandemic wildlife by disrupting the hidden, generally illegal supply chains that destroy wild populations, including those that fuel the wildlife trade.
- People are also wildlife in unexpected places, such as in large cities and commercial harbors. The increased number of animals in urban environments is likely due to reductions in human presence, air and water pollution levels, and noise pollution.
- For example, people have spotted pumas wandering in downtown Santiago, Chile, and dolphins swimming in the usually choppy waters of the Port of Trieste, Italy.

Wildlife and COVID-19: The Bad

- However, many of the immediate positive effects of the pandemic on wildlife — such as reduced road, air, and ship deaths or disruption — will likely reverse if the world goes back to business as usual.
- In many cases, it will take generations of change to help thousands of species around the world recover from the impact of humankind. For example, it may take 10–15 years of sustained reduced fishing to allow the world's depleted fish populations to recover.
- Some studies have also found that the pandemic may actually be causing harm to wildlife.
- In researchers found that reduced human disturbance relating to lockdown has benefitted invasive alien species by interrupting the actions that people were taking to control them. The authors also claim that

pandemic restrictions have reduced the work of conservation and law enforcement organizations that care for wildlife and protected areas.

- And this is a global trend, as the staff of preserves, game parks, sanctuaries, and other wildlife facilities are unable to perform their normal activities.
- Also, the reduction in law enforcement may cause a sudden increase in illegal wildlife killing — in particular, that of endangered animals liable to persecution or poaching.
- Some experts also worry that economic hardship in low income countries may lead to an increase in natural resource exploitation, such as unlicensed logging and the illegal wildlife market, as people run out of ways to earn a living.
- According to satellite images, a surge Trusted Source in deforestation is taking place in several hotspots. Also, illegal fishing rates are on the rise in Brazil and the Philippines.
- The changes in human activity that the pandemic has necessitated may also be having some negative effects. For instance, some species that rely heavily on humans for feeding or scavenging, such as monkeys, gulls, and rats, may be struggling Trusted Source during the pandemic.
- People may also be using outdoor spaces such as parks and nature reserves more during lockdown, which could disturb resident wildlife unaccustomed to human interactions.
- On the flip side, reduced ecotourism rates are crippling many organizations worldwide that rely on human visitors to feed and care for their animals.
- Meanwhile, plastic pollution from improperly disposed-of single-use COVID-19 protective gear also seems to be increasing the global plastic pollution problem and causing wildlife deaths, as animals can ingest plastic items or become entangled or trapped in them.
- According to one estimate, people are throwing away as many as 3.4 billion Trusted Source single-use face masks and face shields daily worldwide.

How Might Climate Change Affect the Spread of Viruses?

- In the coming decades, ecological degradation, rising temperatures, and extreme weather events could intensify the threats to human health posed by viruses.
- All data and statistics are based on publicly available data at the time of publication. Some information may be out of date. Visit our corona virus hub and follow our live updates page for the most recent information on the COVID-19 pandemic.
- We know from past epidemics that changes in temperature, rainfall, and humidity can have profound effects on the spread of infectious disease.

- In the summer of 1878, for example, the southern United States was struck by a catastrophic outbreak of yellow fever, a viral disease indirectly transmitted between people via the mosquito *Aedes aegypti*.
- Around 100,000 people contracted the disease, and up to 20,000 people lost their lives. Some estimates put the economic cost as high as $200m.
- Yellow fever was a regular scourge of cities in the lower Mississippi River basin in the 18th and 19th centuries.
- "During its brief reign — July to October — its ravages were such as to completely paralyze both the social and commercial interests of a given city, and even an entire section of our country," a U.S. Senate Report noted in 1911.

COVID-19 and its Effects on Environment

- Due to the unusual outbreak of COVID-19, almost every big and small city and village in the affected countries like China, Taiwan, Italy, USA, France, Spain, Turkey, Iran, Germany, S Korea, U.K, India, Australia and many more, is under partial or total lockdown for a long period of time ranging from a few weeks up to a few months.
- The major sectors contributing to air pollution are transport, industries, power plants, construction activities, biomass burning, road dust resuspension and residential activities. In addition, certain activities such as operation of DG sets, restaurant, landfill fires, *etc.* also contribute to air pollution.

The global outbreak of corona virus disease 2019 (COVID-19) is affecting every part of human lives, including the physical world. The measures taken to control the spread of the virus and the slowdown of economic activities have significant effects on the environment. Therefore, this study intends to explore the positive and negative environmental impacts of the COVID-19 pandemic, by reviewing the available scientific literatures. This study indicates that, the pandemic situation significantly improves air quality in different cities across the world, reduces GHGs emission, lessens water pollution and noise, and reduces the pressure on the tourist destinations, which may assist with the restoration of the ecological system. In addition, there are also some negative consequences of COVID-19, such as increase of medical waste, haphazard use and disposal of disinfectants, mask, and gloves; and burden of untreated wastes continuously endangering the environment. It seems that, economic activities will return soon after the pandemic, and the situation might change. Hence, this study also outlines possible ways to achieve long-term environmental benefits. It is expected that the proper implementation of the proposed strategies might be helpful for the global environmental sustainability.

- ✰ Under the nationwide lockdown, all transport services – road, air and rail were suspended with exceptions for essential services. Educational institutions, industrial establishments and hospitality services were also suspended. As a result, air quality improvement has been noted in many towns and cities across the world.
- ✰ Due to non-functioning of industries, industrial waste emission has decreased to a large extent. Vehicles are hardly found on the roads resulting in almost zero emission of green-house gases and toxic tiny suspended particles to the environment.
- ✰ Due to lesser demand of power in industries, use of fossil fuels or conventional energy sources have been lowered considerably.
- ✰ Ecosystems are being greatly recovered. In many big cities, the inhabitants are experiencing a clear sky for the first time in their lives.
- ✰ The pollution level in tourist spots such as forests, sea beaches, hill areas, *etc.* is also shrinking largely. Ozone layer has been found to have revived to some extent.
- ✰ The pandemic has displayed its contrasting consequence on human civilization, in the sense that, on one hand, it has caused worldwide panic situation, but created a very positive impact on the world environment on the other.

References

1. Impact of lockdown on air quality, CPCB, M/o Environment, Forest and Climate Change, Govt. of India. http://www.indiaenvironmentportal.org.in/files/file/Impact-Of-Lockdown-On-AirQuality.pdf
2. Website of World Health Organisation. https://www.who.int/emergencies/diseases/novel-coronavirus-2019
3. The Indian Express, dated 08 May, 2020. https://indianexpress.com/article/opinion/web-edits/rediscovering-sustainabledevelopment-through-a-covid-19-lockdown-lens-6399986/

Index

D

E

F

G

Q

R

S

T